T0231327

Analysis of
INDUSTRIAL CLUSTERS in CHINA

Analysis of
INDUSTRIAL CLUSTERS in CHINA

Zhu Yingming

科学出版社
Science Press

CRC CRC Press
Taylor & Francis Group
Boca Raton   London   New York

CRC Press is an imprint of the
Taylor & Francis Group, an **informa** business

CRC Press
Taylor & Francis Group
6000 Broken Sound Parkway NW, Suite 300
Boca Raton, FL 33487-2742

© 2010 by Science Press
CRC Press is an imprint of Taylor & Francis Group, an Informa business

No claim to original U.S. Government works

Printed in the United States of America on acid-free paper
10 9 8 7 6 5 4 3 2 1

International Standard Book Number: 978-1-4200-8919-6 (Hardback)

Library of Congress Cataloging-in-Publication Data

Yingming, Zhu.
  Analysis of industrial clusters in China / Zhu Yingming.
      p. cm.
  Includes bibliographical references and index.
  ISBN 978-1-4200-8919-6 (hardcover : alk. paper)
  1. Industrial clusters--China. I. Title.

HC79.D5.Y56 2009
338.8'7--dc22
                                                                          2009024577

Visit the Taylor & Francis Web site at
http://www.taylorandfrancis.com

and the CRC Press Web site at
http://www.crcpress.com

Preface

At present, economic localization tendency is more remarkable, competition among countries is more intense, and the tendency for localization development over the national economic system within the state stands out especially under the macroscopic context of further development of the economic globalization tidal wave. Each country actively and initiatively takes all kinds of policy steps to improve the international competitiveness of its industries so as to improve national integrated competitiveness for the purpose of gaining more economic benefits under the context of economic globalization. Studies on industrial clusters in national economic systems on the national level are the theoretic foundation of industrial structural adjustment and optimizing upgrade under the new situation, and are also the basic work of study on national innovative industrial clusters. Early in 1960s, western developed countries had already conducted studies on industrial clusters in the national economic system. In 1990s, OECD countries had conducted studies on innovative industrial clusters. China conducts urgent studies on industrial clusters in the national economic system.

In a national economic system, there exists extensive, complicated, and close economic links among industries, and industries exist where close economic links come into being as industrial clusters. Complication and diversification of economic links among industries generate external economies, which are the base of regional growth and development. When any industry with close economic links forms industrial clusters with each other, a favorable development environment is created for a regional self-support growth process. Regional economic activities can attract an inflow of production factors from outside regions so that new productive activities can be introduced, production chains can be extended, and multiplier effects of regional productive activities remarkably heightened. With further development of regional industrial clusters, multiplier effects become quite strong, and rapid economic growth process accompanies the change in economic structures.

Although academe in China has conducted large numbers of theoretical and empirical studies about industrial clusters, so far it has not created an acknowledged concept of industrial cluster, and it has not conducted systemic study on industrial clusters in the national economy. The terminology concerning industrial clusters are: industrial cluster, industrial pile-up, industrial clustering, industrial agglomeration, industrial combo, and regional production complex. The confusion of terms relative to industrial clusters has influenced the research level of the area.

From the point of view of spatial-temporal links of industrial development, passing through the forming process and correlation of regional industries, we can see the spatial centralization of different industrial sectors in a specific location, and this comes into being as regional industrial groups take one or more industrial sectors as the core of close economic links. So industrial clusters can be summarized as relying on a definite social economic context, taking one or more industrial sectors as the core of industrial development in a specific location. With the help of close economic links(functional links) among industrial sectors in a national economy, industrial sectors of different sizes and types are formed and develop economic links among industrial sectors forming a relatively integrated industrial 'aggregation' together.

Because industrial clusters are industrial aggregations formed by industrial sectors with close economic links in a specific territory, close economic links are not only the basic conditions, but also important characteristics of industrial clusters. Economic links among industries within industrial clusters not only include input-output relationships, but also reflect external economic relationships.

There are two basic characteristics of industrial clusters, namely, functional links and spatial links. Functional links are the material bases for industrial clusters being formed and further developed. Functional links are actually input-output links among industries and exhibit technical economic links among industries with input and output goods as linked ties. Because input and output goods can be tangible as well as intangible representing physical goods and value, functional links of industrial clusters can take the form of physical goods and value; the links in the form of goods flow and service flow. Spatial links of industrial clusters are spatial bases for industrial clusters being formed and further developed. Actually, spatial links of industrial clusters are special links in the process of external economies among industries; these exhibit a path dependency of industrial clusters on the social economic environment in given regions and are the functions of given regions. Spatial links of industrial clusters under different spatial ranks have different characteristics because space bears characteristics of rank and stratification. In a given region, we can refer to it as a point in the space, spatial links among industrial clusters can be viewed as vested, and characteristics of economic links among industries in industrial clusters depend mostly on characteristics of functional links. Industrial clusters in given regions can be defined as industrial aggregations formed by different industrial sectors through functional links under the hypothesis condition. As opposed to functional links of industrial clusters, spatial links do not concern technical economic links with all kinds of input and output goods among industries as linked ties.

The study of industrial clusters in the Chinese national economy is conducted in the spatial scope of China. We can abstract the area as a spatial point. Spatial links can be viewed as invariable, economic links among industries in industrial clusters, which give priority to functional links. Identifying problems of industrial clusters in China concerns industrial clusters with close functional links. So, identifying industrial clusters in China can use input-output analyzing methods. Using an input coefficient matrix (b_{ij} and a_{ji})(124×124) and an output coefficient matrix (a_{ij} and b_{ji}) (124×124) of economic links among industries in 1997, and making use of the trianglization method, we can identify industrial clusters in China. Because the industrial clusters identified by these methods are still incomplete (quasi-industrial clusters), we need to determine identifying principles of industrial clusters, and these principles include the lowest threshold principle, the shortest linkage path principle, and the dominant principle.

According to the identifying methods and basic principles mentioned above, we identify four types of industrial clusters in China in 1997: light manufacturing industrial clusters, heavy manufacturing industrial clusters, construction industrial clusters, and services industrial clusters. These are further divided into 11 sub-types of industrial clusters: planting, electronic, textile, iron and steel, petroleum, nonferrous metals, chemistry, construction, business, diet, and administration industrial clusters.

From the point of view of system theory, an industrial cluster is actually a system composed of many industries. In order to study the system—an industrial cluster, first we should know the relationships among industries; that is, we should understand the inner structure of the system because the functions of a system are mainly determined by the structures of the system. Structures of industrial clusters concern the correlation and interdependent relationship among basic integral parts, namely, among industries of industrial clusters. Structures of industrial clusters are determined by the functional and spatial links among industries and, in turn, deeply influence these links and then influence the functional characteristics of industrial clusters. Because of location, structural characteristics of industrial clusters mostly depend on the functional links among industries. In this book, general structures of industrial clusters are analyzed using directional linkage diagrams, recursion structural relationships among industries in Chinese industrial clusters are analyzed using the ISM method, and regression analyses are conducted using the relationship between topological indexes and total output of industrial clusters.

Functions of industrial clusters concern the effects possessed by the industrial system and its internal integral part. The industrial clusters system of China is a huge open industrial system composed of 11 industrial clusters, having external and internal functions. External functions relate to the contacts that one industrial cluster has with other clusters (including other industrial or non-industrial clusters) concerning the internal demands in order to facilitate that industrial clusters, development, and growth. Internal functions are the links among internal industries of industrial clusters. Functions of industrial clusters concern all kinds of flow input, transition, and output among industries in and out of the system. Studying the functions of industrial clusters is actually studying these flows. Basic functions such as scale distributions, function types, central industries, and nodal industries of industrial clusters in China are analyzed in this book. The technological economic structural relationship status of industrial clusters is analyzed, and the technological economic benefit of industrial clusters in China is studied. In order to maintain stable and ordered development of industrial clusters in China, and to realize adjustment and optimizing upgrade of industrial structures in China, we must macro-control all kinds of flow among industries, make them harmonious and unimpeded, make the system develop towards a more ordered advanced phase, and offer decision-making bases for government decision-making sectors.

There were 96 industries in 11 industrial clusters in China in 1997, accounting for 78.2% of all industries. Total output of these clusters is up to 16,603.591 billions RMB, accounting for 83.1% of total output of all industries. Industrial clusters have been important subsystems in the Chinese national economic system, and their working status directly affects the healthy development of the Chinese national economy. Structural and functional analyses of industrial clusters provide abundant raw data and new ideas. During the period of the 11th five-year planning, in order to carry out scientific development, propel the transition of economic growth, and facilitate healthy and sustainable development of the national economy, the policies of industrial clusters should be stressed in industrial development policies in China.

Zhu Yingming

Contents

CHAPTER 1

Conception Definition of Industrial Clusters

In a national economic system, extensive, complicated, and close economic links exist among industries. In industrial clusters, industries with close economic links come into being. The complication and diversification of economic links among industries generates external economies, which are the bases of regional growth and development. When many industries with close economic links form industrial clusters with each other, it provides a favorable development environment for a regional self-support growth process. Regional economic activities can attract the inflow of production factors from outside regions so that new productive activities can be introduced, production chains can be extended, and multiplier effects of regional productive activities can be remarkably heightened. With further development of regional industrial clusters, multiplier effects become quite strong, and rapid economic growth process accompanies the change in economic structures.

Although academe in China has conducted large numbers of theoretical and empirical studies about industrial clusters, so far there is not an acknowledged concept of industrial clusters, and a systemic study about industrial clusters in the national economy has not been conducted. The terminology relative to industrial clusters is industrial cluster, industrial pile-up, industrial clustering, industrial agglomeration, industrial combo, and regional production complex. The confusion of terms relative to industrial clusters has impeded its research.

1.1 BASIC CONCEPT OF INDUSTRIAL CLUSTERS

Industries are a phenomenon of the social division of labor and are generated and developed along with this social division of labor. The result of three major divisions of labor has been forming industrial sectors such as agriculture, stockbreeding, the handicraft industry, and business. Economic links among industries are from inexistence to existence, from the links among sectors of intra-industries to the links among sectors of inter-industries, from simple to complicated, from unidirectional and linear to multidirectional and network. This accelerates the development and change of industrial organization, industrial structure, and spatial structure.

With the development of social productive forces and the progress of production technologies, the internal division of labor of industries is further built up, and new industrial sectors are constantly emerging. In the 1760s, Industrial Revolution I began in Britain and mainly facilitated the development of light and the textile industrial sectors. Industrial Revolution II in the early 1900s saw unprecedented development with heavy industries beginning with iron and steel, petroleum, and chemical industries. Industrial Revolution III in the 1950s placed information technology within realms such as agriculture, industry, and the service industry to be extensively used, and many industrial sectors rapidly came forth. Links among industries developed from unidirectional to multidirectional, and from linear to network, and the extent and depth of functional links among industries are continually increasing.

In the process of industrial development, each industry as the main body of social economic activities follows the basic laws of economics, having lower transaction costs and the maximization of market benefits as the main goals. Each industry develops from relying mainly on internal scale and scope economies to external scale and scope economies so as to achieve maximization of economic benefits. Pursuing external economy is the spatial concentrating process of different industries in the same location, that is, the spatial linking process of different industries. Therefore, the development process of industries is the process of decision-making and spatial centralization in the most optimizing location, in which the extent and depth of spatial links among industries are continually strengthened.

From the point of view of spatial-temporal links of industrial development passing through the forming process and the correlation of regional industries, we can see the spatial centralization of different industrial sectors in a specific location, and they come into being as regional industrial groups take one or more industrial sectors as the core of close economic links. Industrial clusters can be summarized as relying on a definite social economic context, taking one or more industrial sectors as the core of industrial development in a specific location. With the help of close economic links (functional links) among industrial sectors in a national economy, industrial sectors of different sizes and types incur, are formed, and develop economic links among industrial sectors and form a relatively integrated industrial 'aggregation' together.

1.2 BASIC CHARACTERISTICS OF INDUSTRIAL CLUSTERS

Because industrial clusters are industrial aggregations formed by industrial sectors with close economic links in a specific territory, close economic links are not only the basic condition for forming industrial clusters, but are also important characteristics of industrial clusters. Economic links among industries within industrial clusters not only include input-output relationships, but also reflect external economic relationships. Therefore, industrial clusters have two basic characteristics—functional links and spatial links.

1.2.1 Functional links

Making an on-the-spot investigation on the formation and development process of industries, it is not difficult for us to see that functional links are material bases for industrial clusters' formation and further development. Functional links of industrial clusters are actually input-output links among industries and exhibit technical economic links among industries with input goods and output goods as linked ties. Because input and output goods can be tangible or intangible as well as representing physical goods and value, functional links of industrial clusters can take the form of physical goods and value and can be the links in the form of goods and service flows. Because economic links in the form of physical goods go against conducted quantitive studies on industrial clusters and economic links in the form of value are convenient for quantitive study on industrial clusters, we can focus even more particularly on economic technological links in the form of value in the study of industrial clusters.

Functional links of industrial clusters can be divided into three types (Su, 2000):

1.2.1.1 Links of products and services

In the forming and developing process of industries, some industrial sectors provide products or services for others or for each other. Links of products and services are the most basic functional links of industrial clusters. After industrial clusters are well established, quantities and ratios of products and services provided among industries play a very important role in the stable structure, perfect function, and efficiency enhancement of industrial clusters. Productive technological links, price links, and labor employment links in industrial clusters are the economic links derived from products and services.

1.2.1.2 Productive technological links

Productive technological links are those links which different industrial sectors have with related industries around manufacturing techniques and operating technologies according to their specific requirements of productive technologies in the formation and development of industries. For manufacturing techniques of different industrial sectors having different requirements, the performances of their product structures are also different. In the formation and development process of industrial clusters, an industrial sector does not passively accept products or services of other related industrial sectors, but has specific requirements according to the characteristics of manufacturing techniques and specific properties of product structures to ensure product qualities and technical characteristics for the industrial sector. Therefore, manufacturing techniques and operating technologies among industries have inevitable links, and the links of manufacturing techniques are closely related to the supply-demand links with products and services among industries. As an important support of links among industries, the development and changes of manufacturing techniques not only directly affect supply-demand proportion relations of products and services among industries, but also make a certain industry exchange with industries whose products and labor services are related in the process of production, and then proceed to affect the sector compositions and structural relations of industrial clusters. For example, in the initial stage of development of industrial clusters, there are a few types of industrial sectors, and structural relations among sectors are simple. With the continual development of industrial clusters and the strengthening of manufacturing technique links among industries, the types of industrial sectors increase and structural relations among industrial clusters become complicated.

1.2.1.3   Price links

Price links in industrial clusters concern the monetary expression of the magnitude of value of products and services among industries in industrial clusters. Input-output links of products and services among industries necessarily exhibit equivalent exchange relations taking money as the media. Price links among industries have important roles. First, heterogeneous links of products and services among different industries can be uniformly measured in the form of price, thereby establishing foundations for type identification of industrial clusters using input-output tables. Besides, links among industries in industrial clusters measured in the form of price provide effective metric means for structural change analyses, functional characteristic analyses, beneficial characteristic analyses, and evolution laws of industrial clusters.

1.2.2   Spatial links

Spatial links of industrial clusters are specific links in the process of obtaining external agglomeration economies among industries. They exhibit path dependency of industrial clusters on the social economic environment in given regions and are the functions of those regions. Spatial links of industrial clusters under different spatial ranks have different characteristics because space has characteristics of rank and stratification. In a given region, we can abstract it as a point in space, and spatial links among industrial clusters can be viewed as vested. Characteristics of economic links among industries in industrial clusters mostly depend on characteristics of functional links. Industrial clusters in given regions can be defined as industrial aggregations formed by different industrial sectors through functional links under the hypothesis condition. As opposed to functional links of industrial clusters, spatial links of industrial clusters do not concern technical economic links taking all kinds of input goods and output goods among industries as linked ties.

Spatial links of industrial clusters can be divided into three types (Zhu, 2003).

1.2.2.1   Localization economy links

Localization economy links are spatial links generated by different enterprises of each industrial sector for obtaining localization economy in the formation and development of industrial clusters. Localization economy as one of the agglomeration economy types applies to the economy that is internal for the industries but external for the enterprises when many enterprises of a certain industry are located in one place. Localization links result in the development of 'common supply sources' of highly specialization production factors, which can be shared by many enterprises of this industry, thereby reducing inventory levels (production factors). Further development of localization links induce scale expansion of the primary industrial sectors within industries, generate new industrial sectors, give birth to new economic links with other industrial sectors, and further accelerate the development of industrial clusters.

1.2.2.2   Urbanization economy links

Urbanization economy links are spatial links generated by each industrial sector for obtaining urbanization economy in the formation and development of industrial clusters. Urbanization economy as another agglomeration economy type concerns the economy that is internal for the industries, but external for the region when many industries are located in one place. Urbanization economy reflects the economic benefits of the bigger urban environments, in which there are greater labor markets, larger service sectors, and more potential inputs of common middle goods. Different ranks of central cities of the region attract numerous similar industrial sectors to concentrate in the region and generate close spatial links for obtaining urbanization economies.

1.2.2.3   Learning and innovation links

Learning and innovation links concern the particular learning and innovation milieus found in the formation and development of industrial clusters, namely 'Marshal industrial atmosphere,' in which links among industries exhibit intangible knowledge, technologies, and informal links. The long-term environments are convenient for while 'collective learning processes,' information, knowledge, and practice spread quickly in the whole specific area, and can't be copied and imitated by other areas, it further improves learning and innovative abilities of enterprises and decision-making organizations in specific areas. Large numbers of new enterprises can spin-off and develop, economic links among industries are further strengthened, and industrial clusters are further developed under the advantageous environment.

1.3   INFLUENCING FACTORS OF GROWTH AND DEVELOPMENT OF INDUSTRIAL CLUSTERS

Generally speaking, the growth and development of regional industrial clusters are closely linked with the regional economic system and industrial structure status, which are deeply affected by multi-factors such as the regional economic developing level, the industrial base and the extent of opening to the outside world, market developing extent, and historic tradition. As mentioned above, in given areas, spatial links in industrial clusters can be seen as vested, and economic links in industrial clusters mostly depend on functional links among industries. The growth and development of industrial clusters are influenced by structural changes of links among industries, that is, the input and output structures among industries. As time goes by, many factors may cause changes within links among industries, and then proceed to affect growth and development of industrial clusters. Here, the main factors affecting the growth and development of industrial clusters refer only to the factors leading to structural changes of links among industries.

1.3.1 Technical changes

When technology changes, the input and output structures of regional industries change so as to adapt to the requirements for technology change, and the functional link structures of industrial clusters change profoundly. For example, because of technical progress, electric power inputs of aluminum per ton decrease in the process of production. This change not only leads to the change of electric power input coefficients of aluminum, but also causes the change of other input coefficients of aluminum. Because the summation of input coefficients of aluminum production is equal to 1, relative importance of electric power in the process of producing aluminum decreases and consequentially causes an increase in the relative importance of other inputs.

Technical progress leads to the generation of new industrial sectors and new input goods, and input structures and output structures of industrial sectors tend to be diversified, which causes changes in input and output coefficients among industrial sectors. When the number of industrial sectors using products of given industrial sectors increases, output coefficients of this industrial sector become accordingly larger, while the output coefficient value of each industrial sector used becomes relatively smaller. When the input combination used in given industrial sectors becomes more diversified, each input coefficient often becomes smaller with time. It is clear that when patterns of links among industries of given industrial sectors become more complicated, relative intensity of functional links taking input and output links as features tends to weaken with time.

Of course, there is a great difference in the diversification extent of input-output structures among industrial sectors. For instance, the diversification extent of basic capital goods production sectors (such as the engineering industry) is much higher than that of raw materials production sectors (such as the coal exploitation industry). If an industrial cluster is composed of diversified industrial sectors greatly affected by changes in technology, the cluster will eventually break up, instead of forming a giant cluster with many industrial sectors and close interrelationships.

From the point of view of innovation, the motive power causing technical changes introduces continuous innovation waves, which bring a quick increase of regional labor productivity and give rise to the further development of regional specialization. One of the results of regional specialization is the spin-off of new industries and the exit of old industries in the region. Also, functional links of the region present new characteristics. Since reform and opening, the Chinese economy has presented continual and high speed growth. One of the reasons is that the effects of scale economies are based on technical change influence and the substitute effect of new input goods for old input goods in the process of middle input. New industries or industries producing new products possess important status with the functional links of industrial clusters, which can give rise to strong forward, backward and side links, and can come into being with a stronger multiplier effect in the national economy and thereby induce fast and continual development of the regional economy.

1.3.2 Changes of output levels

Changes of output levels lead to changes of relative importance about supply and purchase relations among industries and to changes of functional links among industries, and then proceed to influence growth and development of industrial clusters.

If a_{ij} denotes supply links among industries, b_{ij} denotes purchase links in industries, then

$$a_{ij}^t = \frac{x_{ij}^t}{\sum_i x_{ij}^t}$$

$$b_{ij}^t = \frac{x_{ij}^t}{\displaystyle\sum_j x_{ij}^t}$$

When the formulas above are transformed, then

$$a_{ij}^t = b_{ij}^t \frac{\displaystyle\sum_j x_{ij}^t}{\displaystyle\sum_i x_{ij}^t}$$

$$b_{ij}^t = a_{ij}^t \frac{\displaystyle\sum_i x_{ij}^t}{\displaystyle\sum_j x_{ij}^t}$$

So, the relationship between a_{ij} and b_{ij} changes with the relationship between total input quantities among industries for industry j and total output quantities among industries for industry i. Even though total input quantities among industries for industry j and total output quantities among industries for industry i increase proportionally, there exist changes between a_{ij} and b_{ij} along with changes of one side or the other. So, relative importance about any pair of supply (a_{ij}) and purchase (b_{ij}) relations of industries may change as time goes by, depending on the difference in increasing speed between two industries. In other words, if input growth of industry j is faster than output growth of industry i, then from the point of view of industry i, industry j becomes a relative important buyer. On the other hand, if input growth of industry i is faster than output growth of industry j, then from the point of view of industry i, industry j becomes a relative unimportant buyer. Therefore, relative changes about input and output scale among industries will change relative to the importance of functional links among the industries, thereby influencing the growth and development of industrial clusters as time goes by.

1.3.3   Changes of input goods prices

Changes in relative prices of input goods lead to changes in quantities and proportions of input goods, and functional links among industrial sectors change, thereby influencing growth and development of industrial clusters. Using the indifference curve analyzing method in the utility theory of western economics, we analyze the effect of changes in relative price levels on the growth and development of industrial clusters.

When the price of one kind of input goods changes, there are two effects on functional links among industries: one is that changes in the input proportions of two goods give rise to changes in their relative price; the other is that regional aggregate expenditure levels of input goods purchasing will change. Therefore, the effects of functional links among industries caused by price changes in one kind of input goods can be broken down into income effect and substituted effect. Income effect involves changes in aggregate expenditure levels of purchased input goods caused by price changes of those goods and leads to changes of links among industries. Substituted effect concerns changes in the relative price of input goods caused by price changes and leads to changes in quantities of input goods. Income effects lead to changes in levels of functional links among industries, while substituted effects do not.

Changes in relative prices among input goods will result in substitutions among input goods under the existing technical level conditions. Substitution not only takes place directly among the same kinds of input goods, but also is realized indirectly by saving raw materials due to increasing labor inputs. The effects of changes in relative prices of input goods on the growth and development of industrial clusters differ on account of the different types of input goods. The changes in functional links among industries in industrial clusters using a completely substituted effect among input goods will become complicated and diversified.

CHAPTER 2

Measurement and Identification of Economic Links of Industrial Clusters

The basic characteristics of industrial clusters are functional and spatial links among industries in industrial clusters; type identification of industrial clusters depends on the measurement of functional and spatial links among industries in industrial clusters. For this reason, correct identifying methods of industrial clusters are needed. In order to identify types of industrial clusters accurately, measuring criteria of functional and spatial links for industrial clusters are needed.

2.1 MEASUREMENT OF ECONOMIC LINKS OF INDUSTRIAL CLUSTERS

2.1.1 Measurement of spatial links

2.1.1.1 The coefficient of geographic links

Early criterion for measuring spatial links of industrial activities is introduced comparing the distributed status of the regional percentage in industrial activities. 'The coefficient of geographic links' of Paul Sargant Florence measures spatial association among industries by summing the absolute value of the difference between the regional employment proportions of any two industries within regions. His formula is as follows:

$$G_{ij} = \sum_{k=1}^{n} \left| \frac{x_{ik}}{x_i} - \frac{x_{jk}}{x_j} \right| \tag{2.1}$$

where G_{ij} = the coefficient of geographic links between industry i and industry j; x_{ik} = the number of employees of industry i in region k; x_{jk} = the number of employees of industry j in region k; x_i = the total number of employees of industry i in all n regions; x_j = the total number of employees of industry j in all n regions.

The range of the coefficient is $0 - 2$. If the two industries are not located in the same region, G_{ij} is exactly equal to 2. If all industries are located in the same region, G_{ij} is 0.

Although the coefficient can measure spatial association between industries to a certain degree, because different pairs of industries can yield the same coefficient, and the coefficient of spatial links between the same industries may be different due to different regions divided, the coefficient does not consider the localization patterns among different industries or the effect of the number and size of regions on the coefficient, which causes the coefficient to have larger limit in the practical application.

2.1.1.2 Linear zero order correlation coefficients

The linear zero order correlation coefficient is used as the measure of spatial association of a pair of industrial activities:

$$r_{ij} = \frac{\text{cov}(x_{ig}, x_{jg})}{\sigma_{x_{ig}} \sigma_{x_{jg}}} \tag{2.2}$$

where r_{ij} = the correlation coefficient between industry i and industry j; x_{ig} = the number of employees of industry i in region g; x_{jg} = the number of employees of industry j in region

g; $\sigma_{x_{ig}}$ = the standard deviation of industry i in region g; $\sigma_{x_{jg}}$ = the standard deviation of industry j in region g.

Measuring spatial links between industries using the linear zero order correlation coefficient applies to link directions and link extents. Link directions can be divided into two types: positive correlation and negative correlation. Positive correlation indicates links between industries that change in the same direction; negative correlation denotes links between industries that change in the opposite direction. As far as spatial links between industries are concerned, the range of the coefficient is -1 to $+1$, the larger the absolute value of correlation coefficients are, the closer the spatial links between two industries are (McCarty et al., 1956).

2.1.1.3   Linear regression gravitation models

The logarithm transformation of gravitation models can produce a solvable linear equation using linear regression, so the linear regression gravitation model of spatial interactions is the technique widely used to analyze spatial interactions. According to the principle of maximized personal utility under the restriction of time or budgets, gravitational models have a theoretical base concerned with human behavior, and they become a gravitational model based on human spatial movement. The general formal model is

$$\log I_{ij} = \log a + \sum_{i=1}^{r} b_k \log V_{kj} + b_{r+1} \log P_i + b_{r+2} \log P_j - \lambda \log D_{ij} \qquad (2.3)$$

where I_{ij} is the volume of business from region i to region j; V_{kj} is the score of industry k at site j; P_i and P_j are, respectively, the industrial scale of regions i and j; D_{ij} is the distance from i to j; and a, λ and b_k are the weight.

Because economic geography is concerned with the value of λ generally, therefore, without losing universality, a simplified equation is

$$\log p_{ij} = \log a - \lambda \log D_{ij} \qquad (2.4)$$

where p_{ij} is the probability of trade between i and j. If the multiplier effects of are starting point and end point are excluded, then

$$p_{ij} = I_{ij}/P_i P_j \qquad (2.5)$$

On the assumption that Formula (2.4) denotes that the probability of trade between i and j is a function of the distance between them, then Formula (2.5) has nothing to do with relative distance from selectable regions and also interactions with any regions are not the results of selectable regions acting together.

The common interaction model explains movement as the final results of common actions of selectable end points, so if distance is the only variable influencing movement, the distance from starting point to selective end point is used to define the probability of moving towards any potential end point. The concrete model is

$$p_{ij} = D_{ij}^{\lambda}/\sum_{k=1}^{n} D_{ik}^{\lambda} \qquad (2.6)$$

where p_{ij} is the probability of trade between i and j; n is the number of end points. Because D_{ik} varies with starting point i, there is not a steady relationship between p_{ij} and D_{ik}.

The intervention opportunity model has been widely used by economic geographers and sociologists; its formula is

$$\log I_{ij} = \log a + c_1 \log p_i + c_2 \log p_j - \lambda \log O_{ij} \qquad (2.7)$$

where O_{ij} is the greater number of closer selectable regions from i than from j.

The spatial selection model that has the same universality as Formula (2.3) is

$$p_i(j/T) = \prod_{k=1}^{r} V_{jk}^{bk}/D_{ij}^{\lambda} / \sum_{j=1}^{n} \left(\prod_{k=1}^{r} V_{jk}^{bk}/D_{ij}^{\lambda} \right) \tag{2.8}$$

where $p_i(j/T)$ is the probability that place i may choose end point j among possible T groups of choices. Spatial selective models aren't easy to solve equations like linear regression equations, but they can be solved with the help of the computer (Zhu, 2004).

2.1.2 Measurement of functional links

2.1.2.1 Industry flow models

Referring to W. Alonso's interaction model, The industry flow model is expressed by the following six equations (Zhang et al., 1992):

$$O_i = V_i A_i^{\alpha} \tag{2.9}$$

$$I_j = W_j B_j^{\beta} \tag{2.10}$$

$$E_{ij} = K V_i A_i^{\alpha-1} W_j B_j^{\beta-1} t_{ij} \tag{2.11}$$

$$A_i = K \sum_{j=1}^{n} W_j B_j^{\beta-1} t_{ij} \tag{2.12}$$

$$B_j = K \sum_{i=1}^{n} V_i A_i^{\alpha-1} t_{ij} \tag{2.13}$$

$$t_{ij} = (1 + u_{ij}) \delta c_{ij}^{-\gamma} \exp(\mu c_{ij}) \tag{2.14}$$

where O_i, I_j and E_{ij} are, respectively, the number of outputs, inputs and interflows for each industrial sector; V_i is the interpretative variable of outputs of each industry and is called an industrial thrust; W_j is the interpretative variable of inputs of each industry and is called an industrial pull; A_i is the total gravitation that all industries attract industry i; B_j is the total pressure that all industries flow to industry j; α, β, γ, δ, μ are constants; t_{ij} denotes the spatial relation among industries which is determined by the 'common relation' c_{ij} and the 'special relation' u_{ij}; K is constant.

The characteristic of the model is its generality and universality, when $\alpha = \beta = 1$, it becomes the traditional gravitation model; when $\alpha - \beta = 0$, it is the maximal entropy model with double restrictions; when $\alpha = 0$, $\beta = 1$, it is the distribution model sending restriction; when $\alpha=1$, $\beta = 0$ and $V_i = 1$, it becomes the input-output model; when $\alpha = 0$, $\beta = 1$ and $W_j =1$, it becomes the Markov model.

2.1.2.2 Link intensity of a pair of industries

According to Leontief's point of view, if one industrial sector sells $1/n$ or more of its outputs to another industrial sector, or if one industrial sector purchases $1/n$ or more of its inputs from the another industrial sector, then the two industrial sectors are thought to have functional links. The former functional link is called as a supply link; the latter is called a demand link (Streit, 1969).

The intensity of economic links between a pair of industries is denoted by L_{ij} and is measured by the average of four coefficients derived from the input-output tables:

$$L_{ij} = L_{ji} = \frac{1}{4}\left\{ O_{ij}\left(\frac{1}{O_i} + \frac{1}{I_i}\right) + O_{ji}\left(\frac{1}{O_j} + \frac{1}{I_j}\right) \right\} \qquad (2.15)$$

where O_{ij} stands for output flows from industry i to industry j; O_{ji} stands for output flows from industry j to industry i; O_i denotes total outputs of industry i; I_i refers to total intermediate inputs of industry i; O_j denotes total outputs of industry j; and I_j refers to total intermediate inputs of industry j.

Economic links between a pair of industries L_{ij} must be larger than the average of all links, that is

$$L_{ij} > \frac{1}{n}\sum_i L_{ij} \quad \text{or} \quad L_{ij} > \frac{1}{n}\sum_j L_{ij} \qquad (2.16)$$

2.2 IDENTIFYING METHODS OF INDUSTRIAL CLUSTERS

At present, there are mainly two methods of identifying industrial clusters: multivariable statistical analysis methods and input-output analysis methods. Industrial clusters relying mainly on spatial links are identified using multivariable statistical analysis methods, industrial clusters giving priority to functional links are identified using input-output analysis methods. Simultaneously identifying industrial clusters with functional and spatial links is necessary to integrate the two methods. The two methods are briefly introduced in the section.

2.2.1 Multivariable statistical analysis methods

In multivariable statistical analysis of identifying industrial clusters, the most commonly used methods are factor analysis and cluster analysis. Identifying industrial clusters is accounted for below using factor analysis as an example (Zhu, 2003). Concrete steps are as follows:

First, using the data of investment share of industrial sectors of foreign-invested enterprises in the Yangtze River Delta Area (YRDA), we can obtain 15 cities' investment shares matrix in YRDA (Table 2.1).

Second, factor analysis is carried out using the SPSS software package.

Third, 14 factors where industrial sectors' variance is larger than 1% in factor analysis are extracted, which can explain one hundred percentage of total variance of all sectors, thereby obtaining 14 industrial clusters of foreign-invested enterprises.

Fourth, analyses of economic links in industrial clusters are done. In every industrial cluster of foreign-invested enterprises, the strength of links between industrial sectors and each factor can be denoted by the value of factor load, the larger the value, the more intensive the links between industrial sectors and factors, thereby forming industrial clusters with close links among industrial clusters, and reflecting the status of functional links among industries in the industrial clusters (Table 2.2). The values of factor scores show the strength of linkage between cities and factors; the higher the value of scores is, the more intensive links between cities and factors are, thereby forming the spatial distribution of industrial clusters, reflecting the status of spatial links among industries in the industrial clusters (Table 2.3).

2.2.2 Input-output analysis methods

For two industries i and j, functional links between them are shown in Figure 2.1. It can be seen in Figure 2.1(a) that the link flows from industry i to industry j and contains the total output flows of industry i and the total input flows of industry j. It can be seen in Figure 2.1(b) that the link flows from industry j to industry i and contains the total output flows of industry j and the total input flows of industry i. The status of functional links between industry i and industry j depends upon the output and input coefficients between them. In Figure 2.1(a), output and input coefficients are, respectively, a_{ij} and b_{ij}, and in

Table 2.1   Investment share of industrial sectors of foreign-invested enterprises in YRDA (Units:%)

Location	Farming and forestry herd fishery	Food processing industry	Food manufacturing industry	Drink manufacturing industry	Textile industry	Clothing textile fiber industry	Leather fur industry	Lumber processing industry	Furniture manufacturing industry
	1	2	3	4	5	6	7	8	9
Nanjing	0.0047	0.0060	0.0087	0.0252	0.0206	0.0071	0.0000	0.0000	0.0055
Changzhou	0.0000	0.0261	0.0137	0.0077	0.1146	0.0443	0.0062	0.0000	0.0130
Suzhou	0.0038	0.0129	0.0140	0.0213	0.0406	0.0257	0.0051	0.0064	0.0044
Wuxi	0.0000	0.0040	0.0058	0.0032	0.0785	0.0154	0.0056	0.0062	0.0010
Nantong	0.0312	0.0236	0.0153	0.0304	0.1138	0.0211	0.0022	0.0170	0.0000
Yangzhou	0.0000	0.0000	0.0603	0.0000	0.2326	0.0000	0.0000	0.0000	0.0000
Zhenjiang	0.0023	0.0045	0.0201	0.0164	0.0189	0.0119	0.0000	0.0000	0.0000
Taizhou	0.0000	0.0339	0.0000	0.0380	0.0800	0.0000	0.0000	0.0000	0.0000
Hangzhou	0.0000	0.0031	0.0748	0.0409	0.0907	0.0138	0.0073	0.0000	0.0073
Ningbo	0.0010	0.0225	0.0071	0.0047	0.0532	0.0474	0.0057	0.0000	0.0015
Jiaxing	0.0000	0.0268	0.0000	0.0000	0.0865	0.1281	0.0230	0.0117	0.0000
Huzhou	0.0000	0.0000	0.0000	0.0000	0.0875	0.0742	0.0228	0.0000	0.0780
Shaoxing	0.0000	0.0000	0.0000	0.0053	0.4814	0.0338	0.0062	0.0000	0.0000
Zhoushan	0.0000	0.0000	0.0000	0.0000	0.3257	0.0000	0.0000	0.0000	0.0000
Shanghai	0.0016	0.0080	0.0189	0.0169	0.0201	0.0134	0.0029	0.0032	0.0018

Location	Paper making industry	Printing record industry	Sports industry	Petroleum coking	Chemistry raw material industry	Medicine manufacturing industry	Chemistry textile fiber industry	Rubber product industry	Plastic product industry
	10	11	12	13	14	15	16	17	18
Nanjing	0.0157	0.0071	0.0065	0.0228	0.1274	0.0103	0.0000	0.0000	0.0000
Changzhou	0.0271	0.0081	0.0000	0.0055	0.0558	0.0184	0.0134	0.0156	0.0507
Suzhou	0.1493	0.0015	0.0081	0.0036	0.0461	0.0377	0.0071	0.0217	0.0194
Wuxi	0.0082	0.0156	0.0000	0.0000	0.0715	0.0308	0.0194	0.0051	0.0348
Nantong	0.0000	0.0000	0.0000	0.0480	0.0935	0.0000	0.2225	0.0000	0.0335
Yangzhou	0.0000	0.0000	0.0000	0.0232	0.0000	0.0000	0.0000	0.0000	0.0000
Zhenjiang	0.5906	0.0000	0.0000	0.0110	0.0217	0.0000	0.0000	0.0037	0.0120
Taizhou	0.0000	0.0000	0.0000	0.0000	0.0137	0.0825	0.0000	0.0000	0.0000
Hangzhou	0.0474	0.0258	0.0118	0.0000	0.0506	0.0386	0.0068	0.0452	0.0182
Ningbo	0.1256	0.0028	0.0070	0.1839	0.0281	0.0013	0.0000	0.0013	0.0179
Jiaxing	0.0580	0.0314	0.0000	0.0262	0.0540	0.0000	0.0000	0.1269	0.0043
Huzhou	0.0785	0.0712	0.0000	0.0000	0.0898	0.0245	0.0000	0.0000	0.0000
Shaoxing	0.0000	0.0000	0.0000	0.0000	0.0036	0.0042	0.0883	0.0071	0.0247
Zhoushan	0.0000	0.0000	0.0000	0.0000	0.0000	0.0000	0.0000	0.0000	0.3070
Shanghai	0.0090	0.0102	0.0088	0.0036	0.0497	0.0125	0.0033	0.0032	0.0145

Continued

Location	Nonmetallic product industry 19	Metal smelting industry 20	Metal product industry 21	Ordinary machinery industry 22	Special-purpose equipment manufacturing industry 23	Transportation shipping industry 24	Electricity machinery industry 25	Electron correspondence industry 26	Instrument measuring appliance industry 27
Nanjing	0.1226	0.0065	0.0100	0.0362	0.0047	0.0989	0.0381	0.1977	0.0025
Changzhou	0.0261	0.0494	0.0449	0.0419	0.0275	0.0666	0.0814	0.0572	0.0000
Suzhou	0.0560	0.0255	0.0497	0.0393	0.0187	0.0213	0.0648	0.1306	0.0074
Wuxi	0.0195	0.1173	0.0602	0.0462	0.0357	0.0236	0.0882	0.0603	0.0139
Nantong	0.0205	0.0334	0.0106	0.0127	0.0024	0.0871	0.0091	0.0303	0.0000
Yangzhou	0.0281	0.0000	0.0513	0.1359	0.0000	0.0565	0.0848	0.0727	0.0315
Zhenjiang	0.1564	0.0078	0.0160	0.0092	0.0091	0.0111	0.0081	0.0098	0.0020
Taizhou	0.0191	0.0161	0.0000	0.0347	0.0000	0.1685	0.3204	0.0000	0.0078
Hangzhou	0.0271	0.0013	0.0180	0.0337	0.0416	0.0451	0.0643	0.0451	0.0018
Ningbo	0.0143	0.0029	0.0829	0.0192	0.0198	0.0373	0.0405	0.0075	0.0000
Jiaxing	0.0356	0.0149	0.0067	0.0577	0.0000	0.0000	0.0418	0.0176	0.0000
Huzhou	0.2310	0.0000	0.0159	0.0000	0.0000	0.0000	0.0341	0.0000	0.0000
Shaoxing	0.0000	0.0035	0.0135	0.0329	0.0340	0.0102	0.0000	0.0101	0.0000
Zhoushan	0.0000	0.0000	0.0000	0.0000	0.0000	0.0000	0.0000	0.0000	0.0000
Shanghai	0.0232	0.0248	0.0281	0.0386	0.0260	0.0605	0.0280	0.0754	0.0082

Location	Other manufacturing industry 28	Electric power coal gas 29	Architecture industry 30	Transportation stores in a storehouse 31	Wholesale retail trade 32	Finance insurance industry 33	Real estate 34	Society service industry 35	Health sports industry 36	Other industry 37
Nanjing	0.0000	0.0230	0.0011	0.0061	0.0173	0.0000	0.0959	0.0663	0.0028	0.0025
Changzhou	0.0111	0.0329	0.0211	0.0341	0.0000	0.0000	0.0785	0.0069	0.0000	0.0000
Suzhou	0.0101	0.0162	0.0045	0.0117	0.0018	0.0000	0.0597	0.0383	0.0000	0.0157
Wuxi	0.0008	0.0892	0.0071	0.0300	0.0116	0.0000	0.0427	0.0356	0.0000	0.0126
Nantong	0.0028	0.0606	0.0000	0.0305	0.0059	0.0000	0.0213	0.0160	0.0000	0.0048
Yangzhou	0.0000	0.1890	0.0000	0.0000	0.0000	0.0000	0.0000	0.0340	0.0000	0.0000
Zhenjiang	0.0041	0.0189	0.0073	0.0032	0.0000	0.0000	0.0099	0.0139	0.0000	0.0000
Taizhou	0.0000	0.1430	0.0000	0.0000	0.0269	0.0000	0.0000	0.0156	0.0000	0.0000
Hangzhou	0.0096	0.0314	0.0090	0.0258	0.0000	0.0000	0.0633	0.0995	0.0011	0.0000
Ningbo	0.0035	0.0320	0.0166	0.0520	0.0285	0.0218	0.0377	0.0533	0.0000	0.0192
Jiaxing	0.0000	0.1769	0.0000	0.0253	0.0000	0.0000	0.0309	0.0159	0.0000	0.0000
Huzhou	0.0000	0.0344	0.1047	0.0000	0.0128	0.0000	0.0000	0.0000	0.0408	0.0000
Shaoxing	0.0098	0.1377	0.0100	0.0038	0.0000	0.0000	0.0248	0.0551	0.0000	0.0000
Zhoushan	0.0000	0.0000	0.0000	0.0000	0.1517	0.0000	0.0000	0.2157	0.0000	0.0000
Shanghai	0.0054	0.0053	0.0128	0.0432	0.0283	0.0029	0.2127	0.0420	0.0015	0.1314

Table 2.2 The status of functional links of industrial clusters of foreign-invested enterprises in YRDA

Types of industrial clusters	The status of functional links within industrial clusters
1. Industrial cluster of furniture manufacturing industry, health sports industry, social welfare industry and architecture industry	9(0.988), 36(0.978), 30(0.976), 11(0.870), 19(0.698), 7(0.619), 14(0.380), 6(0.339)
2. Industrial cluster of medicine manufacturing industry, electric machinery and equipment manufacturing industry	15(0.917), 25(0.904), 24(0.729), 4(0.632), 2(0.599), 5(-0.377)
3. Industrial cluster of production and supply of electric power, coal, gas and water	29(0.439), 22(0.327), 35(-0.840), 32(-0.908), 18(-0.919)
4. Industrial cluster of rubber product industry, clothing and other textile fiber products industry	17(0.971), 6(0.829), 7(0.737), 8(0.443), 11(0.402), 2(0.362), 29(0.302), 24(-0.343)
5. Industrial cluster of finance insurance industry and petroleum coking	33(0.969), 13(0.964), 21(0.697), 31(0.619), 2(0.308)
6. Industrial cluster of agricultural product and chemistry textile fiber industry	1(0.958), 16(0.894), 8(0.826), 14(0.341)
7. Industrial cluster of other industry and real estate	37(0.984), 34(0.878), 31(0.489), 12(0.464), 29(-0.313), 5(-0.319)
8. Industrial cluster of instrument measuring appliance industry, culture office supplies and ordinary machinery industry	27(0.866), 22(0.790), 21(0.446), 29(0.418), 3(0.348), 14(-0.301),4(-0.328)
9. Industrial cluster of electron correspondence industry, chemistry raw material industry and chemical product manufacturing industry	26(0.922), 14(0.692), 12(0.443), 34(0.362), 29(-0.329), 5(-0.416)
10. Industrial cluster of other manufacturing industry	28(0.954), 23(0.648), 12(0.333)
11. Industrial cluster of textile industry	5(0.566), 29(0.404), 19(-0.464), 10(-0.895)
12. Industrial cluster of black and nonferrous metal smelting and rolling processing industry	20(0.949), 23(0.464), 21(0.387), 31(0.382)
13. Industrial cluster of food manufacturing industry	3(0.891), 12(0.439), 4(0.403)
14. Industrial cluster of culture and education sporting goods manufacturing industry	12(0.322), 2(-0.411)

Notes: the figure in the bracket is factor load, the figure before bracket is industrial sector codes

Table 2.3 The status of spatial links of industrial clusters of foreign-invested enterprises in YRDA

Types of industrial clusters	The status of spatial links of industrial clusters	
	The city with the strongest link	The city with the weakest link
1. Industrial cluster of furniture manufacturing industry, health sports industry, social welfare industry and architecture industry	Huzhou	Shaoxing
2. Industial cluster of medicine manufacturing industry, electricity machinery and equipment manufacturing industry	Taizhou	Shaoxing
3. Industrial cluster of production and supply of electric power, coal, gas and water	Shaoxing	Zhoushan
4. Industrial cluster of rubber product industry, clothing and other textile fiber products industry	Jiaxing	Zhenjiang
5. Industrial cluster of finance insurance industry and petroleum coking	Ningbo	Shaoxing
6. Industrial cluster of agricultural product and chemistry textile fiber industry	Nantong	Nanjing
7. Industrial cluster of other industry and real estate	Shanghai	Yangzhou
8. Industrial cluster of instrument measuring appliance industry, culture office supplies and ordinary machinery industry	Yangzhou	Nanjing
9. Industrial cluster of electron correspondence industry, chemistry raw material industry and chemical product manufacturing industry	Nanjing	Shaoxing
10. Industrial cluster of other manufacturing industry	Suzhou	Nanjing
11. Industrial cluster of textile industry	Shaoxing	Zhenjiang

Continued

Types of industrial clusters	The status of spatial links of industrial clusters	
	The city with the strongest link	The city with the weakest link
12. Industrial cluster of black and nonferrous metal smelting and rolling processing industry	Wuxi	Yangzhou
13. Industrial cluster of food manufacturing industry	Hangzhou	Suzhou
14. Industrial cluster of culture and education sporting goods manufacturing industry	Suzhou	Changzhou

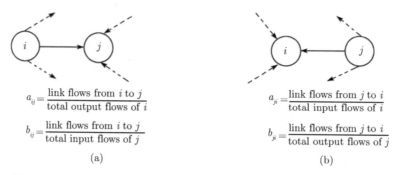

$$a_{ij} = \frac{\text{link flows from } i \text{ to } j}{\text{total output flows of } i}$$ $\qquad$ $$a_{ji} = \frac{\text{link flows from } j \text{ to } i}{\text{total input flows of } i}$$

$$b_{ij} = \frac{\text{link flows from } i \text{ to } j}{\text{total input flows of } j}$$ $\qquad$ $$b_{ji} = \frac{\text{link flows from } j \text{ to } i}{\text{total output flows of } j}$$

(a) $\qquad\qquad$ (b)

Figure 2.1   The modes of functional links between a pair of industries

Figure 2.1(b), output and input coefficients are, respectively, b_{ji} and a_{ji}; then

$$a_{ij} = \frac{x_{ij}}{\sum_j x_{ij}} \tag{2.17}$$

$$a_{ji} = \frac{x_{ji}}{\sum_i x_{ji}} \tag{2.18}$$

$$b_{ij} = \frac{x_{ij}}{\sum_i x_{ij}} \tag{2.19}$$

$$b_{ji} = \frac{x_{ji}}{\sum_j x_{ji}} \tag{2.20}$$

Next, $n \times n$ relationship matrixes, $n \times n$ covariance matrixes and correlation coefficient matrixes are built up by selecting the highest value of the four coefficients for each pair of industries.

In order to identify the types of industrial clusters in a national economy, the relative strength of the links of all industries can be used to evaluate them. Calculations are as follows:

$$(\boldsymbol{R}a - Ia\lambda) = 0$$

$$(\boldsymbol{R} - I\lambda)a = 0$$

$$|\boldsymbol{R} - I\lambda| = 0$$

where a = the eigenvectors; λ = eigenvalues or characteristic roots. The ratio of the characteristic roots to the traces of the $\boldsymbol{R}$ matrix define indexes of links:

$$C_i = \frac{\lambda_i}{tr\boldsymbol{R}} \times 100\%$$

The size and strength of functional links among industries, C_i, can determine whether industrial clusters exist. Generally speaking, larger C_i values indicate the existence of identified industrial clusters (Roepke et al., 1974).

In identified industrial clusters, if two industries lack direct links, k and l, there may exist indirect functional links between them. We can use four types of correlation coefficients between them and industry i to describe input-output structures between them:

$$r\left(a_{ik}, a_{il}\right) \tag{2.21}$$

$$r\left(b_{ki}, a_{li}\right) \tag{2.22}$$

$$r\left(a_{ik}, b_{li}\right) \tag{2.23}$$

$$r\left(b_{ki}, a_{il}\right) \tag{2.24}$$

A high $r\left(a_{ik}, a_{il}\right)$ coefficient indicates that the two industries, k and l, have similar input structures or use supplies from the same producers. A high $r\left(b_{ki}, a_{li}\right)$ coefficient signifies that the two industries, k and l, supply their products to similar sets of users. A high $r\left(a_{ik}, b_{li}\right)$ coefficient implies that suppliers of industry k are users of products of the industry l. A high $r\left(b_{ki}, a_{il}\right)$ coefficient denotes that the suppliers of industry l are users of the products of industry k (Czamanski, 1971).

2.2.3   Graph theory and network flow theory methods

2.2.3.1   Graph theory methods

Identifying industrial clusters using the graph theory (Wang, 1997) method is carried out with the data of input-output tables. Applying the results of an empirical study by Champbell (1970), we present the basic steps of the method:

First, establishing adjoining matrixes. To translate the middle-exchange parts in input-output tables into adjoining matrixes is: if there exist goods or service flows between industries i and j, that is, $x_{ij} > 0$, then there exists $x_{ij} = x_{ji} = 1$ between them, otherwise there exists $x_{ij} = x_{ji} = 0$ (the value in main diagonal being equal to 0).

Second, constructing directed graphs. Every industry in input-output tables is denoted as a node, so goods or service flows from i to j with a unit element in input-output tables are shown by an arc (ring or side) between two nodes connecting the direction from i to j. The extent of functional links and network status of structures among industries in a region can be indicated by the ratios of the number of arcs to the number of maximal possible arcs in the directed graphs, of which n is the number of industries. The ratios are used to depict connectivities as strong or weak. The more connectivities, the more close functional links among the industries, and the higher networking levels of structures among the industries.

Third, constructing distance matrixes. As long as directed graphs from i to j exist, node i is accessible for node j. Distance matrixes are constructed according to the number of the shortest steps (distances) from industry i to industry j in adjoining matrixes. For example, if industry i sold its output to industry k, industry k to industry l and industry l to industry j, then the number of distances between i and j is 3. The ratios of the number of all distances in distance matrixes to the number of steps of each row (industry) are usually used to measure the relative centricity of an industry. The higher the relative centricity, the higher the central status of the industry in its system.

Fourth, identifying industrial clusters. In a special directed graph, if eliminating a given node splits the directed graph into many subgraphs, the node is called nexus and the subgraphs are called connecting branches or strong branches. Therefore, directed graphs are divided into many subgraphs composed of nodes accessible to one another. An industrial aggregation composed of the subgraphs is indentified as an industrial cluster.

Industrial clusters with close functional links can be quickly identified by this method. But defects of the method also exist because much useful information is lost in the process of translating input-output tables into directed graphs, especially that the values of zero and nonzero are considered in the adjoining matrixes and that the size of actual functional links among industries is not considered.

2.2.3.2   Network flow theory methods

Methods of identifying industrial clusters using the algorithm of maximal flow and minimum cut in network flow theory (Wang, 1997) can refer to the empirical study of Slater (1977). The method is as follows:

Functional links among industries can be denoted by matrixes of goods and service flows (called flow matrixes for short). Flow matrixes among industries can be regarded as a network in which each industry is expressed by a point (node). If industry j purchases any output from industry i, an arc section is extended from the node of industry i to that of industry j. Each section of the arc is denoted as trades among industries, that is x_{ij}, which is regarded as the ability flowing through the arc. If every section of arc has a direction, the nodes from which goods and service flow are called sources (industry i). If each section of the arc points to the same nodes, the nodes are called sinks (industry j). Therefore, sources are regarded as generators of the goods and service flows, and sinks as absorbers of them.

If there is only one source (industry i) and one sink (industry j) in a certain network, the objective of problem of maximizing flow is making total flows from sources to sinks through the network in order to confirm feasible patterns. If the arc sections used to traffic capacities are deleted, the network is decomposed into two types of self-complementing nodes: one including sources (industry i), the other including sinks (industry j). The sum of flows of those arc sections is equal to the sum of maximal flows from the sources to the sinks. The cuts decomposing network are the aggregate of those arc sections used to traffic capacities, numerical value of corresponding cuts is equal to maximal flows. One of the two types of self-complementing node is usually composed of source nodes or sink nodes and all other residual and important nodes [$(n-1)$ industries].

If the maximal flows from sources (i) to sinks (j) is lower than the total traffic capacities of all arc sections emitted from industry i (total outputs of industry i) and is also lower than the total traffic capacities of all arc sections pointed to industry j (total consumptions of industry j), then the nodes are usually grouped into productive industrial clusters of sources (i) or consumptive industrial clusters of sinks (j).

Therefore, productive industrial clusters composed of m members which have total outputs of P among industries for source node i sell the outputs which are lower than P to other $n-m$ industries. Consumptive industrial clusters composed of $n-m$ members which have total inputs of C among industries for sink node j purchase the outputs which are lower than C from productive industrial clusters.

We demonstrate the method for identifying for industrial clusters using network flow theory below. In Table 2.4, industry 6 is specified as the source node and industry 2 is specified as the sink node. Industries 1, 3, 5 and 6 form a productive cluster because the quantity, that is, 2160, sold to industries 2 and 4 is lower than the total output 2583, of industry 6 which is the node of the productive industrial cluster. On the other hand, industries 2 and 4 form a consumptive cluster because the quantity, that is, 2160, purchased from industries 1, 3, 5 and 6 is lower than the total input of 2677, of industry 2 which is the node of the consumptive industrial cluster (Shirasu, 1980).

We demonstrate the method to identify the industry cluster that uses the network flow theory in the following example. In Table 2.4, industry 6 is specified as the source node and industry 2 is specified as the sink node. Industries 1, 3, 5 and 6 form a productive cluster because the production that they sold to industries 2 and 4 is less than the sum of production from the source node (industry 6). On the other hand, industries 2 and 4 compose a consumptive cluster, because the consumption that they purchase from industries 1, 3, 5 and 6 is less than the sum of consumption to the sink node (industry 2).

Table 2.4   Demonstrating identifying industrial clusters by network flow theory using simple input–output table

Middle Input	Middle Demand						
	Agriculture, fishery and food industry	Metal and architecture industry	Textile and clothing industry	Mining, chemistry and public utilities	Trade industry	Service industry	Total
Agriculture, fishery and food industry		0	39	175	1	352	567
Metal and architecture industry	422		62	399	163	856	1902
Textile and clothing industry	25	38		55	21	39	178
Mining, chemistry and public utilities	700	1438	322		448	704	3612
Trade industry	404	685	131	274		89	1583
Service industry	268	516	141	417	1241		2583
Total	1819	2677	695	1320	1874	2020	

CHAPTER 3

Type Identification of Industrial Clusters in China

In Chapter 2, we briefly introduced measurements and identified methods of functional and spatial links in general industrial clusters. To identify industrial clusters in China elaborate studies are needed, not only because circumstances of growth and development of industrial clusters in China are particular, but also because identifying types of industrial clusters in China should be especially concerned with the acquirability, comparability and tractability of data. This chapter discusses identifying methods of quasi-industrial clusters in China, identifying principles of industrial clusters in China, and basic types of industrial clusters in China.

3.1 IDENTIFYING METHODS OF QUASI-INDUSTRIAL CLUSTERS IN CHINA

Studying industrial clusters of the Chinese national economy is conducted in the spatial scope of China. We can abstract the area as a spatial point, spatial links among industries can be viewed as invariable, economic links among industries in industrial clusters give priority to functional links, and identifying the problems of industrial clusters in China then becomes identifying the problems of industrial clusters with close functional links. Identifying industrial clusters in China can use input-output analyzing methods. We can use the measuring criteria of inter-industry links, input coefficients b_{ij} and a_{ji} and output coefficient a_{ij} and b_{ji}, to identify industrial clusters in China. Industrial clusters identified by these methods are still incomplete industrial clusters, and we require other principles to complement and modify them, so we refer to industrial clusters identified by this method as quasi-industrial clusters. The Chinese input-output table from 1997 is taken as the basic data for identifying quasi-industrial clusters, using an example in this section to explain the method of identifying quasi-industrial clusters in China.

3.1.1 Data of identifying quasi-industrial clusters in China

We will use the 'middle use' and 'middle input' sections of the Chinese input-output table from 1997 (Appendix 2) as the basic data for identifying quasi-industrial clusters in China. Measuring criteria for identifying industrial clusters are the inter-industrial, economic links (e), namely, input coefficients (b_{ij} and a_{ji}) and output coefficients (a_{ij} and b_{ji}):

$$a_{ij} = \frac{x_{ij}}{\sum_j x_{ij}}, a_{ji} = \frac{x_{ji}}{\sum_i x_{ji}}, b_{ij} = \frac{x_{ij}}{\sum_i x_{ij}}, b_{ji} = \frac{x_{ji}}{\sum_j x_{ji}}$$

Using the data of Appendix 2, we can work out the input coefficients (b_{ij} and a_{ji}) of the inter-industrial economic link matrix (124×124) (Appendix 3) and the output coefficients (a_{ij} and b_{ji}) of the inter-industrial economic link matrix (124×124) (Appendix 4) from 1997.

3.1.2 The trianglization method of identifying quasi-industrial clusters in China

Referring to research findings of former scholars (Roepke et al., 1974; Czamanski, 1974; Shirasu, 1980; Malmberg et al., 1996; DeBresson et al.,1999), and using an

Table 3.1   The trianglization method of identifying quasi-industrial cluster 3

Nos.	003	016	014	001	039	015	024	The average of e_{ij}
003	0							
016	**0.80735**	0						0.40368
014	**0.44815**	0.01507	0					0.42352
001	0.41670	0.00140	**0.72326**	0				0.60298
039	0.00021	0	0.00094	**0.85779**	0			0.65417
015	0	0.00278	0.01163	**0.68966**	0.00006	0		0.66250
024	0.08885	0.00006	0.01270	**0.50132**	0.00080	0.00161	0	0.65433

example for quasi-industrial cluster 3 (Table 3.1), the steps explaining the trianglization method of identifying quasi-industrial clusters in China are as follows:

First, we begin with a random industry of the 124 industries in the Chinese input-output table from 1997. Suppose that industry k is selected as an initial industry and is placed in column 1 row 2 and row 1 column 2 in the lower triangular matrix $\boldsymbol{E}^t$. For example, in the identification of quasi-industrial cluster 3, livestock and livestock products (003) is first placed in column 1 row 2 and row 1 column 2.

Second, industry l coefficients corresponding to the maximal value of input and output coefficients of all industries which have close links with k (e $\geqslant$ 0.20 in input coefficient and output coefficient matrixes) are placed in column 1 row 3 and row 1 column 3. For example, slaughtering, meat processing, eggs and dairy products (016) which are close links with livestock and livestock products (003) (the maximum of coefficient b_{ij} is 0.80735) are placed in column 1 row 3 and row 1 column 3.

Third, searching the maximum of input and output coefficients of all industries (except industry k) having close links with industry l (e $\geqslant$ 0.20) and comparing the coefficient with the second maximum of input and output coefficient of all industries (except for industry l) having close links with industry k, we find that industry g corresponds to the larger of the two coefficients and this is placed in column 1 row 4 and row 1 column 4. Industry k, l and g form a quasi-industrial cluster. For example, the industry which has the closest links with slaughtering, meat processing, eggs and dairy products (016) is leather, furs, down and related products (029) and its maximal coefficient a_{ij} is 0.44034, while the industry which has the sub-closest links with livestock and livestock products (003) is grain mill products, vegetable oil and forage (014) and its maximal coefficient b_{ji} is 0.44815. Comparing the two coefficients, the industry forming a new quasi-industrial cluster with industries 003 and 016 is industry 014.

Fourth, we search the maximum of input and output coefficients of all industries (except industries k, l and g) having close links with industry g (e $\geqslant$ 0.20) and compare the coefficient with the input or output coefficient having the closest links with other industries (except industries k and l) in quasi-industrial clusters and all industry members of non-quasi-industrial clusters; then industry h, corresponding to the maximal input or output coefficient, is placed in column 1 row 5 and row 1 column 5. For example, the industry which has the closest links with industry 014 is crop cultivation (001), and its maximal coefficient a_{ji} is 0.72326. The non-industrial cluster member industry which has the closest links with livestock and livestock products (003) is wool textiles (023), and its maximal coefficient b_{ij} is 0.28701. The non-industrial cluster member industry which has the closest links with slaughtering, meat processing, eggs and dairy products (016) is leather, furs, down and related products (029), and its maximal coefficient a_{ij} is 0.44034. When we compare the three coefficients, we find that the industry that can be a new member of the quasi-industrial cluster is crop cultivation (001).

Fifth, repeat the above processes until the average of inter-industrial links contained in the quasi-industrial cluster begins to drop. The average is defined as

$$\sum_i \sum_j e_{ij}^t / m$$

where $i, j = 1, 2, \ldots, m$; m is the number of industries in industrial clusters. Here, the quasi-industrial cluster is composed of m industries.

Sixth, in the process of selecting industries of quasi-industrial clusters, the values of inter-industrial linkage e_{ij} in the lower triangular matrix $\boldsymbol{E}^t$ are filled. Similarly, these link values are the maximum in input and output coefficients. The value of e_{ij} between the same industries is 0, namely, the value of primary diagonal is 0. In Table 3.1, the values e_{ij} with panes are the maximal e_{ij} values during the processes of selecting quasi-industrial clusters and other values of e_{ij} are the maximal values during the processes of filling.

Seventh, when the method is applied to residual $(124 - m)$ industries by selecting different initiatial industries, we can identify other types of quasi-industrial clusters.

3.1.3 Basic types of quasi-industrial clusters in China

Using the above triangular method of identifying quasi-industrial clusters in China, we can identify 41 types of quasi-industrial clusters (Tables 3.2–3.42). We can see from these quasi-industrial clusters that there are similar industrial compositions for some quasi-industrial clusters, while some industrial clusters do not contain all industries with close links. Therefore, some basic principles are needed to synthetically induce types of quasi-industrial clusters in China and subjoin or delete industrial compositions of industrial clusters in order to form industrial clusters in China.

Table 3.2   Quasi-industrial cluster 1 in China

Nos.	001	039	014	015	122	040	024	Average e_{ij}
001	0							
039	0.8578	0						0.4289
014	0.7233	0.0009	0					0.5273
015	0.6897	0.0001	0.0116	0				0.5708
122	0.5832	0.0017	0.1071	0	0			0.5951
040	0.5808	0.0033	0.0058	0.0001	0.0079	0		0.5955
024	0.5013	0.0008	0.0127	0.0016	0.0003	0	0	0.5843

Table 3.3   Quasi-industrial cluster 2 in China

Nos.	002	013	046	032	033	100	077	078	076	114	120	020	Average e_{ij}
002	0												
013	0.3459	0											0.1730
046	0.2948	0.0572	0										0.2326
032	0.0275	0.2870	0.0101	0									0.2556
033	0.0017	0.0005	0.0071	0.5073	0								0.3078
100	0.0510	0.0577	0.0572	0.0689	0.2616	0							0.3393
077	0.0001	0.0006	0.0017	0.0145	0.0038	0.5216	0						0.3683
078	0	0.0002	0.0114	0.0170	0.0066	0.0636	0.4865	0					0.3954
076	0.0001	0.0008	0.0049	0.0052	0.0007	0.0478	0.0010	0.4586	0				0.4092
114	0.0108	0.0033	0.0016	0.0390	0.0882	0.1549	0.0134	0.0213	0.4251	0			0.4440
120	0.0038	0.0001	0.0024	0.0088	0.0118	0.0532	0.0104	0.0649	0.0230	0.3651	0		0.4530
020	0.0005	0.0001	0.0015	0.0444	0.0160	0.2962	0.0006	0.0001	0.0001	0.0437	0.0024	0	0.4491

Table 3.4   Quasi-industrial cluster 3 in China

Nos.	003	016	014	001	039	015	024	Average e_{ij}
003	0							
016	0.8074	0						0.4037
014	0.4481	0.0151	0					0.4235
001	0.4167	0.0014	0.7233	0				0.6030
039	0.0002	0	0.0009	0.8578	0			0.6542
015	0	0.0028	0.0116	0.6897	0.0001	0		0.6625
024	0.0888	0.0001	0.0127	0.5013	0.0008	0.0016	0	0.6543

Table 3.5   Quasi-industrial cluster 4 in China

Nos.	004	017	069	101	014	001	039	015	122	040	Average e_{ij}
004	0										
017	0.8030	0									0.4015
069	0.4700	0	0								0.4243
101	0.2464	0.3721	0.0093	0							0.4752
014	0.2520	0.3260	0	0.0584	0						0.5074
001	0.1354	0.0003	0.0045	0.0451	0.7233	0					0.5743
039	0	0	0	0.0054	0.0009	0.8578	0				0.6157
015	0	0.0005	0	0.0412	0.0116	0.6897	0.0001	0			0.6316
122	0.0966	0.0011	0	0.0291	0.1071	0.5832	0.0017	0	0		0.6524
040	0	0	0	0.0110	0.0058	0.5808	0.0033	0.0001	0.0079	0	0.6481

Table 3.6   Quasi-industrial cluster 5 in China

Nos.	005	001	039	014	015	122	040	024	028	Average e_{ij}
005	0									
001	0.2976	0								0.1488
039	0.0001	0.8578	0							0.3851
014	0.0082	0.7233	0.0009	0						0.4719
015	0.0005	0.6897	0.0001	0.0116	0					0.5179
122	0.0231	0.5832	0.0017	0.1071	0	0				0.5508
040	0.0008	0.5808	0.0033	0.0058	0.0001	0.0079	0			0.5576
024	0.0723	0.5013	0.0008	0.0127	0.0016	0.0003	0	0		0.5616
028	0.0114	0.0028	0.0058	0.0062	0.0012	0.0047	0.0019	0.5029	0	0.5588

Table 3.7   Quasi-industrial cluster 6 in China

Nos.	006	037	087	055	056	057	009	Average e_{ij}
006	0							
037	0.4607	0						0.2303
087	0.3677	0.0098	0					0.2794
055	0.0388	0.3591	0.0019	0				0.3095
056	0.0133	0.0682	0.0015	0.6146	0			0.3871
057	0.0898	0.1828	0.0156	0.1303	0.3901	0		0.4574
009	0.0060	0.0054	0.0004	0.1724	0.0750	0.3648	0	0.4812

Table 3.8   Quasi-industrial cluster 7 in China

Nos.	007	036	093	104	Average e_{ij}
007	0				
036	0.8186	0			0.4093
093	0.0050	0.4909	0		0.4382
104	0.0001	0.3740	0.0059	0	0.4236

Table 3.9   Quasi-industrial cluster 8 in China

Nos.	008	039	001	014	015	122	040	024	Average e_{ij}
008	0								
039	0.3798	0							0.1899
001	0	0.8578	0						0.4125
014	0	0.0009	0.7233	0					0.4904
015	0.0001	0.0001	0.6897	0.0116	0				0.5326
122	0	0.0017	0.5832	0.1071	0	0			0.5592
040	0.0013	0.0033	0.5808	0.0058	0.0001	0.0079	0		0.5649
024	0.0002	0.0008	0.5013	0.0127	0.0016	0.0003	0	0	0.5589

Table 3.10   Quasi-industrial cluster 9 in China

Nos.	010	059	060	075	099	Average e_{ij}
010	0					
059	0.4539	0				0.2269
060	0.1079	0.6258	0			0.3958
075	0.0095	0.1870	0.4359	0		0.4550
099	0.0025	0.0287	0.0021	0.2559	0	0.4218

Table 3.11   Quasi-industrial cluster 10 in China

Nos.	012	052	090	049	050	048	123	053	Average e_{ij}
012	0								
052	0.2664	0							0.1332
090	0.2496	0.5918	0						0.3692
049	0.0656	0.0062	0.8580	0					0.5094
050	0.1284	0.0083	0.7365	0.0115	0				0.5844
048	0.0969	0.0112	0.6900	0.2906	0.0327	0			0.6739
123	0.0162	0.0017	0.5965	0.0428	0.0054	0.0317	0		0.6768
053	0.1205	0.0217	0.3536	0.0146	0.0343	0.0604	0.0002	0	0.6679

Table 3.12   Quasi-industrial cluster 11 in China

Nos.	018	015	001	039	014	122	040	Average e_{ij}
018	0							
015	0.4734	0						0.2367
001	0.2261	0.6897	0					0.4630
039	0.0014	0.0001	0.8578	0				0.5621
014	0.1680	0.0116	0.7233	0.0009	0			0.6304
122	0.0011	0	0.5832	0.0017	0.1071	0		0.6409
040	0.0198	0.0001	0.5808	0.0033	0.0058	0.0079	0	0.6376

Table 3.13   Quasi-industrial cluster 12 in China

Nos.	019	001	039	014	015	122	040	024	Average e_{ij}
019	0								
001	0.3926	0							0.1963
039	0.0019	0.8578	0						0.4174
014	0.0902	0.7233	0.0009	0					0.5167
015	0.0388	0.6897	0.0001	0.0116	0				0.5614
122	0.0009	0.5832	0.0017	0.1071	0	0			0.5833
040	0.0046	0.5808	0.0033	0.0058	0.0001	0.0079	0		0.5860
024	0.0027	0.5013	0.0008	0.0127	0.0016	0.0003	0	0	0.5777

Table 3.14   Quasi-industrial cluster 13 in China

Nos.	020	001	039	014	015	122	040	Average e_{ij}
020	0							
001	0.3270	0						0.1635
039	0	0.8578	0					0.3949
014	0.0449	0.7233	0.0009	0				0.4885
015	0.2553	0.6897	0.0001	0.0116	0			0.5821
122	0.0007	0.5832	0.0017	0.1071	0	0		0.6005
040	0	0.5808	0.0033	0.0058	0.0001	0.0079	0	0.6002

Table 3.15   Quasi-industrial cluster 14 in China

Nos.	021	001	039	014	015	122	040	024	Average e_{ij}
021	0								
001	0.3365	0							0.1682
039	0	0.8578	0						0.3981
014	0	0.7233	0.0009	0					0.4796
015	0	0.6897	0.0001	0.0116	0				0.5240
122	0.0008	0.5832	0.0017	0.1071	0	0			0.5521
040	0	0.5808	0.0033	0.0058	0.0001	0.0079	0		0.5586
024	0.0035	0.5013	0.0008	0.0127	0.0016	0.0003	0	0	0.5538

Table 3.16   Quasi-industrial cluster 15 in China

Nos.	022	027	001	039	014	015	122	040	024	028	018	Average e_{ij}
022	0											
027	0.3740	0										0.1870
001	0.0835	0.4961	0									0.3179
039	0.0079	0	0.8578	0								0.4548
014	0.0016	0	0.7233	0.0009	0							0.5090
015	0.0006	0	0.6897	0.0001	0.0116	0						0.5412
122	0.0003	0	0.5832	0.0017	0.1071	0	0					0.5628
040	0.0005	0	0.5808	0.0033	0.0058	0.0001	0.0079	0				0.5672
024	0.0262	0.1647	0.5013	0.0008	0.0127	0.0016	0.0003	0	0			0.5828
028	0.2742	0.1901	0.0028	0.0058	0.0062	0.0012	0.0047	0.0019	0.5029	0		0.6235
018	0.0045	0	0.2261	0.0014	0.1680	0.4734	0.0011	0.0198	0.0041	0.0039	0	0.6489
003	0.0004	0.0977	0.4167	0.0002	0.4481	0	0.1891	0	0.0888	0.0014	0.1004	0.7067
016	0	0	0.0014	0	0.0151	0.0028	0	0	0	0.0010	0.0810	0.7222
029	0.0875	0.0163	0.0017	0.0007	0.0005	0.0003	0.0004	0.0004	0.0118	0.0338	0.0006	0.7220

Table 3.17   Quasi-industrial cluster 16 in China

Nos.	023	028	024	001	039	014	015	122	040	027	018	003	016	029	Average e_{ij}
023	0														0.1803
028	0.3606	0													0.2953
024	0.0224	0.5029	0												0.3476
001	0.0003	0.0028	0.5013	0											0.4522
039	0.0062	0.0058	0.0008	0.8578	0										0.5007
014	0.0001	0.0062	0.0127	0.7233	0.0009	0									0.5298
015	0.0002	0.0012	0.0016	0.6897	0.0001	0.0116	0								0.5507
122	0.0001	0.0047	0.0003	0.5832	0.0017	0.1071	0	0							0.5562
040	0	0.0019	0	0.5808	0.0033	0.0058	0.0001	0.0079	0						0.5977
027	0.1203	0.1901	0.1647	0.4961	0	0	0	0	0	0					0.6249
018	0.0001	0.0039	0.0041	0.2261	0.0014	0.1680	0.4734	0.0011	0.0198	0	0				0.7087
003	0.2870	0.0014	0.0888	0.4167	0.0002	0.4481	0	0.1891	0	0.0977	0.1004	0			0.7240
016	0	0.0010	0	0.0014	0	0.0151	0.0028	0	0	0	0.0810	0.8074	0		0.7180
029	0.0080	0.0338	0.0118	0.0017	0.0007	0.0005	0.0003	0.0004	0.0004	0.0163	0.0006	0.1249	0.4403	0	

Table 3.18   Quasi-industrial cluster 17 in China

Nos.	025	028	024	001	039	014	015	122	040	027	018	003	016	029	Average e_{ij}
025	0														0.1206
028	0.2411	0													0.2489
024	0.0027	0.5029	0												0.3156
001	0.0117	0.0028	0.5013	0											0.4254
039	0.0001	0.0058	0.0008	0.8578	0										0.4784
014	0.0002	0.0062	0.0127	0.7233	0.0009	0									0.5106
015	0.0002	0.0012	0.0016	0.6897	0.0001	0.0116	0								0.5339
122	0	0.0047	0.0003	0.5832	0.0017	0.1071	0	0							0.5413
040	0	0.0019	0	0.5808	0.0033	0.0058	0.0001	0.0079	0						0.5728
027	0.0056	0.1901	0.1647	0.4961	0.0014	0.1680	0.4734	0.0011	0.0198	0					0.6023
018	0.0001	0.0039	0.0041	0.2261	0.0002	0.4481	0	0.1891	0	0.0977	0				0.6734
003	0.1122	0.0014	0.0888	0.4167	0	0.0151	0.0028	0	0	0	0.1004	0.8074	0		0.6915
016	0	0.0010	0	0.0014	0.0007	0.0005	0.0003	0.0004	0.0004	0.0163	0.0810	0.1249	0.4403	0	0.6901
029	0.0405	0.0338	0.0118	0.0017	0.0007	0.0005	0.0003	0.0004	0.0004	0.0163	0.0006	0.1249	0.4403	0	

Table 3.19   Quasi-industrial cluster 18 in China

Nos.	026	022	027	001	039	014	015	122	040	024	028	018	003	016	029	019	Average e_{ij}
026	0																
022	0.3095	0															0.1548
027	0.0063	0.3740	0														0.2299
001	0.0028	0.0835	0.4961	0													0.3180
039	0.0006	0.0079	0	0.8578	0												0.4277
014	0.0006	0.0016	0	0.7233	0.0009	0											0.4775
015	0.0005	0.0006	0	0.6897	0.0001	0.0116	0										0.5096
122	0.0004	0.0003	0	0.5832	0.0017	0.1071	0	0									0.5325
040	0.0001	0.0005	0	0.5808	0.0033	0.0058	0.0001	0.0079	0								0.5398
024	0.0155	0.0262	0.1647	0.5013	0.0008	0.0127	0.0016	0.0003	0	0							0.5582
028	0.1132	0.2742	0.1901	0.0028	0.0058	0.0062	0.0012	0.0047	0.0019	0.5029	0						0.6077
018	0.0004	0.0045	0	0.2261	0.0014	0.1680	0.4734	0.0011	0.0198	0.0041	0.0039	0					0.6323
003	0.0074	0.0004	0.0977	0.4167	0.0002	0.4481	0	0.1891	0	0.0888	0.0014	0.1004	0				0.6875
016	0.0001	0	0	0.0014	0	0.0151	0.0028	0	0	0	0.0010	0.0810	0.8074	0			0.7033
029	0.1050	0.0875	0.0163	0.0017	0.0007	0.0005	0.0003	0.0004	0.0004	0.0118	0.0338	0.0006	0.1249	0.4403	0		0.7114
019	0.0005	0.0002	0	0.3926	0.0019	0.0902	0.0388	0.0009	0.0046	0.0027	0.0026	0.0508	0.0018	0.0007	0.0007	0	0.7037

Table 3.20   Quasi-industrial cluster 19 in China

Nos.	030	031	090	049	050	048	123	052	053	Average e_{ij}
030	0									
031	0.3115	0								0.1557
090	0.3043	0.1540	0							0.2566
049	0.0034	0.0006	0.8580	0						0.4080
050	0.0024	0.0083	0.7365	0.0115	0					0.4781
048	0.0058	0.0009	0.6900	0.2906	0.0327	0				0.5684
123	0.0004	0.0034	0.5965	0.0428	0.0054	0.0317	0			0.5844
052	0.0025	0.0046	0.5918	0.0062	0.0083	0.0112	0.0017	0		0.5896
053	0.0021	0.0044	0.3536	0.0146	0.0343	0.0604	0.0002	0.0217	0	0.5786

Table 3.21   Quasi-industrial cluster 20 in China

Nos.	034	032	033	013	002	100	077	078	076	114	120	020	Average e_{ij}
034	0												0.1264
032	0.2529	0											0.2632
033	0.0294	0.5073	0										0.2693
013	0.0002	0.2870	0.0005	0									0.2911
002	0.0030	0.0275	0.0017	0.3459	0								0.3301
100	0.0857	0.0689	0.2616	0.0577	0.0510	0							0.3602
077	0.0006	0.0145	0.0038	0.0006	0.0001	0.5216	0						0.3877
078	0.0065	0.0170	0.0066	0.0002	0	0.0636	0.4865	0					0.4033
076	0.0133	0.0052	0.0007	0.0008	0.0001	0.0478	0.0010	0.4586	0				0.4474
114	0.0885	0.0390	0.0882	0.0033	0.0108	0.1549	0.0134	0.0213	0.4251	0			0.4565
120	0.0068	0.0088	0.0118	0.0001	0.0038	0.0532	0.0104	0.0649	0.0230	0.3651	0		0.4522
020	0.0001	0.0444	0.0160	0.0001	0.0005	0.2962	0.0006	0.0001	0.0001	0.0437	0.0024	0	

Table 3.22   Quasi-industrial cluster 21 in China

Nos.	043	047	084	045	022	027	001	039	014	015	122	040	024	028	018	003	016	029	019	Average e_{ij}
043	0																			0.1767
047	0.3533	0																		0.2347
084	0.3234	0.0275	0																	0.2651
045	0.2981	0.0100	0.0479	0																0.2819
022	0.0185	0.0510	0.0651	0.2144	0															0.3131
027	0.0007	0.0119	0.0107	0.0722	0.3740	0														0.3590
001	0.0036	0.0373	0.0069	0.0066	0.0835	0.4961	0													0.4434
039	0.0133	0.1480	0.0063	0.0016	0.0079	0	0.8578	0												0.4774
014	0.0107	0.0089	0.0033	0	0.0016	0	0.7233	0.0009	0											0.5012
015	0.0014	0.0100	0.0023	0	0.0006	0	0.6897	0.0001	0.0116	0										0.5194
122	0.0004	0.0078	0.0002	0.0007	0.0003	0	0.5832	0.0017	0.1071	0	0									0.5305
040	0.0154	0.0323	0.0066	0.0004	0.0005	0	0.5808	0.0033	0.0058	0.0001	0.0079	0								0.5496
024	0.0010	0.0031	0.0090	0.0578	0.0262	0.1647	0.5013	0.0008	0.0127	0.0016	0.0003	0.0019	0							0.5978
028	0.0048	0.0110	0.0408	0.1781	0.2742	0.1901	0.0028	0.0058	0.0062	0.0012	0.0047	0.0198	0.5029	0						0.6219
018	0.0062	0.0366	0.0125	0.0020	0.0045	0	0.2261	0.0014	0.1680	0.4734	0.0011	0.0041	0.0041	0.0039	0					0.6716
003	0.0009	0.0010	0.0718	0.0004	0.0004	0.0977	0.4167	0.0002	0.4481	0.1680	0.1891	0.0004	0.0888	0.0014	0.1004	0				0.6862
016	0.0005	0.0077	0.0026	0	0	0	0.0014	0.0014	0.0151	0.0005	0.0028	0.0004	0	0.0010	0.0810	0.8074	0			0.6901
029	0.0066	0.0166	0.0105	0.0036	0.0875	0.0163	0.0017	0.0007	0.0005	0.0003	0.0003	0.0004	0.0118	0.0338	0.0006	0.1249	0.4403	0		
019	0.0180	0.0067	0.0082	0.0007	0.0002	0	0.3926	0.0019	0.0902	0.0388	0.0388	0.0046	0.0027	0.0026	0.0508	0.0018	0.0007	0.0007	0	0.6865

Table 3.23   Quasi-industrial cluster 22 in China

Nos.	051	078	077	100	076	114	120	020	Average e_{ij}
051	0								
078	0.2253	0							0.1126
077	0.0116	0.4865	0						0.2411
100	0.0743	0.0636	0.5216	0					0.3457
076	0.0111	0.4586	0.0010	0.0478	0				0.3803
114	0.0059	0.0213	0.0134	0.1549	0.4251	0			0.4203
120	0.0027	0.0649	0.0104	0.0532	0.0230	0.3651	0		0.4345
020	0.0050	0.0001	0.0006	0.2962	0.0001	0.0437	0.0024	0	0.4237

Table 3.24   Quasi-industrial cluster 23 in China

Nos.	058	057	056	055	009	037	006	087	Average e_{ij}
058	0								
057	0.3474	0							0.1737
056	0.0835	0.3901	0						0.2737
055	0.0016	0.1303	0.6146	0					0.3919
009	0.0933	0.3648	0.0750	0.1724	0				0.4546
037	0.0225	0.1828	0.0682	0.3591	0.0054	0			0.4851
006	0.0135	0.0898	0.0133	0.0388	0.0060	0.4607	0		0.5047
087	0.0010	0.0156	0.0015	0.0019	0.0004	0.0098	0.3677	0	0.4914

Table 3.25   Quasi-industrial cluster 24 in China

Nos.	061	057	056	055	009	037	006	087	Average e_{ij}
061	0								
057	0.2607	0							0.1304
056	0.1209	0.3901	0						0.2572
055	0.1066	0.1303	0.5496	0					0.3896
009	0.1540	0.3648	0.0750	0.1724	0				0.4649
037	0.0283	0.1828	0.0682	0.3591	0.0054	0			0.4947
006	0.0425	0.0898	0.0133	0.0388	0.0060	0.4607	0		0.5170
087	0.0138	0.0156	0.0015	0.0019	0.0004	0.0098	0.3677	0	0.5038

Table 3.26   Quasi-industrial cluster 25 in China

Nos.	065	001	039	014	015	122	040	024	Average e_{ij}
065	0								
001	0.3444	0							0.1722
039	0	0.8578	0						0.4007
014	0	0.7233	0.0009	0					0.4816
015	0	0.6897	0.0001	0.0116	0				0.5255
122	0.0026	0.5832	0.0017	0.1071	0	0			0.5537
040	0	0.5808	0.0033	0.0058	0.0001	0.0079	0		0.5600
024	0.0002	0.5013	0.0008	0.0127	0.0016	0.0003	0	0	0.5547

Table 3.27   Quasi-industrial cluster 26 in China

Nos.	066	057	056	055	009	037	006	087	Average e_{ij}
066	0								
057	0.2020	0							0.1010
056	0.0561	0.3901	0						0.2161
055	0.0136	0.1303	0.5496	0					0.3354
009	0.1320	0.3648	0.0750	0.1724	0				0.4172
037	0.0293	0.1828	0.0682	0.3591	0.0054	0			0.4551
006	0.0348	0.0898	0.0133	0.0388	0.0060	0.4607	0		0.4820
087	0.0088	0.0156	0.0015	0.0019	0.0004	0.0098	0.3677	0	0.4725

Table 3.28 Quasi-industrial cluster 27 in China

Nos.	067	091	117	103	036	007	093	104	Average e_{ij}
067	0								
091	0.4209	0							0.2105
117	0	0.3569	0						0.2593
103	0	0.0023	0.4325	0					0.3031
036	0.0143	0.2259	0.0214	0.3116	0				0.3571
007	0.0030	0.0116	0	0	0.8186	0			0.4365
093	0.0004	0.0077	0.0007	0.0052	0.4909	0.0050	0		0.4470
104	0	0.0025	0.0037	0.0028	0.3740	0.0001	0.0059	0	0.4397

Table 3.29 Quasi-industrial cluster 28 in China

Nos.	070	105	124	111	098	102	Average e_{ij}
070	0						
105	0.5302	0					0.2651
124	0	0.3188	0				0.2830
111	0.0015	0.0187	0.3619	0			0.3078
098	0	0.0004	0.3115	0.0101	0		0.3106
102	0.0001	0.0004	0.2669	0.0167	0	0	0.3062

Table 3.30 Quasi-industrial cluster 29 in China

Nos.	071	057	056	055	009	037	006	087	Average e_{ij}
071	0								
057	0.2008	0							0.1004
056	0.0019	0.3901	0						0.1976
055	0.0001	0.1303	0.6146	0					0.3345
009	0.0003	0.3648	0.0750	0.1724	0				0.3901
037	0.0001	0.1828	0.0682	0.3591	0.0054	0			0.4277
006	0.0021	0.0898	0.0133	0.0388	0.0060	0.4607	0		0.4538
087	0.0011	0.0156	0.0015	0.0019	0.0004	0.0098	0.3677	0	0.4468

Table 3.31 Quasi-industrial cluster 30 in China

Nos.	079	078	077	100	076	114	120	020	Average e_{ij}
079	0								
078	0.2761	0							0.1380
077	0.0173	0.4865	0						0.2599
100	0.0578	0.0636	0.5216	0					0.3557
076	0.0356	0.4586	0.0010	0.0478	0				0.3932
114	0.0769	0.0213	0.0134	0.1549	0.4251	0			0.4429
120	0.1530	0.0649	0.0104	0.0532	0.0230	0.3651	0		0.4753
020	0.0002	0.0001	0.0006	0.2962	0.0001	0.0437	0.0024	0	0.4588

Table 3.32 Quasi-industrial cluster 31 in China

Nos.	081	078	077	100	076	114	Average e_{ij}
081	0						
078	0.2295	0					0.1148
077	0.0003	0.4865	0				0.2388
100	0.1332	0.0636	0.5216	0			0.3587
076	0.0004	0.4586	0.0010	0.0478	0		0.3885
114	0.1376	0.0213	0.0134	0.1549	0.4251	0	0.4491

Table 3.33   Quasi-industrial cluster 32 in China

Nos.	085	059	060	075	010	099	Average e_{ij}
085	0						
059	0.2136	0					0.1068
060	0.0052	0.6258					0.2815
075	0.0010	0.1870	0.4359	0			0.3671
010	0.0001	0.3490	0.1079	0.0095	0		0.3870
099	0.0038	0.0287	0.0021	0.2559	0.0025	0	0.3713

Table 3.34   Quasi-industrial cluster 33 in China

Nos.	088	006	037	087	055	056	057	009	058	086	089	Average e_{ij}
088	0											
006	0.3279	0										0.1640
037	0.0162	0.4607	0									0.2683
087	0.0041	0.3677	0.0098	0								0.2966
055	0.0026	0.0388	0.3591	0.0019	0							0.3178
056	0.0104	0.0133	0.0682	0.0015	0.6146	0						0.3828
057	0.0998	0.0898	0.1828	0.0156	0.1303	0.3901	0					0.4579
009	0.0002	0.0060	0.0054	0.0004	0.1724	0.0750	0.3648	0				0.4787
058	0.0047	0.0135	0.0225	0.0010	0.0016	0.0835	0.3474	0.0933				0.4885
086	0.0719	0.3087	0.0879	0.0466	0.0492	0.0498	0.0506	0.1509	0.1210	0		0.5333
089	0.0038	0.0104	0.0016	0.0139	0.0026	0.0085	0.0221	0.0039	0.0013	0.4003	0	0.5274

Table 3.35   Quasi-industrial cluster 34 in China

Nos.	092	036	007	093	104	Average e_{ij}
092	0					
036	0.2897	0				0.1449
007	0.0067	0.8186	0			0.3717
093	0.0233	0.4909	0.0050	0		0.4086
104	0.0074	0.3740	0.0001	0.0059	0	0.4043

Table 3.36   Quasi-industrial cluster 35 in China

Nos.	094	036	007	093	104	Average e_{ij}
094	0					
036	0.3717	0				0.1858
007	0.0032	0.8186	0			0.3978
093	0.0056	0.4909	0.0050	0		0.4237
104	0.0026	0.3740	0.0001	0.0059	0	0.4155

Table 3.37   Quasi-industrial cluster 36 in China

Nos.	097	100	077	078	076	114	120	020	Average e_{ij}
097	0								
100	0.2853	0							0.1426
077	0.0016	0.5216	0						0.2695
078	0.0013	0.0636	0.4865	0					0.3400
076	0.0005	0.0478	0.0010	0.4586	0				0.3736
114	0.0160	0.1549	0.0134	0.0213	0.4251	0			0.4164
120	0.0004	0.0532	0.0104	0.0649	0.0230	0.3651	0		0.4308
020	0.0006	0.2962	0.0006	0.0001	0.0001	0.0437	0.0024	0	0.4199

Table 3.38 Quasi-industrial cluster 37 in China

Nos.	108	100	077	078	076	114	120	020	Average e_{ij}
108	0								
100	0.2577	0							0.1289
077	0.0033	0.5216	0						0.2609
078	0.0073	0.0636	0.4865	0					0.3350
076	0.0036	0.0478	0.0010	0.4586	0				0.3702
114	0.0586	0.1549	0.0134	0.0213	0.4251	0			0.4207
120	0.0029	0.0532	0.0104	0.0649	0.0230	0.3651	0		0.4349
020	0.0031	0.2962	0.0006	0.0001	0.0001	0.0437	0.0024	0	0.4238

Table 3.39 Quasi-industrial cluster 38 in China

Nos.	109	036	007	093	104	094	Average e_{ij}
109	0						
036	0.2217	0					0.1108
007	0.0015	0.8186	0				0.3473
093	0.0139	0.4909	0.0050	0			0.3879
104	0.0130	0.3740	0.0001	0.0059	0		0.3889
094	0.0047	0.3717	0.0032	0.0056	0.0026	0	0.3887

Table 3.40 Quasi-industrial cluster 39 in China

Nos.	112	105	070	124	111	098	102	033	032	013	Average e_{ij}
112	0										
105	0.2236	0									0.1118
070	0	0.5302	0								0.2513
124	0.0835	0.3188	0	0							0.2890
111	0.0897	0.0187	0.0015	0.3619	0						0.3256
098	0.0033	0.0004	0	0.3115	0.0101	0					0.3255
102	0.1605	0.0004	0.0001	0.2669	0.0167	0	0				0.3426
033	0.0004	0.0021	0.0004	0.1687	0.0102	0.2173	0.0026	0			0.3499
032	0.0002	0.0011	0.0013	0.0142	0.0114	0.0340	0.0015	0.5073	0		0.3745
013	0	0.0001	0.0008	0	0.0008	0	0.0005	0.0005	0.2870	0	0.3660

Table 3.41 Quasi-industrial cluster 40 in China

Nos.	113	100	077	078	076	114	120	020	001	Average e_{ij}
113	0									
100	0.2070	0								0.1035
077	0.0024	0.5216	0							0.2437
078	0.0044	0.0636	0.4865	0						0.3214
076	0.0031	0.0478	0.0010	0.4586	0					0.3592
114	0.0171	0.1549	0.0134	0.0213	0.4251	0				0.4046
120	0.0003	0.0532	0.0104	0.0649	0.0230	0.3651	0			0.4207
020	0.0994	0.2962	0.0006	0.0001	0.0001	0.0437	0.0024	0		0.4234
001	0.0316	0.0397	0.0008	0	0.0001	0.0118	0.0034	0.3270	0	0.4224

Table 3.42   Quasi-industrial cluster 41 in China

Nos.	119	032	033	013	002	100	077	078	076	114	120	020	Average e_{ij}
119	0												
032	0.2466	0											0.1233
033	0.1028	0.5073	0										0.2856
013	0.0001	0.2870	0.0005	0									0.2861
002	0.0019	0.0275	0.0017	0.3459	0								0.3043
100	0.0614	0.0689	0.2616	0.0577	0.0510	0							0.3370
077	0.0031	0.0145	0.0038	0.0006	0.0001	0.5216	0						0.3665
078	0.0023	0.0170	0.0066	0.0002	0	0.0636	0.4865	0					0.3927
076	0.0031	0.0052	0.0007	0.0008	0.0001	0.0478	0.0010	0.4586	0				0.4066
114	0.0952	0.0390	0.0882	0.0033	0.0108	0.1549	0.0134	0.0213	0.4251	0			0.4510
120	0.0042	0.0088	0.0118	0.0001	0.0038	0.0532	0.0104	0.0649	0.0230	0.3651	0		0.4596
020	0.0017	0.0444	0.0160	0.0001	0.0005	0.2962	0.0006	0.0001	0.0001	0.0437	0.0024	0	0.4551

3.2   IDENTIFYING PRINCIPLES OF INDUSTRIAL CLUSTERS IN CHINA

Although we can confirm quasi-industrial clusters in China using the above method, there exist some problems with the method. So, we need to establish some basic principle, synthetically induce quasi-industrial clusters in China, complement or delete industries of quasi-industrial clusters on the basis of it, and accurately identify different types of industrial clusters in China.

3.2.1   The lowest doorsill principle

The closest economic links are among industries of quasi-industrial clusters determined by the above method, so the number of industries contained in quasi-industrial clusters is relatively fewer. Besides inter-industry $e_{ij} \geqslant 0.20$, basic conditions of whether an industry becomes the member of quasi-industrial clusters still depend on lowering of the average value of e_{ij} when adding a new member. The algorithm sometimes doesn't find industries which have such intense links, but $e_{ij} \geqslant 0.20$ after quasi-industrial clusters contain most members with the closest links. This is because the average value of inter-industrial linkage has begun to drop. This industry cannot become a member of the quasi-industrial cluster.

Therefore, for a certain industry to be a member of one industrial cluster, it must satisfy one of the following conditions: First, more than 20% of the output of the industry is absorbed by the industries of the quasi-industrial cluster; second, more than 20% of the output of a certain industry in the quasi-industrial cluster is absorbed by the industry; third, more than 20% of the input of the industry comes from a certain industry of the quasi-industrial cluster; fourth, more than 20% of the input of a certain industry in quasi-industrial cluster comes from the industry. This is the lowest doorsill principle of industrial clusters.

For example, as far as quasi-industrial cluster 1 is concerned, according to the identifying method of quasi-industrial clusters in China, the quasi-industrial cluster titled the planting quasi-industrial cluster is made up of 7 industries: crop cultivation (001), chemical fertilizers (039), grain mill products, vegetable oil and forage (014), sugar refining (015), farming, forestry, herd, fishing service industry (122), chemical pesticides (040), and hemp textiles (024). According to this principle, the industries linked closely with crop cultivation (001) ($e_{ij} >20\%$) are industries 024, 027, 019, 021, 020, 005, 018, 097, and 065; the industry linked closely with chemical fertilizers (039) ($e_{ij} >20\%$) is industry 008; the industries linked closely with grain mill products, vegetable oil and forage (014) ($e_{ij} >20\%$) are industries 097, 004, 003, and 017; the industries linked closely with sugar refining (015) ($e_{ij} >20\%$) are industries 018 and 020; and the industry linked closely with hemp textiles (024) ($e_{ij} >20\%$) is industry 028. These industries may become members of the planting industrial cluster.

3.2.2 The shortest link path principle

Determined by the lowest doorsill principle, an industry may become the member of one industrial cluster, but whether the industry does become a member of the industrial cluster depends on the shortest link path principle.

Making use of graph theory and network theory, we can draw network graphs of industrial clusters which have been identified. If a certain industry is determined by the lowest doorsill principle to belong to two industrial clusters, which industrial cluster the industry becomes a member of depends on its shortest path amounts linked with the central industries of the industrial clusters (underneath). If the industry's shortest path amounts linked with the central industries of industrial cluster A are smaller than those of industrial cluster B, the industry becomes a member of industrial cluster A; otherwise the industry might become a member of cluster B.

For example, industry 093 has close links with member industry 086 of the iron and steel industrial cluster ($e_{ij} > 20\%$); therefore industry 093 may become a member of the iron and steel industrial cluster according to the lowest doorsill principle. Industry 086 has close links with member industry 093 of the petroleum industrial cluster ($e_{ij} > 20\%$), so industry 093 may become a member of the petroleum industrial cluster also according to the lowest doorsill principle. Because the number of shortest link paths between industry 086 and the main members of the petroleum industrial cluster is 2, while that between industry 093 and the main members of the iron and steel industrial cluster is 4, industry 086 becomes a member of petroleum industrial cluster. Industry 093 cannot become a member of iron and steel industrial cluster according to the shortest connection path principle.

3.2.3 The dominant principle

For multiple member industries which become industrial cluster C, they may also become the member industries of industrial cluster D in terms of the lowest doorsill principle or the shortest link path principle. In such a case, those industries which have very small economic links (e_{ij}) with most member industries are deleted from industrial cluster D. This is called the dominance principle.

For example, in the chemistry quasi-industrial cluster, industries 001, 039, 014, 015, 122, 040, and 024 are all members of the planting industrial cluster. Because industries 014, 015, 122, 040, and 024 have slender links with the main members of the chemistry industrial cluster, these five industries are deleted from the chemistry industrial cluster according to the dominance principle.

3.3 BASIC TYPES OF INDUSTRIAL CLUSTERS IN CHINA

According to the above identifying methods of quasi-industrial clusters and basic principles in China, we identify four types of industrial clusters: light manufacturing industrial clusters, heavy manufacturing industrial clusters, construction industrial clusters and service industrial clusters in China in 1997. These are further divided into 11 sub-types of industrial clusters (Table 3.43).

3.3.1 Light manufacturing industrial clusters

The light manufacturing industrial clusters in China in 1997 have 3 sub-types: the planting industrial cluster, the electronics industrial cluster, and the textile industrial cluster.

3.3.1.1 Planting industrial cluster

The planting industrial cluster is integrated to form through quasi-industrial clusters 1, 3, 5, 8, 11, 12, 13, 14 and 25. The industry types forming the planting industrial cluster are

Table 3.43   Types of industrial clusters in China

Types of industrial clusters	Sub-types of industrial clusters	Industries component(codes)
Light manufacturing industrial clusters	Planting industrial cluster	001, 039, 014, 015, 122, 040, 024, 027, 003, 019, 065, 021, 020, 005, 018, 097, 016, 008
	Electronics industrial cluster	078, 077, 100, 076, 114, 120, 079, 081, 051
	Textile industrial cluster	022, 027, 026, 028, 046, 045, 024, 023, 025, 001
Heavy manufacturing industrial clusters	Iron and steel industrial cluster	088, 006, 037, 087, 055, 056, 057, 009, 058, 086, 089, 061, 090, 066, 071
	Petroleum industrial cluster	036, 007, 093, 104, 103, 092, 091, 102, 067, 109, 086, 117, 094
	Nonferrous metal industrial cluster	010, 059, 060, 075, 099, 085
	Chemistry industrial cluster	043, 047, 084, 045
Construction industrial clusters	Construction industrial cluster	090, 049, 050, 048, 123, 052, 053, 030, 031, 012, 075, 061, 013, 097, 099, 057, 080
Service industrial clusters	Commerce industrial cluster	100, 077, 020, 097, 108, 021, 113, 040, 033, 032, 034, 119, 013, 002, 046
	Diet industrial cluster	101, 020, 016, 019, 018, 021, 017, 004, 014, 001, 122
	Administration industrial cluster	124, 111, 098, 102, 105, 070, 112, 117

the most, and they are crop cultivation (001), chemical fertilizers (039), grain mill products, vegetable oil and forage (014), sugar refining (015), technical services for agriculture, forestry, livestock and fishing (122), chemical pesticides (040), hemp textiles (024), other textiles (027), livestock and livestock products (003), wines, spirits and liquors (019), agriculture, forestry, animal husbandry and fishing machinery (065), tobacco products (021), non-alcoholic beverage products (020), other agriculture products (005), other food products (018), storage and warehousing (097), slaughtering, meat processing, eggs and dairy products (016), and natural gas products (008).

3.3.1.2   Electronics industrial cluster

The electronics industrial cluster is integrated to form through quasi-industrial clusters 2, 22, 30, 31 and 6. The 9 industries forming the industrial cluster are electronic element and device (078), electronic appliances (077), wholesale and retail trade (100), electronic computer (076), other social services (114), scientific research (120), other electronic and communication equipment (079), cultural and office equipment (081) and glass and glass products (051).

3.3.1.3   Textile industrial cluster

The textile industrial cluster is integrated to form through quasi-industrial clusters 15, 16, 17 and 18. The 10 industries forming the industrial cluster are cotton textiles (022), other textiles (027), knitted mills (026), wearing apparel (028), rubber products (046), chemical fibers (045), hemp textiles (024), wool textiles (023), silk textiles (025), and crop cultivation (001).

3.3.2   Heavy manufacturing industrial clusters

The heavy manufacturing industrial cluster in China in 1997 have 4 sub-types: the iron and steel industrial cluster, the petroleum industrial cluster, the nonferrous metal industrial cluster and the chemistry industrial cluster.

3.3.2.1   Iron and steel industrial cluster

The iron and steel industrial cluster is integrated to form through quasi-industrial clusters 6, 23, 24, 26, 29 and 33. This cluster is composed of 15 industries: gas production and supply (088), coal mining and processing (006), coking (037), steam and hot water production and supply (087), iron-smelting (055), steel-smelting (056), steel processing (057), ferrous ore mining (009), alloy iron smelting (058), electricity production and supply (086), water production and supply (089), metal products (061), construction (090), other special industrial equipment (066), and bicycle (071).

3.3.2.2   Petroleum industrial cluster

The petroleum industrial cluster is integrated to form through quasi-industrial clusters 7, 27, 34, 35 and 38. This cluster is composed of 13 industries: petroleum refining (036), crude petroleum products (007), pipeline transport (093), water passenger transport (104), highway passenger transport (103), highway freight transport (092), railway freight transport (091), railway passenger transport (102), railroad transport equipment (067), public services (109), electricity production and supply (086), social welfare (117), and water freight transport (094).

3.3.2.3   Nonferrous metal industrial cluster

The nonferrous metal industrial cluster is integrated to form through quasi-industrial cluster 9 and 32. This cluster is composed of 6 industries: nonferrous metal mining (010), nonferrous metal smelting (059), nonferrous metal processing(060), other electric machinery and equipment (075), telecommunication (099), and scrap and waste (085).

3.3.2.4   Chemistry industrial cluster

The chemistry cluster is obtained from quasi-industrial cluster 21. This cluster is composed of 4 industries: other chemical products (043), plastic products (047), other manufacturing products (084), and chemical fibers (045).

3.3.3   Construction industrial clusters

The construction industrial clusters in China in 1997 has only 1 sub-type, that is, the construction industrial cluster.

The construction cluster is integrated to form through quasi- industrial cluster 10 and 19. This cluster is composed of 17 industries: construction (090), cement and asbestos products (049), bricks, tiles, lime and light-weight building materials (050), cement (048), geological prospecting and water conservancy (123), pottery, china and earthenware (052), fireproof products (053), sawmills and fibreboard (030), furniture and products of wood, bamboo, cane, palm, straw, etc. (031), nonmetal minerals and other ore mining (012), other electric machinery and equipment (075), metal products (061), logging and transport of timber and bamboo (013), storage and warehousing (097), telecommunication (099), steel processing (057), instruments, meters and other measuring equipment (080).

3.3.4   Service industrial clusters

The service industrial clusters in China in 1997 have 3 sub-types: the commerce industrial cluster, the diet industrial cluster and the administration industrial cluster.

3.3.4.1   Commerce industrial cluster

The commerce industrial cluster is integrated to form through quasi-industrial cluster 2, 20, 22, 37 and 40. This cluster is composed of 15 industries: wholesale and retail trade (100), electronic appliances (077), non-alcoholic beverage (020), storage and warehousing (097), real estate (108), tobacco products (021), recreational services (113), chemical pesticides (040), printing and record medium reproduction (033), paper and products (032), cultural goods (034), culture and arts, radio, film and television services (119), logging and transport of timber and bamboo (013), forestry (002), and rubber products (046).

3.3.4.2   Diet industrial cluster

The diet industrial cluster is obtained form quasi-industrial cluster 4. This cluster is composed of 11 industries: eating and drinking places (101), non-alcoholic beverages (020), slaughtering, meat processing, eggs and dairy products (016), wines, spirits and liquors (019), other food products (018), tobacco products (021), prepared fish and seafood (017), fishery (004), grain mill products, vegetable oil and forage (014), crop cultivation (001), and technical services for agriculture, forestry, livestock and fishing (122).

3.3.4.3   Administration industrial cluster

The administration industrial cluster is integrated to form through quasi-industrial cluster 28 and 39. This cluster is composed of 8 industries: public administration and other sectors (124), hotels (111), post services (098), railway passenger transport (102), air passenger transport (105), aircraft (070), tourism (112), and social welfare (117).

CHAPTER 4

Structure Analysis of Industrial Clusters in China

Industrial clusters are composed of multiple industries, among which there exist definite relationships. From the point of view of system theory, an industrial cluster is actually a system composed of many industries. In order to study the system—an industrial cluster, first, we should know relationships among industries; that is, we should understand the inner structure of the system because the functions of a system are mainly determined by the structures of the system. Structures of industrial clusters have correlation and interdependent relationships among basic integral parts, namely among industries. Structures of industrial clusters are determined by the functional links and spatial links among industries and, in turn, deeply influence functional links and spatial links of industrial clusters and then proceed to influence functional characteristics of industrial clusters. Under the condition of location, structural characteristics of industrial clusters mostly depend on the ways of functional links among industries.

4.1 GENERAL STRUCTURAL MODELS OF INDUSTRIAL CLUSTERS IN CHINA

In the study of structures of industrial clusters, we need to know not only the comprising industries of industrial clusters, but also the directions of functional links among industries. Therefore, the study of directed connecting graphs is one of important contents of the study on structural models of industrial clusters.

Generally speaking, the structural model of a system reveals mainly the qualitative structures of geometry or the topology of the system, instead of expounding in detail on the algebra descriptions or quantity property of the model. It is an objective model between a conceptual model and a mathematical model with charts and graphs used to make clear the correlations among parts and elements of the system. As far as the particular system of

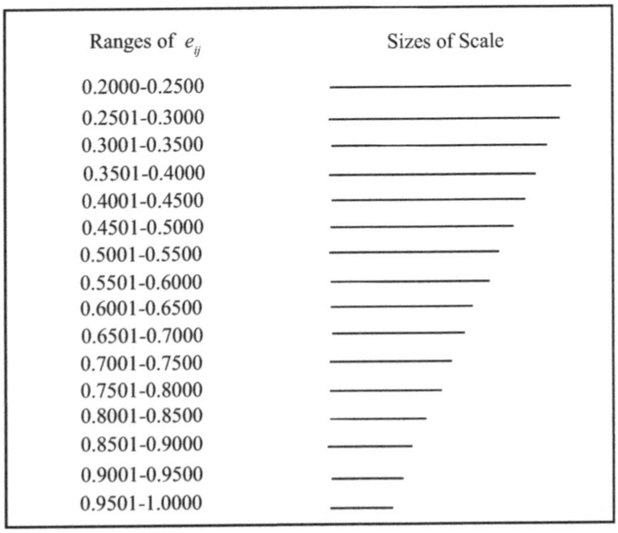

Ranges of e_{ij}	Sizes of Scale
0.2000-0.2500	
0.2501-0.3000	
0.3001-0.3500	
0.3501-0.4000	
0.4001-0.4500	
0.4501-0.5000	
0.5001-0.5500	
0.5501-0.6000	
0.6001-0.6500	
0.6501-0.7000	
0.7001-0.7500	
0.7501-0.8000	
0.8001-0.8500	
0.8501-0.9000	
0.9001-0.9500	
0.9501-1.0000	

Figure 4.1   The legend of general structural model figure of industrial clusters in China

industrial clusters is concerned, the general structural model of industrial clusters reflects the relationship among industries forming the industrial clusters, does not requiring complicated quantity relations, and can simply and exactly show the essential characteristics of industrial clusters, that is, economic links among industries, especially functional links among industries.

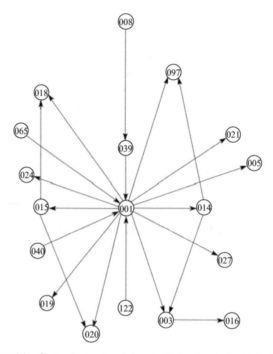

Figure 4.2   General structural figure of planting industrial cluster

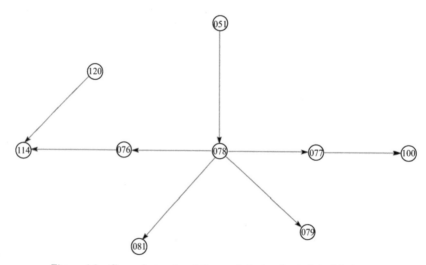

Figure 4.3   General structural figure of electronics industrial cluster

Suppose that each node denotes every industry of the industrial clusters, arrowheads denote the status of functional links among industries, the directions of arrowheads denote the input-output relationship among industries, the length of line segments denotes the strength

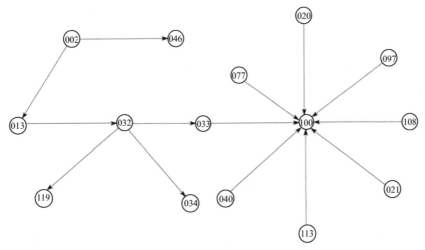

Figure 4.4   General structural figure of commerce industrial cluster

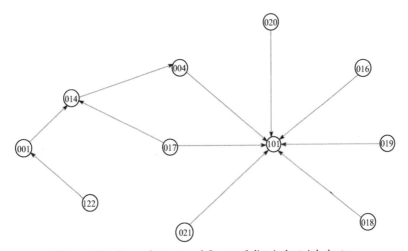

Figure 4.5   General structural figure of diet industrial cluster

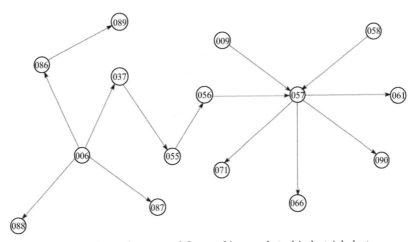

Figure 4.6   General structural figure of iron and steel industrial cluster

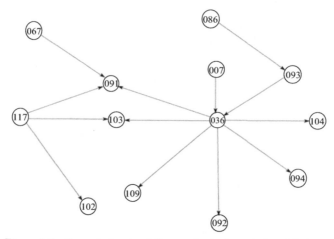

Figure 4.7   General structural figure of petroleum industrial cluster

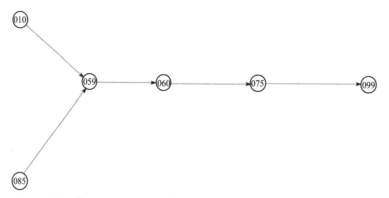

Figure 4.8   General structural figure of nonferrous metal industial cluster

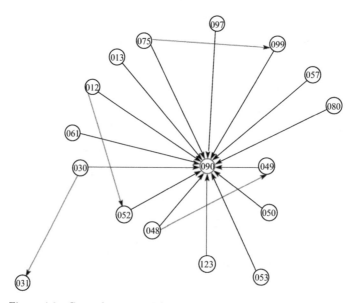

Figure 4.9   General structural figure of construction industrial cluster

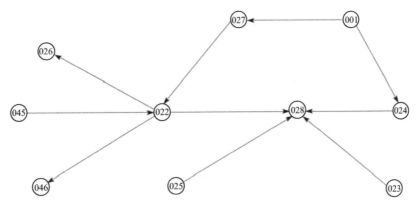

Figure 4.10   General structural figure of textile industrial cluster

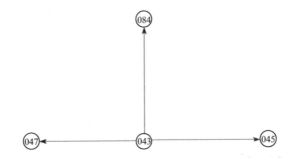

Figure 4.11   General structural figure of chemistry industrial cluster

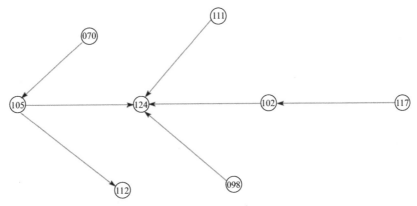

Figure 4.12   General structural figure of administration industrial cluster

of the functional links e_{ij} (the maximum among the input coefficients b_{ij} and a_{ji} and the output coefficients a_{ij} and b_{ji}), and the length of line segments is inversely proportional to the strength of the functional links among industries (Figure 4.1). On the basis of this, we draw the general structural model diagrams of industrial clusters in China (Figures 4.2-4.12) (numerals in the circles in the figures are the industrial codes, see also Appendix 1).

4.2   RECURSION STRUCTURAL MODELS OF INDUSTRIAL CLUSTERS IN CHINA

As the method of researching structural models of industrial clusters in China, Interpretive Structural Modeling (ISM) is discussed in this section. The method of ISM is one method used by Research Institute of Bottelle in U.S.A. to analyze complicated social problems. It decomposes complicated problems into many subproblems. The orders of these subproblems are synthetically and systemically determined with the help of computers (Cheng, 2001). On the basis of studying general structural models of industrial clusters in China, recursion relationships among industries within industrial clusters can be further explained using the method of ISM.

The steps for analyzing recursion structural relationships among industries within industrial clusters in China are as follows: corresponding accessible matrixes M are constructed according to directed connection graphs; decomposition among ranks is carried through (accessible aggregations $R \cap$ antecedent aggregations $S =$ accessible aggregations R); accessible matrixes are translated into structural matrixes J according to decomposition among ranks; recursion structural figures are drawn in accordance with structural matrixes J.

4.2.1　Light manufacturing industrial clusters

4.2.1.1　Planting industrial cluster

The accessible matrix M of the planting industrial cluster is shown in Table 4.1. Its structural matrix J (Table 4.2) is determined according to the accessible matrix, thereby drawing its recursion structural model figure (Figure 4.13).

Table 4.1　The accessible matrix of planting industrial cluster

Nos.	014	027	003	016	122	020	019	040	015	024	065	018	039	008	097	021	005	001
014	1	0	1	1	0	0	0	0	0	0	0	0	0	0	1	0	0	0
027	0	1	0	0	0	0	0	0	0	0	0	0	0	0	0	0	0	0
003	0	0	1	1	0	0	0	0	0	0	0	0	0	0	0	0	0	0
016	0	0	0	1	0	0	0	0	0	0	0	0	0	0	0	0	0	0
122	1	1	1	1	1	1	1	0	1	1	0	1	0	0	1	1	1	1
020	0	0	0	0	0	1	0	0	0	0	0	0	0	0	0	0	0	0
019	0	0	0	0	0	0	1	0	0	0	0	0	0	0	0	0	0	0
040	1	1	1	1	0	1	1	1	1	1	0	1	0	0	1	1	1	1
015	0	0	0	0	0	1	0	0	1	0	0	1	0	0	0	0	0	0
024	0	0	0	0	0	0	0	0	0	1	0	0	0	0	0	0	0	0
065	1	1	1	1	0	1	1	0	1	1	1	1	0	0	1	1	1	1
018	0	0	0	0	0	0	0	0	0	0	0	1	0	0	0	0	0	0
039	1	1	1	1	0	1	1	0	1	1	0	1	1	0	1	1	1	1
008	1	1	1	1	0	1	1	0	1	1	0	1	1	1	1	1	1	1
097	0	0	0	0	0	0	0	0	0	0	0	0	0	0	1	0	0	0
021	0	0	0	0	0	0	0	0	0	0	0	0	0	0	0	1	0	0
005	0	0	0	0	0	0	0	0	0	0	0	0	0	0	0	0	1	0
001	1	1	1	1	0	1	1	0	1	1	0	1	0	0	1	1	1	1

Table 4.2　The structural matrix of planting industrial cluster

Nos.	027	016	020	019	024	018	097	021	005	003	015	014	001	122	040	065	039	008
027	0	0	0	0	0	0	0	0	0	0	0	0	0	0	0	0	0	0
016	0	0	0	0	0	0	0	0	0	0	0	0	0	0	0	0	0	0
020	0	0	0	0	0	0	0	0	0	0	0	0	0	0	0	0	0	0
019	0	0	0	0	0	0	0	0	0	0	0	0	0	0	0	0	0	0
024	0	0	0	0	0	0	0	0	0	0	0	0	0	0	0	0	0	0
018	0	0	0	0	0	0	0	0	0	0	0	0	0	0	0	0	0	0
097	0	0	0	0	0	0	0	0	0	0	0	0	0	0	0	0	0	0
021	0	0	0	0	0	0	0	0	0	0	0	0	0	0	0	0	0	0
005	0	0	0	0	0	0	0	0	0	0	0	0	0	0	0	0	0	0
003	0	1	0	0	0	0	0	0	0	0	0	0	0	0	0	0	0	0
015	0	0	1	0	0	1	0	0	0	0	0	0	0	0	0	0	0	0
014	0	0	0	0	0	0	1	0	0	1	0	0	0	0	0	0	0	0

Continued

Nos.	027	016	020	019	024	018	097	021	005	003	015	014	001	122	040	065	039	008
001	1	0	1	1	1	1	1	1	1	1	1	1	0	0	0	0	0	0
122	0	0	0	0	0	0	0	0	0	0	0	0	1	0	0	0	0	0
040	0	0	0	0	0	0	0	0	0	0	0	0	1	0	0	0	0	0
065	0	0	0	0	0	0	0	0	0	0	0	0	1	0	0	0	0	0
039	0	0	0	0	0	0	0	0	0	0	0	0	1	0	0	0	0	0
008	0	0	0	0	0	0	0	0	0	0	0	0	0	0	0	0	1	0

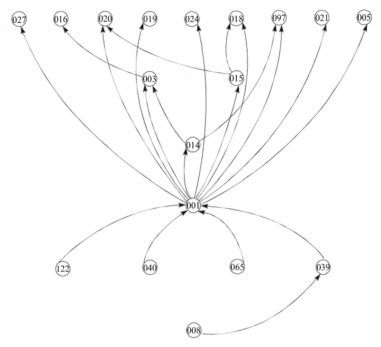

Figure 4.13   The recursion structural figure of planting industrial cluster

4.2.1.2   Electronics industrial cluster

The accessible matrix M of the electronics industrial cluster is shown in Table 4.3. Its structural matrix J (Table 4.4) is determined according to the accessible matrix, thereby drawing its recursion structural model figure (Figure 4.14).

Table 4.3   The accessible matrix of electronics industrial cluster

Nos.	120	114	076	078	051	081	079	077	100
120	1	1	0	0	0	0	0	0	0
114	0	1	0	0	0	0	0	0	0
076	0	1	1	0	0	0	0	0	0
078	0	1	1	1	0	1	1	1	1
051	0	1	1	1	1	1	1	1	1
081	0	0	0	0	0	1	0	0	0
079	0	0	0	0	0	0	1	0	0
077	0	0	0	0	0	0	0	1	1
100	0	0	0	0	0	0	0	0	1

Table 4.4   The structural matrix of electronics industrial cluster

Nos.	114	081	079	100	120	076	077	078	051
114	0	0	0	0	0	0	0	0	0
081	0	0	0	0	0	0	0	0	0
079	0	0	0	0	0	0	0	0	0
100	0	0	0	0	0	0	0	0	0

Continued

Nos.	114	081	079	100	120	076	077	078	051
120	1	0	0	0	0	0	0	0	0
076	1	0	0	0	0	0	0	0	0
077	0	0	0	1	0	0	0	0	0
078	0	1	1	0	0	1	1	0	0
051	0	0	0	0	0	0	0	1	0

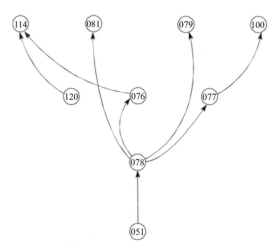

Figure 4.14   The recursion structural figure of electronics industrial cluster

4.2.1.3   Textile industrial cluster

The accessible matrix M of the textile industrial cluster is shown in Table 4.5. Its structural matrix J (Table 4.6) is determined according to the accessible matrix, thereby drawing its recursion structural model figure (Figure 4.15).

Table 4.5   The accessible matrix of textile industrial cluster

Nos.	046	045	026	022	027	001	024	028	023	025
046	1	0	0	0	0	0	0	0	0	0
045	1	1	1	1	0	0	0	1	0	0
026	0	0	1	0	0	0	0	0	0	0
022	1	0	1	1	0	0	0	1	0	0
027	1	0	1	1	1	0	0	1	0	0
001	1	0	1	1	1	1	1	1	0	0
024	0	0	0	0	0	0	1	1	0	0
028	0	0	0	0	0	0	0	1	0	0
023	0	0	0	0	0	0	0	1	1	0
025	0	0	0	0	0	0	0	1	0	1

Table 4.6   The structural matrix of textile industrial cluster

Nos.	046	026	028	022	024	023	025	045	027	001
046	0	0	0	0	0	0	0	0	0	0
026	0	0	0	0	0	0	0	0	0	0
028	0	0	0	0	0	0	0	0	0	0
022	1	1	1	0	0	0	0	0	0	0
024	0	0	1	0	0	0	0	0	0	0
023	0	0	1	0	0	0	0	0	0	0
025	0	0	1	0	0	0	0	0	0	0
045	0	0	0	1	0	0	0	0	0	0
027	0	0	0	1	0	0	0	0	0	0
001	0	0	0	0	1	0	0	0	1	0

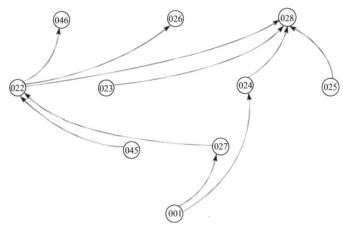

Figure 4.15   The recursion structural figure of textile industrial cluster

4.2.2   Heavy manufacturing industrial clusters

4.2.2.1   Iron and steel industrial cluster

The accessible matrix M of the iron and steel industrial cluster is shown in Table 4.7. Its structural matrix J (Table 4.8) is determined according to the accessible matrix, thereby drawing its recursion structural model figure (Figure 4.16).

Table 4.7   The accessible matrix of iron and steel industrial cluster

Nos.	088	087	037	086	006	089	055	056	009	058	061	090	066	071	057
088	1	0	0	0	0	0	0	0	0	0	0	0	0	0	0
087	0	1	0	0	0	0	0	0	0	0	0	0	0	0	0
037	0	0	1	0	0	0	1	1	0	0	1	1	1	1	1
086	0	0	0	1	0	1	0	0	0	0	0	0	0	0	0
006	1	1	1	1	1	1	1	1	0	0	1	1	1	1	1
089	0	0	0	0	0	1	0	0	0	0	0	0	0	0	0
055	0	0	0	0	0	0	1	1	0	0	1	1	1	1	1
056	0	0	0	0	0	0	0	1	0	0	1	1	1	1	1
009	0	0	0	0	0	0	0	0	1	0	1	1	1	1	1
058	0	0	0	0	0	0	0	0	0	1	1	1	1	1	1
061	0	0	0	0	0	0	0	0	0	0	1	0	0	0	0
090	0	0	0	0	0	0	0	0	0	0	0	1	0	0	0
066	0	0	0	0	0	0	0	0	0	0	0	0	1	0	0
071	0	0	0	0	0	0	0	0	0	0	0	0	0	1	0
057	0	0	0	0	0	0	0	0	0	0	1	1	1	1	1

Table 4.8   The structural matrix of iron and steel industrial cluster

Nos.	088	087	089	061	090	066	071	086	057	056	009	058	055	037	006
088	0	0	0	0	0	0	0	0	0	0	0	0	0	0	0
087	0	0	0	0	0	0	0	0	0	0	0	0	0	0	0
089	0	0	0	0	0	0	0	0	0	0	0	0	0	0	0
061	0	0	0	0	0	0	0	0	0	0	0	0	0	0	0
090	0	0	0	0	0	0	0	0	0	0	0	0	0	0	0
066	0	0	0	0	0	0	0	0	0	0	0	0	0	0	0
071	0	0	0	0	0	0	0	0	0	0	0	0	0	0	0
086	0	0	1	0	0	0	0	0	0	0	0	0	0	0	0
057	0	0	0	1	1	1	1	0	0	0	0	0	0	0	0
056	0	0	0	0	0	0	0	0	1	0	0	0	0	0	0
009	0	0	0	0	0	0	0	0	1	0	0	0	0	0	0
058	0	0	0	0	0	0	0	0	1	0	0	0	0	0	0

Continued

Nos.	088	087	089	061	090	066	071	086	057	056	009	058	055	037	006
055	0	0	0	0	0	0	0	0	0	1	0	0	0	0	0
037	0	0	0	0	0	0	0	0	0	0	0	0	1	0	0
006	1	1	0	0	0	0	0	1	0	0	0	0	0	1	0

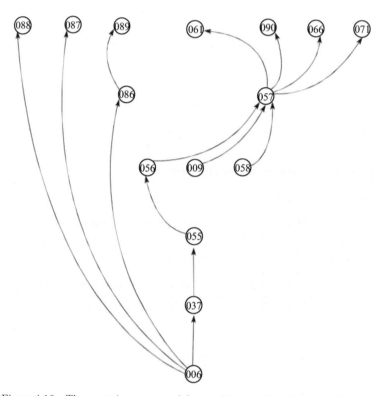

Figure 4.16   The recursion structural figure of iron and steel industrial cluster

4.2.2.2   Petroleum industrial cluster

The accessible matrix M of the petroleum industrial cluster is shown in Table 4.9. Its structural matrix J (Table 4.10) is determined according to the accessible matrix, thereby drawing its recursion structural model figure (Figure 4.17).

Table 4.9   **The accessible matrix of petroleum industrial cluster**

Nos.	067	091	117	102	103	007	086	093	104	094	092	109	036
067	1	1	0	0	0	0	0	0	0	0	0	0	0
091	0	1	0	0	0	0	0	0	0	0	0	0	0
117	0	1	1	1	1	0	0	0	0	0	0	0	0
102	0	0	0	1	0	0	0	0	0	0	0	0	0
103	0	0	0	0	1	0	0	0	0	0	0	0	0
007	0	1	0	0	1	1	0	0	1	1	1	1	1
086	0	1	0	0	1	0	1	1	1	1	1	1	1
093	0	1	0	0	1	0	0	1	1	1	1	1	1
104	0	0	0	0	0	0	0	0	1	0	0	0	0
094	0	0	0	0	0	0	0	0	0	1	0	0	0
092	0	0	0	0	0	0	0	0	0	0	1	0	0
109	0	0	0	0	0	0	0	0	0	0	0	1	0
036	0	1	0	0	1	0	0	0	1	1	1	1	1

Table 4.10 The structural matrix of petroleum industrial cluster

Nos.	091	102	103	104	094	092	109	067	117	036	007	093	086
091	0	0	0	0	0	0	0	0	0	0	0	0	0
102	0	0	0	0	0	0	0	0	0	0	0	0	0
103	0	0	0	0	0	0	0	0	0	0	0	0	0
104	0	0	0	0	0	0	0	0	0	0	0	0	0
094	0	0	0	0	0	0	0	0	0	0	0	0	0
092	0	0	0	0	0	0	0	0	0	0	0	0	0
109	0	0	0	0	0	0	0	0	0	0	0	0	0
067	1	0	0	0	0	0	0	0	0	0	0	0	0
117	1	1	1	0	0	0	0	0	0	0	0	0	0
036	1	0	1	1	1	1	1	0	0	0	0	0	0
007	1	0	0	0	0	0	0	0	0	0	0	0	0
093	1	0	0	0	0	0	0	0	0	0	0	0	0
086	0	0	0	0	0	0	0	0	0	0	0	1	0

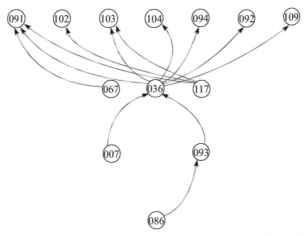

Figure 4.17 The recursion structural figure of petroleum industrial cluster

4.2.2.3 Nonferrous metal industrial cluster

The accessible matrix M of the nonferrous metal industrial cluster is shown in Table 4.11. Its structural matrix J (Table 4.12) is determined according to the accessible matrix, thereby drawing its recursion structural model figure (Figure 4.18).

Table 4.11 The accessible matrix of nonferrous metal industrial cluster

Nos.	010	085	059	060	075	099
010	1	0	1	1	1	1
085	0	1	1	1	1	1
059	0	0	1	1	1	1
060	0	0	0	1	1	1
075	0	0	0	0	1	1
099	0	0	0	0	0	1

Table 4.12 The structural matrix of nonferrous metal industrial cluster

Nos.	099	075	060	059	010	085
099	0	0	0	0	0	0
075	1	0	0	0	0	0
060	0	1	0	0	0	0
059	0	0	1	0	0	0
010	0	0	0	1	0	0
085	0	0	0	1	0	0

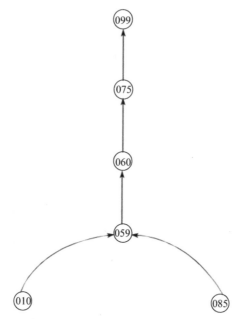

Figure 4.18   The recursion structural figure of the nonferrous metal industrial cluster

4.2.2.4   Chemistry industrial cluster

The accessible matrix M of the chemistry industrial cluster is shown in Table 4.13. Its structural matrix J (Table 4.14) is determined according to the accessible matrix, thereby drawing its recursion structural model figure (Figure 4.19).

Table 4.13   The accessible matrix of chemistry industrial cluster

Nos.	047	084	045	043
047	1	0	0	0
084	0	1	0	0
045	0	0	1	0
043	1	1	1	1

Table 4.14   The structural matrix of chemistry industrial cluster

Nos.	047	084	045	043
047	0	0	0	0
084	0	0	0	0
045	0	0	0	0
043	1	1	1	0

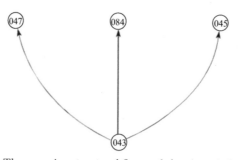

Figure 4.19   The recursion structural figure of chemistry industrial cluster

4.2.3   Construction industrial clusters

The accessible matrix M of the construction industrial cluster is shown in Table 4.15. Its structural matrix J (Table 4.16) is determined according to the accessible matrix, thereby drawing its recursion structural model figure (Figure 4.20).

Table 4.15   The accessible matrix of construction industrial cluster

Nos.	031	030	061	012	013	075	097	099	057	080	049	050	053	123	048	052	090
031	1	0	0	0	0	0	0	0	0	0	0	0	0	0	0	0	0
030	0	1	0	0	0	0	0	0	0	0	0	0	0	0	0	0	1
061	0	0	1	0	0	0	0	0	0	0	0	0	0	0	0	0	1
012	0	0	0	1	0	0	0	0	0	0	0	0	0	0	0	1	1
013	0	0	0	0	1	0	0	0	0	0	0	0	0	0	0	0	1
075	0	0	0	0	0	1	0	1	0	0	0	0	0	0	0	0	1
097	0	0	0	0	0	0	1	0	0	0	0	0	0	0	0	0	1
099	0	0	0	0	0	0	0	1	0	0	0	0	0	0	0	0	1
057	0	0	0	0	0	0	0	0	1	0	0	0	0	0	0	0	1
080	0	0	0	0	0	0	0	0	0	1	0	0	0	0	0	0	1
049	0	0	0	0	0	0	0	0	0	0	1	0	0	0	0	0	1
050	0	0	0	0	0	0	0	0	0	0	0	1	0	0	0	0	1
053	0	0	0	0	0	0	0	0	0	0	0	0	1	0	0	0	1
123	0	0	0	0	0	0	0	0	0	0	0	0	0	1	0	0	1
048	0	0	0	0	0	0	0	0	0	0	1	0	0	0	1	0	1
052	0	0	0	0	0	0	0	0	0	0	0	0	0	0	0	1	1
090	0	0	0	0	0	0	0	0	0	0	0	0	0	0	0	0	1

Table 4.16   The structural matrix of construction industrial cluster

Nos.	031	090	030	061	013	097	099	057	080	049	050	053	123	052	012	075	048
031	0	0	0	0	0	0	0	0	0	0	0	0	0	0	0	0	0
090	0	0	0	0	0	0	0	0	0	0	0	0	0	0	0	0	0
030	1	1	0	0	0	0	0	0	0	0	0	0	0	0	0	0	0
061	0	1	0	0	0	0	0	0	0	0	0	0	0	0	0	0	0
013	0	1	0	0	0	0	0	0	0	0	0	0	0	0	0	0	0
097	0	1	0	0	0	0	0	0	0	0	0	0	0	0	0	0	0
099	0	1	0	0	0	0	0	0	0	0	0	0	0	0	0	0	0
057	0	1	0	0	0	0	0	0	0	0	0	0	0	0	0	0	0
080	0	1	0	0	0	0	0	0	0	0	0	0	0	0	0	0	0
049	0	1	0	0	0	0	0	0	0	0	0	0	0	0	0	0	0
050	0	1	0	0	0	0	0	0	0	0	0	0	0	0	0	0	0
053	0	1	0	0	0	0	0	0	0	0	0	0	0	0	0	0	0
123	0	1	0	0	0	0	0	0	0	0	0	0	0	0	0	0	0
052	0	1	0	0	0	0	0	0	0	0	0	0	0	0	0	0	0
012	0	1	0	0	0	0	0	0	0	0	0	0	0	1	0	0	0
075	0	1	0	0	0	0	1	0	0	0	0	0	0	0	0	0	0
048	0	1	0	0	0	0	0	0	0	1	0	0	0	0	0	0	0

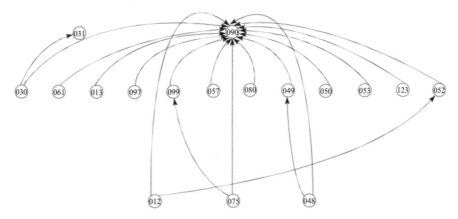

Figure 4.20   The recursion structural figure of construction industrial cluster

4.2.4   Services industrial clusters

4.2.4.1   Commerce industrial cluster

The accessible matrix M of the commerce industrial cluster is shown in Table 4.17. Its structural matrix J (Table 4.18) is determined according to the accessible matrix, thereby drawing its recursion structural model figure (Figure 4.21).

Table 4.17   The accessible matrix of commerce industrial cluster

Nos.	046	002	013	032	119	034	033	077	020	097	108	021	113	040	100
046	1	0	0	0	0	0	0	0	0	0	0	0	0	0	0
002	1	1	1	1	1	1	1	0	0	0	0	0	0	0	1
013	0	0	1	1	1	1	1	0	0	0	0	0	0	0	1
032	0	0	0	1	1	1	1	0	0	0	0	0	0	0	1
119	0	0	0	0	1	0	0	0	0	0	0	0	0	0	0
034	0	0	0	0	0	1	0	0	0	0	0	0	0	0	0
033	0	0	0	0	0	0	1	0	0	0	0	0	0	0	1
077	0	0	0	0	0	0	0	1	0	0	0	0	0	0	1
020	0	0	0	0	0	0	0	0	1	0	0	0	0	0	1
097	0	0	0	0	0	0	0	0	0	1	0	0	0	0	1
108	0	0	0	0	0	0	0	0	0	0	1	0	0	0	1
021	0	0	0	0	0	0	0	0	0	0	0	1	0	0	1
113	0	0	0	0	0	0	0	0	0	0	0	0	1	0	1
040	0	0	0	0	0	0	0	0	0	0	0	0	0	1	1
100	0	0	0	0	0	0	0	0	0	0	0	0	0	0	1

Table 4.18   The structural matrix of commerce industrial cluster

Nos.	046	119	034	100	033	077	020	097	108	021	113	040	032	013	002
046	0	0	0	0	0	0	0	0	0	0	0	0	0	0	0
119	0	0	0	0	0	0	0	0	0	0	0	0	0	0	0
034	0	0	0	0	0	0	0	0	0	0	0	0	0	0	0
100	0	0	0	0	0	0	0	0	0	0	0	0	0	0	0
033	0	0	0	1	0	0	0	0	0	0	0	0	0	0	0
077	0	0	0	1	0	0	0	0	0	0	0	0	0	0	0
020	0	0	0	1	0	0	0	0	0	0	0	0	0	0	0
097	0	0	0	1	0	0	0	0	0	0	0	0	0	0	0
108	0	0	0	1	0	0	0	0	0	0	0	0	0	0	0
021	0	0	0	1	0	0	0	0	0	0	0	0	0	0	0
113	0	0	0	1	0	0	0	0	0	0	0	0	0	0	0
040	0	0	0	1	0	0	0	0	0	0	0	0	0	0	0
032	0	1	1	0	1	0	0	0	0	0	0	0	0	0	0
013	0	0	0	0	0	0	0	0	0	0	0	1	0	0	0
002	1	0	0	0	0	0	0	0	0	0	0	0	0	1	0

4.2.4.2   Diet industrial cluster

The accessible matrix M of the diet industrial cluster is shown in Table 4.19. Its structural matrix J (Table 4.20) is determined according to the accessible matrix, thereby drawing its recursion structural model figure (Figure 4.22).

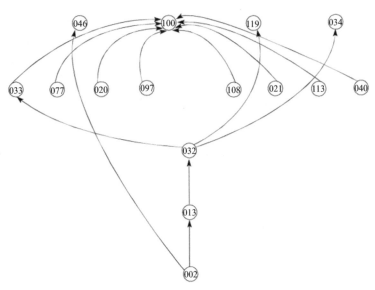

Figure 4.21   The recursion structural figure of commerce industrial cluster

Table 4.19   The accessible matrix of diet industrial cluster

Nos.	122	001	014	004	017	020	016	019	018	021	101
122	1	1	1	1	0	0	0	0	0	0	1
001	0	1	1	1	0	0	0	0	0	0	1
014	0	0	1	1	0	0	0	0	0	0	1
004	0	0	0	1	0	0	0	0	0	0	1
017	0	0	1	1	1	0	0	0	0	0	1
020	0	0	0	0	0	1	0	0	0	0	1
016	0	0	0	0	0	0	1	0	0	0	1
019	0	0	0	0	0	0	0	1	0	0	1
018	0	0	0	0	0	0	0	0	1	0	1
021	0	0	0	0	0	0	0	0	0	1	1
101	0	0	0	0	0	0	0	0	0	0	1

Table 4.20   The structural matrix of diet industrial cluster

Nos.	101	004	020	016	019	018	021	014	001	017	122
101	0	0	0	0	0	0	0	0	0	0	0
004	1	0	0	0	0	0	0	0	0	0	0
020	1	0	0	0	0	0	0	0	0	0	0
016	1	0	0	0	0	0	0	0	0	0	0
019	1	0	0	0	0	0	0	0	0	0	0
018	1	0	0	0	0	0	0	0	0	0	0
021	1	0	0	0	0	0	0	0	0	0	0
014	0	1	0	0	0	0	0	0	0	0	0
001	0	0	0	0	0	0	0	1	0	0	0
017	1	0	0	0	0	0	0	1	0	0	0
122	0	0	0	0	0	0	0	0	1	0	0

4.2.4.3   Administration industrial cluster

The accessible matrix M of the administration industrial cluster is shown in Table 4.21. Its structural matrix J (Table 4.22) is determined according to the accessible matrix, thereby drawing its recursion structural model figure (Figure 4.23).

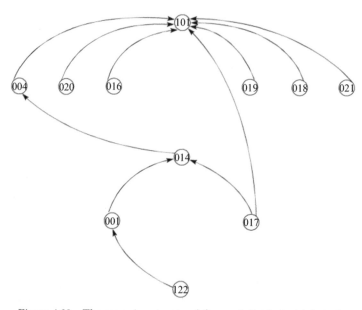

Figure 4.22   The recursion structural figure of diet industrial cluster

Table 4.21   The accessible matrix of administration industrial cluster

Nos.	070	105	112	111	124	098	102	117
070	1	1	1	0	1	0	0	0
105	0	1	1	0	1	0	0	0
112	0	0	1	0	0	0	0	0
111	0	0	0	1	1	0	0	0
124	0	0	0	0	1	0	0	0
098	0	0	0	0	1	1	0	0
102	0	0	0	0	1	0	1	0
117	0	0	0	0	1	0	1	1

Table 4.22   The structural matrix of administration industrial cluster

Nos.	112	124	105	111	098	102	070	117
112	0	0	0	0	0	0	0	0
124	0	0	0	0	0	0	0	0
105	1	1	0	0	0	0	0	0
111	0	1	0	0	0	0	0	0
098	0	1	0	0	0	0	0	0
102	0	1	0	0	0	0	0	0
070	0	0	1	0	0	0	0	0
117	0	0	0	0	0	1	0	0

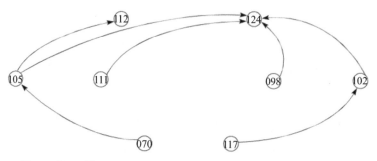

Figure 4.23   The recursion figure of administration industrial cluster

4.3   STRUCTURE-EFFECT RELATIONSHIPS OF INDUSTRIAL CLUSTERS IN CHINA

The properties of industrial clusters depend on the structures of industrial clusters; that is, there is a correlation between the structures and the properties of industrial clusters. As shown in Figure 4.24, the structures of industrial clusters are diagrams, and they are mathematical quantities. If we want to build such correlation, we need to extract the characteristics from the structures, use these characteristics (as independent variables) to construct mathematical models, and then proceed to analyze the properties of industrial clusters using the constructed mathematical models.

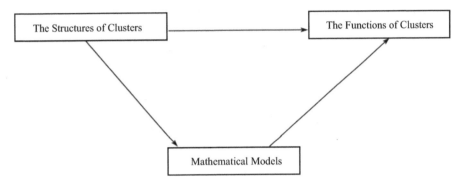

Figure 4.24   Multilateral QSAR/QSPR method

4.3.1   Computing methods of structural topological indexes of industrial clusters in China

When industrial sectors in the structural diagrams of industrial clusters are represented by nodes and the strength of functional links in the structural diagrams of industrial clusters is denoted by sides, structural diagrams of industrial clusters are topological diagrams. Structural topological diagrams are coloured diagrams; some parameters of economic links are needed to characterize the coloured diagrams. Therefore, two columns are added on the basis of distance matrixes, of which one column is the square root of 'the degree of branching' (the number of connected sides of industrial sectors) for each industrial sector and the other is the square root of van der Waals radiuses (the sum of each industry's inputs and outputs divided by the sum of all industries' inputs and outputs); that is, vertexes of structural diagrams are coloured up.

The main types of parameters studied by quantitative structure-quality pertinence (QSAR/QSPR) are topological types of parameters, geometric types of parameters, electronic types of parameters and physical chemical types of parameters. Topological types of parameters directly derive from structural diagrams, whose algorithms are comparatively simple, and better effects can be obtained in the factor analysis at the same time. Therefore, topological types of parameters are chosen in the study of structure-effect relationships of industrial clusters.

Topological indexes are mathematical invariables derived from structural diagrams of industrial clusters, and there are two requirments for structural diagrams of industrial clusters: one is better singularity (topological indexes of different structures being different), namely high selectivity; the other is good correlation with the quality of clusters in order to carry through the study on correlations between structures and qualities of industrial clusters (QSAR/QSPR) (Kier and Hall, 1986).

In this study, we choose the topological index Am adopted by Chinese scholars (Xu and Hu, 2000). The index is built on the base of an augmented distance matrix; Am_1, Am_2, Am_3 are derived from matrixes A, B, C.

$$a_{ij} = \begin{cases} 1, & \text{when the path between vertex } i \text{ and } j \text{ is } P = 1 \\ 0, & \text{others} \end{cases}$$

$$b_{ij} = \begin{cases} 1, & \text{when the path between vertex } i \text{ and } j \text{ is } P = 2 \\ 0, & \text{others} \end{cases}$$

$$c_{ij} = \begin{cases} 1, & \text{when the path between vertex } i \text{ and } j \text{ is } P = 3 \\ 0, & \text{others} \end{cases}$$

After adding two columns on the distance matrix, augmented matrixes are, respectively, G_1, G_2, G_3 and dealing with matrixes G_1-G_3 is as follows:

$$Z_1 = G_1 \times G_1', \quad Z_2 = G_2 \times G_2', \quad Z_3 = G_3 \times G_3'$$

where G_1', G_2' and G_3' are, respectively, transposed matrixes of G_1, G_2 and G_3.

Then the new topological indexes Am_1, Am_2, Am_3 are defined as follows:

$$Am_1 = \frac{\lambda_{\max_1}}{2}, \quad Am_2 = \frac{\lambda_{\max_2}}{2}, \quad Am_3 = \frac{\lambda_{\max_3}}{2}$$

where $\lambda_{\max_1}$, $\lambda_{\max_2}$ and $\lambda_{\max_3}$ are, respectively, the maximal eigenvalues of Z_1, Z_2 and Z_3.

4.3.2   Topological indexes of industrial clusters in China

Translating directed structural diagrams into nondirected structural topological diagrams, using the computing methods of topological indexes above, on the basis of figuring out the augmented matrixes G of each industry, we calculate their matrix Z and finally the topological indexes of each industrial cluster.

4.3.2.1   Planting industrial cluster

Matrixes G and Z corresponding to the planting industrial cluster in China are as follows:

$$G_1 = \begin{bmatrix}
1.73 & 0.16 & 0 & 0 & 1 & 0 & 0 & 0 & 0 & 0 & 0 & 0 & 0 & 0 & 0 & 0 & 1 & 0 & 0 & 1 \\
1 & 0.09 & 0 & 0 & 0 & 0 & 0 & 0 & 0 & 0 & 0 & 0 & 0 & 0 & 0 & 0 & 0 & 0 & 0 & 1 \\
1.73 & 0.16 & 1 & 0 & 0 & 1 & 0 & 0 & 0 & 0 & 0 & 0 & 0 & 0 & 0 & 0 & 0 & 0 & 0 & 1 \\
1 & 0.08 & 0 & 0 & 1 & 0 & 0 & 0 & 0 & 0 & 0 & 0 & 0 & 0 & 0 & 0 & 0 & 0 & 0 & 0 \\
1 & 0.04 & 0 & 0 & 0 & 0 & 0 & 0 & 0 & 0 & 0 & 0 & 0 & 0 & 0 & 0 & 0 & 0 & 0 & 1 \\
1.41 & 0.06 & 0 & 0 & 0 & 0 & 0 & 0 & 0 & 0 & 1 & 0 & 0 & 0 & 0 & 0 & 0 & 0 & 0 & 1 \\
1 & 0.08 & 0 & 0 & 0 & 0 & 0 & 0 & 0 & 0 & 0 & 0 & 0 & 0 & 0 & 0 & 0 & 0 & 0 & 1 \\
1 & 0.05 & 0 & 0 & 0 & 0 & 0 & 0 & 0 & 0 & 0 & 0 & 0 & 0 & 0 & 0 & 0 & 0 & 0 & 1 \\
1.73 & 0.04 & 0 & 0 & 0 & 0 & 0 & 1 & 0 & 0 & 0 & 0 & 0 & 1 & 0 & 0 & 0 & 0 & 0 & 1 \\
1 & 0.03 & 0 & 0 & 0 & 0 & 0 & 0 & 0 & 0 & 0 & 0 & 0 & 0 & 0 & 0 & 0 & 0 & 0 & 1 \\
1 & 0.07 & 0 & 0 & 0 & 0 & 0 & 0 & 0 & 0 & 0 & 0 & 0 & 0 & 0 & 0 & 0 & 0 & 0 & 1 \\
1.41 & 0.1 & 0 & 0 & 0 & 0 & 0 & 0 & 0 & 0 & 1 & 0 & 0 & 0 & 0 & 0 & 0 & 0 & 0 & 1 \\
1.41 & 0.1 & 0 & 0 & 0 & 0 & 0 & 0 & 0 & 0 & 0 & 0 & 0 & 0 & 0 & 1 & 0 & 0 & 0 & 1 \\
1 & 0.02 & 0 & 0 & 0 & 0 & 0 & 0 & 0 & 0 & 0 & 0 & 0 & 0 & 1 & 0 & 0 & 0 & 0 & 0 \\
1.41 & 0.02 & 1 & 0 & 0 & 0 & 0 & 0 & 0 & 0 & 0 & 0 & 0 & 0 & 0 & 0 & 0 & 0 & 0 & 1 \\
1 & 0.06 & 0 & 0 & 0 & 0 & 0 & 0 & 0 & 0 & 0 & 0 & 0 & 0 & 0 & 0 & 0 & 0 & 0 & 1 \\
1 & 0.07 & 0 & 0 & 0 & 0 & 0 & 0 & 0 & 0 & 0 & 0 & 0 & 0 & 0 & 0 & 0 & 0 & 0 & 1 \\
3.87 & 0.22 & 1 & 1 & 1 & 0 & 1 & 1 & 1 & 1 & 1 & 1 & 1 & 1 & 0 & 1 & 1 & 1 & 0
\end{bmatrix}$$

$$G_2 = \begin{bmatrix}
1.73 & 0.16 & 0 & 2 & 2 & 2 & 2 & 2 & 2 & 2 & 2 & 2 & 2 & 2 & 2 & 0 & 2 & 2 & 2 & 2 \\
1 & 0.09 & 2 & 0 & 2 & 0 & 2 & 2 & 2 & 2 & 2 & 2 & 2 & 2 & 2 & 0 & 2 & 2 & 2 & 0 \\
1.73 & 0.16 & 2 & 2 & 0 & 0 & 2 & 2 & 2 & 2 & 2 & 2 & 2 & 2 & 2 & 0 & 2 & 2 & 2 & 2 \\
1 & 0.08 & 2 & 0 & 0 & 0 & 0 & 0 & 0 & 0 & 0 & 0 & 0 & 0 & 0 & 0 & 0 & 0 & 0 & 2 \\
1 & 0.04 & 2 & 2 & 2 & 0 & 0 & 2 & 2 & 2 & 2 & 2 & 2 & 2 & 2 & 0 & 2 & 2 & 2 & 0 \\
1.41 & 0.06 & 2 & 2 & 2 & 0 & 2 & 0 & 2 & 2 & 2 & 2 & 2 & 2 & 2 & 0 & 2 & 2 & 2 & 2 \\
1 & 0.08 & 2 & 2 & 2 & 0 & 2 & 2 & 0 & 2 & 2 & 2 & 2 & 2 & 2 & 0 & 2 & 2 & 2 & 0 \\
1 & 0.05 & 2 & 2 & 2 & 0 & 2 & 2 & 2 & 0 & 2 & 2 & 2 & 2 & 2 & 0 & 2 & 2 & 2 & 0 \\
1.73 & 0.04 & 2 & 2 & 2 & 2 & 2 & 2 & 2 & 2 & 0 & 2 & 2 & 2 & 2 & 0 & 2 & 2 & 2 & 2 \\
1 & 0.03 & 2 & 2 & 2 & 0 & 2 & 2 & 2 & 2 & 2 & 0 & 2 & 2 & 2 & 0 & 2 & 2 & 2 & 0 \\
1 & 0.07 & 2 & 2 & 2 & 0 & 2 & 2 & 2 & 2 & 2 & 2 & 0 & 2 & 2 & 0 & 2 & 2 & 2 & 0 \\
1.41 & 0.1 & 2 & 2 & 2 & 0 & 2 & 2 & 2 & 2 & 2 & 2 & 2 & 0 & 2 & 0 & 2 & 0 & 2 & 2 \\
1.41 & 0.1 & 2 & 2 & 2 & 0 & 2 & 2 & 2 & 2 & 2 & 2 & 2 & 2 & 0 & 0 & 2 & 2 & 2 & 0 \\
1 & 0.02 & 0 & 0 & 0 & 0 & 0 & 0 & 0 & 0 & 0 & 0 & 0 & 0 & 0 & 0 & 0 & 0 & 0 & 2 \\
1.41 & 0.02 & 2 & 2 & 2 & 0 & 2 & 2 & 2 & 2 & 2 & 2 & 2 & 2 & 0 & 0 & 2 & 2 & 2 & 2 \\
1 & 0.06 & 2 & 2 & 2 & 0 & 2 & 2 & 2 & 2 & 2 & 2 & 2 & 2 & 0 & 2 & 0 & 2 & 0 \\
1 & 0.07 & 2 & 2 & 2 & 0 & 2 & 2 & 2 & 2 & 2 & 2 & 2 & 2 & 0 & 2 & 2 & 0 & 0 \\
3.87 & 0.22 & 2 & 0 & 2 & 2 & 0 & 2 & 0 & 0 & 2 & 0 & 0 & 2 & 0 & 2 & 2 & 0 & 0 & 0
\end{bmatrix}$$

$$G_3 = \begin{bmatrix}
1.73 & 0.16 & 0 & 3 & 3 & 3 & 3 & 3 & 0 & 3 & 3 & 3 & 3 & 3 & 3 & 3 & 3 & 3 & 3 \\
1 & 0.09 & 3 & 0 & 0 & 3 & 0 & 3 & 0 & 0 & 3 & 0 & 0 & 3 & 0 & 3 & 3 & 0 & 0 \\
1.73 & 0.16 & 3 & 0 & 0 & 0 & 3 & 3 & 3 & 3 & 3 & 3 & 3 & 3 & 3 & 3 & 3 & 3 & 3 \\
1 & 0.08 & 3 & 3 & 0 & 0 & 3 & 3 & 3 & 3 & 3 & 3 & 3 & 3 & 0 & 3 & 3 & 3 & 3 \\
1 & 0.04 & 3 & 0 & 3 & 3 & 0 & 3 & 0 & 0 & 3 & 0 & 0 & 3 & 0 & 3 & 3 & 0 & 0 \\
1.41 & 0.06 & 3 & 3 & 3 & 3 & 3 & 0 & 3 & 3 & 3 & 3 & 3 & 3 & 3 & 3 & 3 & 3 & 3 \\
1 & 0.08 & 3 & 0 & 3 & 3 & 0 & 3 & 0 & 0 & 3 & 0 & 0 & 3 & 0 & 3 & 3 & 0 & 0 \\
1 & 0.05 & 0 & 0 & 3 & 3 & 0 & 3 & 0 & 0 & 0 & 0 & 0 & 3 & 0 & 3 & 3 & 0 & 0 \\
1.73 & 0.04 & 3 & 3 & 3 & 3 & 3 & 3 & 3 & 0 & 0 & 3 & 3 & 3 & 3 & 3 & 3 & 3 & 3 \\
1 & 0.03 & 3 & 0 & 3 & 3 & 0 & 3 & 0 & 0 & 3 & 0 & 0 & 3 & 0 & 3 & 3 & 0 & 0 \\
1 & 0.07 & 3 & 0 & 3 & 3 & 0 & 3 & 0 & 0 & 3 & 0 & 0 & 3 & 0 & 3 & 3 & 0 & 0 \\
1.41 & 0.1 & 3 & 3 & 3 & 3 & 3 & 3 & 3 & 3 & 3 & 3 & 3 & 0 & 3 & 3 & 3 & 3 & 3 \\
1.41 & 0.1 & 3 & 0 & 3 & 3 & 0 & 3 & 0 & 0 & 3 & 0 & 0 & 3 & 0 & 0 & 3 & 0 & 0 \\
1 & 0.02 & 3 & 3 & 3 & 0 & 3 & 3 & 3 & 3 & 3 & 3 & 3 & 3 & 0 & 0 & 3 & 3 & 3 \\
1.41 & 0.02 & 3 & 3 & 3 & 3 & 3 & 3 & 3 & 3 & 3 & 3 & 3 & 3 & 3 & 3 & 0 & 3 & 3 \\
1 & 0.06 & 3 & 0 & 3 & 3 & 0 & 3 & 0 & 0 & 3 & 0 & 0 & 3 & 0 & 3 & 3 & 0 & 0 \\
1 & 0.07 & 3 & 0 & 3 & 3 & 0 & 3 & 0 & 0 & 3 & 0 & 0 & 3 & 0 & 3 & 3 & 0 & 0 \\
3.87 & 0.22 & 0 & 0 & 3 & 3 & 0 & 3 & 0 & 0 & 0 & 0 & 0 & 3 & 0 & 0 & 3 & 0 & 0
\end{bmatrix}$$

$$z_1 = \begin{bmatrix}
6.03 & 2.75 & 4.03 & 2.75 & 2.74 & 3.46 & 2.74 & 2.74 & 4.01 & 2.74 & 2.74 & 3.46 & 3.47 & 1.74 & 3.45 & 2.74 & 2.74 & 8.74 \\
2.75 & 2.01 & 2.75 & 1.01 & 2 & 2.42 & 2.01 & 2 & 2.74 & 2 & 2.01 & 2.42 & 2.42 & 1 & 2.42 & 2.01 & 2.01 & 3.89 \\
4.03 & 2.75 & 6.03 & 1.75 & 2.74 & 3.46 & 2.75 & 2.74 & 4.01 & 2.74 & 2.74 & 3.47 & 3.47 & 1.74 & 4.45 & 2.74 & 2.74 & 7.74 \\
2.75 & 1.01 & 1.75 & 2.01 & 1 & 1.42 & 1.01 & 1 & 1.74 & 1 & 1.01 & 1.42 & 1.42 & 1 & 1.42 & 1.01 & 1.01 & 4.89 \\
2.74 & 2 & 2.74 & 1 & 2 & 2.42 & 2 & 2 & 2.73 & 2 & 2 & 2.42 & 2.42 & 1 & 2.42 & 2 & 2 & 3.88 \\
3.46 & 2.42 & 3.46 & 1.42 & 2.42 & 4 & 2.42 & 2.42 & 3.45 & 2.42 & 2.42 & 4.01 & 3.01 & 1.42 & 3 & 2.42 & 2.42 & 6.49 \\
2.74 & 2.01 & 2.75 & 1.01 & 2 & 2.42 & 2.01 & 2 & 2.74 & 2 & 2.01 & 2.42 & 2.42 & 1 & 2.42 & 2.01 & 2.01 & 3.89 \\
2.74 & 2 & 2.74 & 1 & 2 & 2.42 & 2 & 2 & 2.73 & 2 & 2 & 2.42 & 2.42 & 1 & 2.42 & 2 & 2 & 3.89 \\
4.01 & 2.74 & 4.01 & 1.74 & 2.73 & 3.45 & 2.74 & 2.73 & 6 & 2.73 & 2.74 & 3.45 & 3.45 & 1.73 & 3.45 & 2.73 & 2.73 & 8.72 \\
2.74 & 2 & 2.74 & 1 & 2 & 2.42 & 2 & 2 & 2.73 & 2 & 2 & 2.42 & 2.42 & 1 & 2.41 & 2 & 2 & 3.88 \\
2.74 & 2.01 & 2.74 & 1.01 & 2 & 2.42 & 2.01 & 2 & 2.74 & 2 & 2.01 & 2.42 & 2.42 & 1 & 2.42 & 2 & 2 & 3.89 \\
3.46 & 2.42 & 3.47 & 1.42 & 2.42 & 4.01 & 2.42 & 2.42 & 3.45 & 2.42 & 2.42 & 4.01 & 3.01 & 1.42 & 3 & 2.42 & 2.42 & 6.5 \\
3.47 & 2.42 & 3.47 & 1.42 & 2.42 & 3.01 & 2.42 & 2.42 & 3.45 & 2.42 & 2.42 & 3.01 & 4.01 & 1.42 & 3 & 2.42 & 2.42 & 5.5 \\
1.74 & 1 & 1.74 & 1 & 1.42 & 1.42 & 1 & 1 & 1.73 & 1 & 1 & 1.42 & 1.42 & 2 & 1.41 & 1 & 1 & 4.88 \\
3.45 & 2.42 & 4.45 & 1.42 & 2.42 & 3 & 2.42 & 2.42 & 3.45 & 2.41 & 2.42 & 3 & 3 & 1.41 & 4 & 2.42 & 2.42 & 6.48 \\
2.74 & 2.01 & 2.74 & 1.01 & 2 & 2.42 & 2.01 & 2 & 2.73 & 2 & 2 & 2.42 & 2.42 & 1 & 2.42 & 2 & 2 & 3.89 \\
2.74 & 2.01 & 2.74 & 1.01 & 2 & 2.42 & 2.01 & 2 & 2.73 & 2 & 2 & 2.42 & 2.42 & 1 & 2.42 & 2 & 2 & 3.89 \\
8.74 & 3.89 & 7.74 & 4.89 & 3.88 & 6.49 & 3.89 & 3.89 & 8.72 & 3.88 & 3.89 & 6.5 & 5.5 & 4.88 & 6.48 & 3.89 & 3.89 & 30.1
\end{bmatrix}$$

$$z_2 = \begin{bmatrix}
67 & 53.7 & 59 & 5.75 & 53.7 & 58.5 & 53.7 & 53.7 & 59 & 53.7 & 53.7 & 58.5 & 54.5 & 5.74 & 58.5 & 53.7 & 53.7 & 30.7 \\
53.7 & 57 & 53.7 & 5.01 & 53 & 53.4 & 53 & 53 & 53.7 & 53 & 53 & 53.4 & 53.4 & 1 & 53.4 & 53 & 53 & 27.9 \\
59 & 53.7 & 63 & 9.75 & 53.7 & 58.5 & 53.7 & 53.7 & 59 & 53.7 & 53.7 & 58.5 & 54.5 & 5.74 & 58.5 & 53.7 & 53.7 & 26.7 \\
5.75 & 5.01 & 9.75 & 9.01 & 5 & 9.42 & 5.01 & 5 & 9.74 & 5 & 5.01 & 9.42 & 5.42 & 5 & 9.42 & 5.01 & 5.01 & 7.89 \\
53.7 & 53 & 53.7 & 5 & 57 & 53.4 & 53 & 53 & 53.7 & 53 & 53 & 53.4 & 53.4 & 1 & 53.4 & 53 & 53 & 27.9 \\
58.5 & 53.4 & 58.5 & 9.42 & 53.4 & 62 & 53.4 & 53.4 & 58.5 & 53.4 & 53.4 & 58 & 54 & 5.42 & 58 & 53.4 & 53.4 & 25.5 \\
53.7 & 53 & 53.7 & 5.01 & 53 & 53.4 & 57 & 53 & 53.7 & 53 & 53 & 53.4 & 53.4 & 1 & 53.4 & 53 & 53 & 27.9 \\
53.7 & 53 & 53.7 & 5 & 53 & 53.4 & 53 & 57 & 53.7 & 53 & 53 & 53.4 & 53.4 & 1 & 53.4 & 53 & 53 & 27.9 \\
59 & 53.7 & 59 & 9.74 & 53.7 & 58.5 & 53.7 & 53.7 & 63 & 53.7 & 53.7 & 58.5 & 54.5 & 5.73 & 58.5 & 53.7 & 53.7 & 26.7 \\
53.7 & 53 & 53.7 & 5 & 53 & 53.4 & 53 & 53 & 53.7 & 57 & 53 & 53.4 & 53.4 & 1 & 53.4 & 53 & 53 & 27.9 \\
53.7 & 53 & 53.7 & 5.01 & 53 & 53.4 & 53 & 53 & 53.7 & 53 & 57 & 53.4 & 53.4 & 1 & 53.4 & 53 & 53 & 27.9 \\
58.5 & 53.4 & 58.5 & 9.42 & 53.4 & 58 & 53.4 & 53.4 & 58.5 & 53.4 & 53.4 & 62 & 54 & 5.42 & 58 & 53.4 & 53.4 & 25.5 \\
54.5 & 53.4 & 54.5 & 5.42 & 53.4 & 54 & 53.4 & 53.4 & 54.5 & 53.4 & 53.4 & 54 & 58 & 1.42 & 54 & 53.4 & 53.4 & 29.5 \\
5.74 & 1 & 5.74 & 5 & 1 & 5.42 & 1 & 1 & 5.73 & 1 & 1 & 5.42 & 1.42 & 5 & 5.41 & 1 & 1 & 3.88 \\
58.5 & 53.4 & 58.5 & 9.42 & 53.4 & 58 & 53.4 & 53.4 & 58.5 & 53.4 & 53.4 & 58 & 54 & 5.41 & 62 & 53.4 & 53.4 & 25.5 \\
53.7 & 53 & 53.7 & 5.01 & 53 & 53.4 & 53 & 53 & 53.7 & 53 & 53 & 53.4 & 53.4 & 1 & 53.4 & 57 & 53 & 27.9 \\
53.7 & 53 & 53.7 & 5.01 & 53 & 53.4 & 53 & 53 & 53.7 & 53 & 53 & 53.4 & 53.4 & 1 & 53.4 & 53 & 57 & 27.9 \\
30.7 & 27.9 & 26.7 & 7.89 & 27.9 & 25.5 & 27.9 & 27.9 & 26.7 & 27.9 & 27.9 & 25.5 & 29.5 & 3.88 & 25.5 & 27.9 & 27.9 & 47.1
\end{bmatrix}$$

$$z_3 = \begin{bmatrix}
138 & 55.7 & 111 & 110 & 64.7 & 128 & 64.7 & 55.7 & 129 & 64.7 & 64.7 & 128 & 56.5 & 110 & 128 & 64.7 & 64.7 & 51.7 \\
55.7 & 64 & 55.7 & 46 & 64 & 55.4 & 64 & 46 & 55.7 & 64 & 64 & 55.4 & 55.4 & 46 & 55.4 & 64 & 64 & 39.9 \\
111 & 55.7 & 138 & 128 & 55.7 & 128 & 55.7 & 37.7 & 111 & 55.7 & 55.7 & 128 & 47.5 & 110 & 128 & 55.7 & 55.7 & 33.7 \\
110 & 46 & 128 & 136 & 46 & 127 & 46 & 28 & 110 & 46 & 46 & 127 & 46.4 & 118 & 127 & 46 & 46 & 30.9 \\
64.7 & 64 & 55.7 & 46 & 73 & 64.4 & 73 & 55 & 64.7 & 73 & 73 & 64.4 & 64.4 & 55 & 64.4 & 73 & 73 & 48.9 \\
128 & 55.4 & 128 & 127 & 64.4 & 155 & 64.4 & 46.4 & 128 & 64.4 & 64.4 & 146 & 56 & 118 & 146 & 64.4 & 64.4 & 41.5 \\
64.7 & 64 & 55.7 & 46 & 73 & 64.4 & 73 & 55 & 64.7 & 73 & 73 & 64.4 & 64.4 & 55 & 64.4 & 73 & 73 & 48.9 \\
55.7 & 46 & 37.7 & 28 & 55 & 46.4 & 55 & 55 & 55.7 & 55 & 55 & 46.4 & 46.4 & 37 & 46.4 & 55 & 55 & 48.9 \\
129 & 55.7 & 111 & 110 & 64.7 & 128 & 64.7 & 55.7 & 138 & 64.7 & 64.7 & 128 & 56.5 & 110 & 128 & 64.7 & 64.7 & 51.7 \\
64.7 & 64 & 55.7 & 46 & 73 & 64.4 & 73 & 55 & 64.7 & 73 & 73 & 64.4 & 64.4 & 55 & 64.4 & 73 & 73 & 48.9 \\
64.7 & 64 & 55.7 & 46 & 73 & 64.4 & 73 & 55 & 64.7 & 73 & 73 & 64.4 & 64.4 & 55 & 64.4 & 73 & 73 & 48.9 \\
128 & 55.4 & 128 & 127 & 64.4 & 146 & 64.4 & 46.4 & 128 & 64.4 & 64.4 & 155 & 56 & 118 & 146 & 64.4 & 64.4 & 41.5 \\
56.5 & 55.4 & 47.5 & 46.4 & 64.4 & 56 & 64.4 & 46.4 & 56.5 & 64.4 & 64.4 & 56 & 65 & 55.4 & 56 & 64.4 & 64.4 & 50.5 \\
110 & 46 & 110 & 118 & 55 & 118 & 55 & 37 & 110 & 55 & 55 & 118 & 55.4 & 127 & 118 & 55 & 55 & 39.9 \\
128 & 55.4 & 128 & 127 & 64.4 & 146 & 64.4 & 46.4 & 128 & 64.4 & 64.4 & 155 & 56 & 118 & 155 & 64.4 & 64.4 & 41.5 \\
64.7 & 64 & 55.7 & 46 & 73 & 64.4 & 73 & 55 & 64.7 & 73 & 73 & 64.4 & 64.4 & 55 & 64.4 & 73 & 73 & 48.9 \\
64.7 & 64 & 55.7 & 46 & 73 & 64.4 & 73 & 55 & 64.7 & 73 & 73 & 64.4 & 64.4 & 55 & 64.4 & 73 & 73 & 48.9 \\
51.7 & 39.9 & 33.7 & 30.9 & 48.9 & 41.5 & 48.9 & 48.9 & 51.7 & 48.9 & 48.9 & 41.5 & 50.5 & 39.9 & 41.5 & 48.9 & 48.9 & 60.1
\end{bmatrix}$$

Maximal eigenvalues of matrix Z of the planting industrial cluster are 16.14230, 15.2763 and 10.9220, and their topological indexes are 8.0715, 7.6382 and 5.4610.

4.3.2.2   Electronics industrial cluster

Matrixes G and Z corresponding to the electronics industrial cluster in China are as follows:

$$G_1 = \begin{bmatrix}
1 & 0.03 & 0 & 1 & 0 & 0 & 0 & 0 & 0 & 0 & 0 \\
1.41 & 0.11 & 1 & 0 & 1 & 0 & 0 & 0 & 0 & 0 & 0 \\
1.41 & 0.07 & 0 & 1 & 0 & 1 & 0 & 0 & 0 & 0 & 0 \\
2.24 & 0.11 & 0 & 0 & 1 & 0 & 1 & 1 & 1 & 1 & 0 \\
1 & 0.07 & 0 & 0 & 0 & 1 & 0 & 0 & 0 & 0 & 0 \\
1 & 0.04 & 0 & 0 & 0 & 1 & 0 & 0 & 0 & 0 & 0 \\
1 & 0.08 & 0 & 0 & 0 & 1 & 0 & 0 & 0 & 0 & 0 \\
1.41 & 0.07 & 0 & 0 & 0 & 1 & 0 & 0 & 0 & 0 & 1 \\
1 & 0.22 & 0 & 0 & 0 & 0 & 0 & 0 & 0 & 1 & 0
\end{bmatrix}$$

$$G_2 = \begin{bmatrix}
1 & 0.03 & 0 & 0 & 2 & 0 & 0 & 0 & 0 & 0 & 0 \\
1.41 & 0.11 & 0 & 0 & 0 & 2 & 0 & 0 & 0 & 0 & 0 \\
1.41 & 0.07 & 2 & 0 & 0 & 0 & 2 & 2 & 2 & 2 & 0 \\
2.24 & 0.11 & 0 & 2 & 0 & 0 & 0 & 0 & 0 & 0 & 2 \\
1 & 0.07 & 0 & 0 & 2 & 0 & 0 & 2 & 2 & 2 & 0 \\
1 & 0.04 & 0 & 0 & 2 & 0 & 2 & 0 & 2 & 2 & 0 \\
1 & 0.08 & 0 & 0 & 2 & 0 & 2 & 2 & 0 & 2 & 0 \\
1.41 & 0.07 & 0 & 0 & 2 & 0 & 2 & 2 & 2 & 0 & 0 \\
1 & 0.22 & 0 & 0 & 0 & 2 & 0 & 0 & 0 & 0 & 0
\end{bmatrix}$$

$$G_3 = \begin{bmatrix}
1 & 0.03 & 0 & 0 & 0 & 3 & 0 & 0 & 0 & 0 & 0 \\
1.41 & 0.11 & 0 & 0 & 0 & 0 & 3 & 3 & 3 & 3 & 0 \\
1.41 & 0.07 & 0 & 0 & 0 & 0 & 0 & 0 & 0 & 0 & 3 \\
2.24 & 0.11 & 3 & 0 & 0 & 0 & 0 & 0 & 0 & 0 & 0 \\
1 & 0.07 & 0 & 3 & 0 & 0 & 0 & 0 & 0 & 0 & 3 \\
1 & 0.04 & 0 & 3 & 0 & 0 & 0 & 0 & 0 & 0 & 3 \\
1 & 0.08 & 0 & 3 & 0 & 0 & 0 & 0 & 0 & 0 & 3 \\
1.41 & 0.07 & 0 & 3 & 0 & 0 & 0 & 0 & 0 & 0 & 0 \\
1 & 0.22 & 0 & 0 & 3 & 0 & 3 & 3 & 3 & 0 & 0
\end{bmatrix}$$

$$Z_1 = \begin{bmatrix}
2 & 1.42 & 2.42 & 2.24 & 1 & 1 & 1 & 1.42 & 1.01 \\
1.42 & 4.01 & 2.01 & 4.17 & 1.42 & 1.42 & 1.42 & 2.01 & 1.44 \\
2.42 & 2.01 & 4.01 & 3.17 & 2.42 & 2.42 & 2.42 & 3.01 & 1.43 \\
2.24 & 4.17 & 3.17 & 10 & 2.24 & 2.24 & 2.24 & 3.17 & 3.26 \\
1 & 1.42 & 2.42 & 2.24 & 2 & 2 & 2.01 & 2.42 & 1.01 \\
1 & 1.42 & 2.42 & 2.24 & 2 & 2 & 2 & 2.42 & 1.01 \\
1 & 1.42 & 2.42 & 2.24 & 2.01 & 2 & 2.01 & 2.42 & 1.02 \\
1.42 & 2.01 & 3.01 & 3.17 & 2.42 & 2.42 & 2.42 & 4.01 & 1.43 \\
1.01 & 1.44 & 1.43 & 3.26 & 1.01 & 1.01 & 1.02 & 1.43 & 2.05
\end{bmatrix}$$

$$Z_2 = \begin{bmatrix}
5 & 1.42 & 1.42 & 2.24 & 5 & 5 & 5 & 5.42 & 1.01 \\
1.42 & 6.01 & 2.01 & 3.17 & 1.42 & 1.42 & 1.42 & 2.01 & 5.44 \\
1.42 & 2.01 & 22 & 3.17 & 13.4 & 13.4 & 13.4 & 14 & 1.43 \\
2.24 & 3.17 & 3.17 & 13 & 2.24 & 2.24 & 2.24 & 3.17 & 2.26 \\
5 & 1.42 & 13.4 & 2.24 & 17 & 13 & 13 & 13.4 & 1.01 \\
5 & 1.42 & 13.4 & 2.24 & 13 & 17 & 13 & 13.4 & 1.01 \\
5 & 1.42 & 13.4 & 2.24 & 13 & 13 & 17 & 13.4 & 1.02 \\
5.42 & 2.01 & 14 & 3.17 & 13.4 & 13.4 & 13.4 & 18 & 1.43 \\
1.01 & 5.44 & 1.43 & 2.26 & 1.01 & 1.01 & 1.02 & 1.43 & 5.05
\end{bmatrix}$$

$$Z_3 = \begin{bmatrix}
10 & 1.42 & 1.42 & 2.24 & 1 & 1 & 1 & 1.42 & 1.01 \\
1.42 & 38 & 2.01 & 3.17 & 1.42 & 1.42 & 1.42 & 2.01 & 28.4 \\
1.42 & 2.01 & 11 & 3.17 & 10.4 & 10.4 & 10.4 & 2.01 & 1.43 \\
2.24 & 3.17 & 3.17 & 14 & 2.24 & 2.24 & 2.24 & 3.17 & 2.26 \\
1 & 1.42 & 10.4 & 2.24 & 19 & 19 & 19 & 10.4 & 1.01 \\
1 & 1.42 & 10.4 & 2.24 & 19 & 19 & 19 & 10.4 & 1.01 \\
1 & 1.42 & 10.4 & 2.24 & 19 & 19 & 19 & 10.4 & 1.02 \\
1.42 & 2.01 & 2.01 & 3.17 & 10.4 & 10.4 & 10.4 & 11 & 1.43 \\
1.01 & 28.4 & 1.43 & 2.26 & 1.01 & 1.01 & 1.02 & 1.43 & 37.1
\end{bmatrix}$$

Maximal eigenvalues of matrix Z of the electronics industrial cluster are 5.0222, 6.6588 and 5.5611, and their topological indexes are 2.5111, 3.3294 and 2.7806.

4.3.2.3   Commerce industrial cluster

Matrixes G and Z corresponding to the commerce industrial cluster in China are as follows:

$$G_1 = \begin{bmatrix}
1 & 0.09 & 0 & 1 & 0 & 0 & 0 & 0 & 0 & 0 & 0 & 0 & 0 & 0 & 0 & 0 & 0 \\
1.41 & 0.06 & 1 & 0 & 1 & 0 & 0 & 0 & 0 & 0 & 0 & 0 & 0 & 0 & 0 & 0 & 0 \\
1.41 & 0.05 & 0 & 1 & 0 & 1 & 0 & 0 & 0 & 0 & 0 & 0 & 0 & 0 & 0 & 0 & 0 \\
2 & 0.13 & 0 & 0 & 1 & 0 & 1 & 1 & 1 & 0 & 0 & 0 & 0 & 0 & 0 & 0 & 0 \\
1 & 0.05 & 0 & 0 & 0 & 1 & 0 & 0 & 0 & 0 & 0 & 0 & 0 & 0 & 0 & 0 & 0 \\
1 & 0.04 & 0 & 0 & 0 & 1 & 0 & 0 & 0 & 0 & 0 & 0 & 0 & 0 & 0 & 0 & 0 \\
1.41 & 0.08 & 0 & 0 & 0 & 1 & 0 & 0 & 0 & 0 & 0 & 0 & 0 & 0 & 0 & 0 & 1 \\
1 & 0.07 & 0 & 0 & 0 & 0 & 0 & 0 & 0 & 0 & 0 & 0 & 0 & 0 & 0 & 0 & 1 \\
1 & 0.06 & 0 & 0 & 0 & 0 & 0 & 0 & 0 & 0 & 0 & 0 & 0 & 0 & 0 & 0 & 1 \\
1 & 0.02 & 0 & 0 & 0 & 0 & 0 & 0 & 0 & 0 & 0 & 0 & 0 & 0 & 0 & 0 & 1 \\
1 & 0.06 & 0 & 0 & 0 & 0 & 0 & 0 & 0 & 0 & 0 & 0 & 0 & 0 & 0 & 0 & 1 \\
1 & 0.06 & 0 & 0 & 0 & 0 & 0 & 0 & 0 & 0 & 0 & 0 & 0 & 0 & 0 & 0 & 1 \\
1 & 0.03 & 0 & 0 & 0 & 0 & 0 & 0 & 0 & 0 & 0 & 0 & 0 & 0 & 0 & 0 & 1 \\
1 & 0.05 & 0 & 0 & 0 & 0 & 0 & 0 & 0 & 0 & 0 & 0 & 0 & 0 & 0 & 0 & 1 \\
2.83 & 0.22 & 0 & 0 & 0 & 0 & 0 & 0 & 1 & 1 & 1 & 1 & 1 & 1 & 1 & 1 & 0
\end{bmatrix}$$

$$
\boldsymbol{G}_2 =
\begin{bmatrix}
1 & 0.09 & 0 & 0 & 2 & 0 & 0 & 0 & 0 & 0 & 0 & 0 & 0 & 0 & 0 & 0 & 0 \\
1.41 & 0.06 & 0 & 0 & 0 & 2 & 0 & 0 & 0 & 0 & 0 & 0 & 0 & 0 & 0 & 0 & 0 \\
1.41 & 0.05 & 2 & 0 & 0 & 0 & 2 & 2 & 2 & 0 & 0 & 0 & 0 & 0 & 0 & 0 & 0 \\
2 & 0.13 & 0 & 2 & 0 & 0 & 0 & 0 & 0 & 0 & 0 & 0 & 0 & 0 & 0 & 0 & 2 \\
1 & 0.05 & 0 & 0 & 2 & 0 & 0 & 2 & 2 & 0 & 0 & 0 & 0 & 0 & 0 & 0 & 0 \\
1 & 0.04 & 0 & 0 & 2 & 0 & 2 & 0 & 2 & 0 & 0 & 0 & 0 & 0 & 0 & 0 & 0 \\
1.41 & 0.08 & 0 & 0 & 2 & 0 & 2 & 2 & 0 & 2 & 2 & 2 & 2 & 2 & 2 & 2 & 0 \\
1 & 0.07 & 0 & 0 & 0 & 0 & 0 & 0 & 2 & 0 & 2 & 2 & 2 & 2 & 2 & 2 & 0 \\
1 & 0.06 & 0 & 0 & 0 & 0 & 0 & 0 & 2 & 2 & 0 & 2 & 2 & 2 & 2 & 2 & 0 \\
1 & 0.02 & 0 & 0 & 0 & 0 & 0 & 0 & 2 & 2 & 2 & 0 & 2 & 2 & 2 & 2 & 0 \\
1 & 0.06 & 0 & 0 & 0 & 0 & 0 & 0 & 2 & 2 & 2 & 2 & 0 & 2 & 2 & 2 & 0 \\
1 & 0.06 & 0 & 0 & 0 & 0 & 0 & 0 & 2 & 2 & 2 & 2 & 2 & 0 & 2 & 2 & 0 \\
1 & 0.03 & 0 & 0 & 0 & 0 & 0 & 0 & 2 & 2 & 2 & 2 & 2 & 2 & 0 & 2 & 0 \\
1 & 0.05 & 0 & 0 & 0 & 0 & 0 & 0 & 2 & 2 & 2 & 2 & 2 & 2 & 2 & 0 & 0 \\
2.83 & 0.22 & 0 & 0 & 0 & 2 & 0 & 0 & 0 & 0 & 0 & 0 & 0 & 0 & 0 & 0 & 0
\end{bmatrix}
$$

$$
\boldsymbol{G}_3 =
\begin{bmatrix}
1 & 0.09 & 0 & 0 & 0 & 3 & 0 & 0 & 0 & 0 & 0 & 0 & 0 & 0 & 0 & 0 & 0 \\
1.41 & 0.06 & 0 & 0 & 0 & 0 & 3 & 3 & 3 & 0 & 0 & 0 & 0 & 0 & 0 & 0 & 0 \\
1.41 & 0.05 & 0 & 0 & 0 & 0 & 0 & 0 & 0 & 0 & 0 & 0 & 0 & 0 & 0 & 0 & 3 \\
2 & 0.13 & 3 & 0 & 0 & 0 & 0 & 0 & 0 & 3 & 3 & 3 & 3 & 3 & 3 & 3 & 0 \\
1 & 0.05 & 0 & 3 & 0 & 0 & 0 & 0 & 0 & 0 & 0 & 0 & 0 & 0 & 0 & 0 & 3 \\
1 & 0.04 & 0 & 3 & 0 & 0 & 0 & 0 & 0 & 0 & 0 & 0 & 0 & 0 & 0 & 0 & 3 \\
1.41 & 0.08 & 0 & 3 & 0 & 0 & 0 & 0 & 0 & 0 & 0 & 0 & 0 & 0 & 0 & 0 & 0 \\
1 & 0.07 & 0 & 0 & 0 & 3 & 0 & 0 & 0 & 0 & 0 & 0 & 0 & 0 & 0 & 0 & 0 \\
1 & 0.06 & 0 & 0 & 0 & 3 & 0 & 0 & 0 & 0 & 0 & 0 & 0 & 0 & 0 & 0 & 0 \\
1 & 0.02 & 0 & 0 & 0 & 3 & 0 & 0 & 0 & 0 & 0 & 0 & 0 & 0 & 0 & 0 & 0 \\
1 & 0.06 & 0 & 0 & 0 & 3 & 0 & 0 & 0 & 0 & 0 & 0 & 0 & 0 & 0 & 0 & 0 \\
1 & 0.06 & 0 & 0 & 0 & 3 & 0 & 0 & 0 & 0 & 0 & 0 & 0 & 0 & 0 & 0 & 0 \\
1 & 0.03 & 0 & 0 & 0 & 3 & 0 & 0 & 0 & 0 & 0 & 0 & 0 & 0 & 0 & 0 & 0 \\
1 & 0.05 & 0 & 0 & 0 & 3 & 0 & 0 & 0 & 0 & 0 & 0 & 0 & 0 & 0 & 0 & 0 \\
2.83 & 0.22 & 0 & 0 & 3 & 0 & 3 & 3 & 0 & 0 & 0 & 0 & 0 & 0 & 0 & 0 & 0
\end{bmatrix}
$$

$$
\boldsymbol{z}_1 =
\begin{bmatrix}
2.01 & 1.42 & 2.42 & 2.01 & 1 & 1 & 1.42 & 1.01 & 1.01 & 1 & 1.01 & 1.01 & 1 & 1 & 2.85 \\
1.42 & 4 & 2 & 3.84 & 1.42 & 1.42 & 2 & 1.42 & 1.42 & 1.42 & 1.42 & 1.42 & 1.42 & 1.42 & 4.01 \\
2.42 & 2 & 4 & 2.83 & 2.42 & 2.42 & 3 & 1.42 & 1.42 & 1.42 & 1.42 & 1.42 & 1.42 & 1.42 & 4.01 \\
2.01 & 3.84 & 2.83 & 8.02 & 2.01 & 2 & 2.84 & 2.01 & 2.01 & 2 & 2.01 & 2.01 & 2 & 2.01 & 6.69 \\
1 & 1.42 & 2.42 & 2.01 & 2 & 2 & 2.42 & 1 & 1 & 1 & 1 & 1 & 1 & 1 & 2.84 \\
1 & 1.42 & 2.42 & 2 & 2 & 2 & 2.42 & 1 & 1 & 1 & 1 & 1 & 1 & 1 & 2.84 \\
1.42 & 2 & 3 & 2.84 & 2.42 & 2.42 & 4.01 & 2.42 & 2.42 & 2.42 & 2.42 & 2.42 & 2.42 & 2.42 & 4.02 \\
1.01 & 1.42 & 1.42 & 2.01 & 1 & 1 & 2.42 & 2.01 & 2 & 2 & 2 & 2 & 2 & 2 & 2.84 \\
1.01 & 1.42 & 1.42 & 2.01 & 1 & 1 & 2.42 & 2 & 2 & 2 & 2 & 2 & 2 & 2 & 2.84 \\
1 & 1.42 & 1.42 & 2 & 1 & 1 & 2.42 & 2 & 2 & 2 & 2 & 2 & 2 & 2 & 2.83 \\
1.01 & 1.42 & 1.42 & 2.01 & 1 & 1 & 2.42 & 2 & 2 & 2 & 2 & 2 & 2 & 2 & 2.84 \\
1.01 & 1.42 & 1.42 & 2.01 & 1 & 1 & 2.42 & 2 & 2 & 2 & 2 & 2 & 2 & 2 & 2.84 \\
1 & 1.42 & 1.42 & 2 & 1 & 1 & 2.42 & 2 & 2 & 2 & 2 & 2 & 2 & 2 & 2.84 \\
1 & 1.42 & 1.42 & 2.01 & 1 & 1 & 2.42 & 2 & 2 & 2 & 2 & 2 & 2 & 2 & 2.84 \\
2.85 & 4.01 & 4.01 & 6.69 & 2.84 & 2.84 & 4.02 & 2.84 & 2.84 & 2.83 & 2.84 & 2.84 & 2.84 & 2.84 & 16.1
\end{bmatrix}
$$

$$
\boldsymbol{z}_2 =
\begin{bmatrix}
5.01 & 1.42 & 1.42 & 2.01 & 5 & 5 & 5.42 & 1.01 & 1.01 & 1 & 1.01 & 1.01 & 1 & 1 & 2.85 \\
1.42 & 6 & 2 & 2.84 & 1.42 & 1.42 & 2 & 1.42 & 1.42 & 1.42 & 1.42 & 1.42 & 1.42 & 1.42 & 8.01 \\
1.42 & 2 & 18 & 2.83 & 9.42 & 9.42 & 10 & 5.42 & 5.42 & 5.42 & 5.42 & 5.42 & 5.42 & 5.42 & 4.01 \\
2.01 & 2.84 & 2.83 & 12 & 2.01 & 2 & 2.84 & 2.01 & 2.01 & 2 & 2.01 & 2.01 & 2 & 2.01 & 5.69 \\
5 & 1.42 & 9.42 & 2.01 & 13 & 9 & 9.42 & 5 & 5 & 5 & 5 & 5 & 5 & 5 & 2.84 \\
5 & 1.42 & 9.42 & 2 & 9 & 13 & 9.42 & 5 & 5 & 5 & 5 & 5 & 5 & 5 & 2.84 \\
5.42 & 2 & 10 & 2.84 & 9.42 & 9.42 & 42 & 25.4 & 25.4 & 25.4 & 25.4 & 25.4 & 25.4 & 25.4 & 4.02 \\
1.01 & 1.42 & 5.42 & 2.01 & 5 & 5 & 25.4 & 29 & 25 & 25 & 25 & 25 & 25 & 25 & 2.84 \\
1.01 & 1.42 & 5.42 & 2.01 & 5 & 5 & 25.4 & 25 & 29 & 25 & 25 & 25 & 25 & 25 & 2.84 \\
1 & 1.42 & 5.42 & 2 & 5 & 5 & 25.4 & 25 & 25 & 29 & 25 & 25 & 25 & 25 & 2.83 \\
1.01 & 1.42 & 5.42 & 2.01 & 5 & 5 & 25.4 & 25 & 25 & 25 & 29 & 25 & 25 & 25 & 2.84 \\
1.01 & 1.42 & 5.42 & 2.01 & 5 & 5 & 25.4 & 25 & 25 & 25 & 25 & 29 & 25 & 25 & 2.84 \\
1 & 1.42 & 5.42 & 2 & 5 & 5 & 25.4 & 25 & 25 & 25 & 25 & 25 & 29 & 25 & 2.84 \\
1 & 1.42 & 5.42 & 2.01 & 5 & 5 & 25.4 & 25 & 25 & 25 & 25 & 25 & 25 & 29 & 2.84 \\
2.85 & 8.01 & 4.01 & 5.69 & 2.84 & 2.84 & 4.02 & 2.84 & 2.84 & 2.83 & 2.84 & 2.84 & 2.84 & 2.84 & 12.1
\end{bmatrix}
$$

$$\mathbf{z}_3 = \begin{bmatrix}
10 & 1.42 & 1.42 & 2.01 & 1 & 1 & 1.42 & 10 & 10 & 10 & 10 & 10 & 10 & 10 & 2.85 \\
1.42 & 29 & 2 & 2.84 & 1.42 & 1.42 & 2 & 1.42 & 1.42 & 1.42 & 1.42 & 1.42 & 1.42 & 1.42 & 22 \\
1.42 & 2 & 11 & 2.83 & 10.4 & 10.4 & 2 & 1.42 & 1.42 & 1.42 & 1.42 & 1.42 & 1.42 & 1.42 & 4.01 \\
2.01 & 2.84 & 2.83 & 76 & 2.01 & 2 & 2.84 & 2.01 & 2.01 & 2 & 2.01 & 2.01 & 2 & 2.01 & 5.69 \\
1 & 1.42 & 10.4 & 2.01 & 19 & 19 & 10.4 & 1 & 1 & 1 & 1 & 1 & 1 & 1 & 2.84 \\
1 & 1.42 & 10.4 & 2 & 19 & 19 & 10.4 & 1 & 1 & 1 & 1 & 1 & 1 & 1 & 2.84 \\
1.42 & 2 & 2 & 2.84 & 10.4 & 10.4 & 11 & 1.42 & 1.42 & 1.42 & 1.42 & 1.42 & 1.42 & 1.42 & 4.02 \\
10 & 1.42 & 1.42 & 2.01 & 1 & 1 & 1.42 & 10 & 10 & 10 & 10 & 10 & 10 & 10 & 2.84 \\
10 & 1.42 & 1.42 & 2.01 & 1 & 1 & 1.42 & 10 & 10 & 10 & 10 & 10 & 10 & 10 & 2.84 \\
10 & 1.42 & 1.42 & 2 & 1 & 1 & 1.42 & 10 & 10 & 10 & 10 & 10 & 10 & 10 & 2.83 \\
10 & 1.42 & 1.42 & 2.01 & 1 & 1 & 1.42 & 10 & 10 & 10 & 10 & 10 & 10 & 10 & 2.84 \\
10 & 1.42 & 1.42 & 2.01 & 1 & 1 & 1.42 & 10 & 10 & 10 & 10 & 10 & 10 & 10 & 2.84 \\
10 & 1.42 & 1.42 & 2 & 1 & 1 & 1.42 & 10 & 10 & 10 & 10 & 10 & 10 & 10 & 2.84 \\
10 & 1.42 & 1.42 & 2.01 & 1 & 1 & 1.42 & 10 & 10 & 10 & 10 & 10 & 10 & 10 & 2.84 \\
2.85 & 22 & 4.01 & 5.69 & 2.84 & 2.84 & 4.02 & 2.84 & 2.84 & 2.83 & 2.84 & 2.84 & 2.84 & 2.84 & 35.1
\end{bmatrix}$$

Maximal eigenvalues of matrix $\mathbf{Z}$ of the commerce industrial cluster are 8.8605, 8.9071 and 10.4991, and their topological indexes are 4.4302, 4.4535 and 5.2496.

4.3.2.4 Diet industrial cluster

Matrixes $\mathbf{G}$ and $\mathbf{Z}$ corresponding to the diet industrial cluster in China are as follows:

$$\mathbf{G}_1 = \begin{bmatrix}
1 & 0.04 & 0 & 1 & 0 & 0 & 0 & 0 & 0 & 0 & 0 & 0 & 0 \\
1.41 & 0.22 & 1 & 0 & 1 & 0 & 0 & 0 & 0 & 0 & 0 & 0 & 0 \\
1.73 & 0.16 & 0 & 1 & 0 & 1 & 1 & 0 & 0 & 0 & 0 & 0 & 0 \\
1.41 & 0.09 & 0 & 0 & 1 & 0 & 0 & 0 & 0 & 0 & 0 & 0 & 1 \\
1.41 & 0.05 & 0 & 0 & 1 & 0 & 0 & 0 & 0 & 0 & 0 & 0 & 1 \\
1 & 0.06 & 0 & 0 & 0 & 0 & 0 & 0 & 0 & 0 & 0 & 0 & 1 \\
1 & 0.08 & 0 & 0 & 0 & 0 & 0 & 0 & 0 & 0 & 0 & 0 & 1 \\
1 & 0.08 & 0 & 0 & 0 & 0 & 0 & 0 & 0 & 0 & 0 & 0 & 1 \\
1 & 0.1 & 0 & 0 & 0 & 0 & 0 & 0 & 0 & 0 & 0 & 0 & 1 \\
1 & 0.06 & 0 & 0 & 0 & 0 & 0 & 0 & 0 & 0 & 0 & 0 & 1 \\
2.65 & 0.1 & 0 & 0 & 0 & 1 & 1 & 1 & 1 & 1 & 1 & 1 & 0
\end{bmatrix}$$

$$\mathbf{G}_2 = \begin{bmatrix}
1 & 0.04 & 0 & 0 & 2 & 0 & 0 & 0 & 0 & 0 & 0 & 0 & 0 \\
1.41 & 0.22 & 0 & 0 & 0 & 2 & 2 & 0 & 0 & 0 & 0 & 0 & 0 \\
1.73 & 0.16 & 2 & 0 & 0 & 0 & 0 & 0 & 0 & 0 & 0 & 0 & 2 \\
1.41 & 0.09 & 0 & 2 & 0 & 0 & 2 & 2 & 2 & 2 & 2 & 2 & 0 \\
1.41 & 0.05 & 0 & 2 & 0 & 2 & 0 & 2 & 2 & 2 & 2 & 2 & 0 \\
1 & 0.06 & 0 & 0 & 0 & 2 & 2 & 0 & 2 & 2 & 2 & 2 & 0 \\
1 & 0.08 & 0 & 0 & 0 & 2 & 2 & 2 & 0 & 2 & 2 & 2 & 0 \\
1 & 0.08 & 0 & 0 & 0 & 2 & 2 & 2 & 2 & 0 & 2 & 2 & 0 \\
1 & 0.1 & 0 & 0 & 0 & 2 & 2 & 2 & 2 & 2 & 0 & 2 & 0 \\
1 & 0.06 & 0 & 0 & 0 & 2 & 2 & 2 & 2 & 2 & 2 & 0 & 0 \\
2.65 & 0.1 & 0 & 0 & 2 & 0 & 0 & 0 & 0 & 0 & 0 & 0 & 0
\end{bmatrix}$$

$$\mathbf{G}_3 = \begin{bmatrix}
1 & 0.04 & 0 & 0 & 0 & 3 & 3 & 0 & 0 & 0 & 0 & 0 & 0 \\
1.41 & 0.22 & 0 & 0 & 0 & 0 & 0 & 0 & 0 & 0 & 0 & 0 & 3 \\
1.73 & 0.16 & 0 & 0 & 0 & 0 & 0 & 3 & 3 & 3 & 3 & 3 & 0 \\
1.41 & 0.09 & 3 & 0 & 0 & 0 & 0 & 0 & 0 & 0 & 0 & 0 & 0 \\
1.41 & 0.05 & 3 & 0 & 0 & 0 & 0 & 0 & 0 & 0 & 0 & 0 & 0 \\
1 & 0.06 & 0 & 0 & 3 & 0 & 0 & 0 & 0 & 0 & 0 & 0 & 0 \\
1 & 0.08 & 0 & 0 & 3 & 0 & 0 & 0 & 0 & 0 & 0 & 0 & 0 \\
1 & 0.08 & 0 & 0 & 3 & 0 & 0 & 0 & 0 & 0 & 0 & 0 & 0 \\
1 & 0.1 & 0 & 0 & 3 & 0 & 0 & 0 & 0 & 0 & 0 & 0 & 0 \\
1 & 0.06 & 0 & 0 & 3 & 0 & 0 & 0 & 0 & 0 & 0 & 0 & 0 \\
2.65 & 0.1 & 0 & 3 & 0 & 0 & 0 & 0 & 0 & 0 & 0 & 0 & 0
\end{bmatrix}$$

$$Z_1 = \begin{bmatrix}
2 & 1.42 & 2.74 & 1.42 & 1.42 & 1 & 1 & 1 & 1 & 1 & 2.65 \\
1.42 & 4.05 & 2.48 & 3.02 & 3.01 & 1.43 & 1.43 & 1.43 & 1.44 & 1.43 & 3.76 \\
2.74 & 2.48 & 6.03 & 2.46 & 2.46 & 1.74 & 1.75 & 1.74 & 1.75 & 1.74 & 6.6 \\
1.42 & 3.02 & 2.46 & 4.01 & 4 & 2.42 & 2.42 & 2.42 & 2.42 & 2.42 & 3.75 \\
1.42 & 3.01 & 2.46 & 4 & 4 & 2.42 & 2.42 & 2.42 & 2.42 & 2.42 & 3.75 \\
1 & 1.43 & 1.74 & 2.42 & 2.42 & 2 & 2 & 2 & 2.01 & 2 & 2.65 \\
1 & 1.43 & 1.75 & 2.42 & 2.42 & 2 & 2.01 & 2.01 & 2.01 & 2.01 & 2.65 \\
1 & 1.43 & 1.74 & 2.42 & 2.42 & 2 & 2.01 & 2.01 & 2.01 & 2.01 & 2.65 \\
1 & 1.44 & 1.75 & 2.42 & 2.42 & 2.01 & 2.01 & 2.01 & 2.01 & 2.01 & 2.66 \\
1 & 1.43 & 1.74 & 2.42 & 2.42 & 2 & 2.01 & 2.01 & 2.01 & 2 & 2.65 \\
2.65 & 3.76 & 6.6 & 3.75 & 3.75 & 2.65 & 2.65 & 2.65 & 2.66 & 2.65 & 14
\end{bmatrix}$$

$$Z_2 = \begin{bmatrix}
5 & 1.42 & 1.74 & 1.42 & 1.42 & 1 & 1 & 1 & 1 & 1 & 6.65 \\
1.42 & 10.1 & 2.48 & 6.02 & 6.01 & 9.43 & 9.43 & 9.43 & 9.44 & 9.43 & 3.76 \\
1.74 & 2.48 & 11 & 2.46 & 2.46 & 1.74 & 1.75 & 1.74 & 1.75 & 1.74 & 4.6 \\
1.42 & 6.02 & 2.46 & 30 & 26 & 21.4 & 21.4 & 21.4 & 21.4 & 21.4 & 3.75 \\
1.42 & 6.01 & 2.46 & 26 & 30 & 21.4 & 21.4 & 21.4 & 21.4 & 21.4 & 3.75 \\
1 & 9.43 & 1.74 & 21.4 & 21.4 & 25 & 21 & 21 & 21 & 21 & 2.65 \\
1 & 9.43 & 1.75 & 21.4 & 21.4 & 21 & 25 & 21 & 21 & 21 & 2.65 \\
1 & 9.43 & 1.74 & 21.4 & 21.4 & 21 & 21 & 25 & 21 & 21 & 2.65 \\
1 & 9.44 & 1.75 & 21.4 & 21.4 & 21 & 21 & 21 & 25 & 21 & 2.66 \\
1 & 9.43 & 1.74 & 21.4 & 21.4 & 21 & 21 & 21 & 21 & 25 & 2.65 \\
6.65 & 3.76 & 4.6 & 3.75 & 3.75 & 2.65 & 2.65 & 2.65 & 2.66 & 2.65 & 11
\end{bmatrix}$$

$$Z_3 = \begin{bmatrix}
19 & 1.42 & 1.74 & 1.42 & 1.42 & 1 & 1 & 1 & 1 & 1 & 2.65 \\
1.42 & 11.1 & 2.48 & 2.02 & 2.01 & 1.43 & 1.43 & 1.43 & 1.44 & 1.43 & 3.76 \\
1.74 & 2.48 & 48 & 2.46 & 2.46 & 1.74 & 1.75 & 1.74 & 1.75 & 1.74 & 4.6 \\
1.42 & 2.02 & 2.46 & 11 & 11 & 1.42 & 1.42 & 1.42 & 1.42 & 1.42 & 3.75 \\
1.42 & 2.01 & 2.46 & 11 & 11 & 1.42 & 1.42 & 1.42 & 1.42 & 1.42 & 3.75 \\
1 & 1.43 & 1.74 & 1.42 & 1.42 & 10 & 10 & 10 & 10 & 10 & 2.65 \\
1 & 1.43 & 1.75 & 1.42 & 1.42 & 10 & 10 & 10 & 10 & 10 & 2.65 \\
1 & 1.43 & 1.74 & 1.42 & 1.42 & 10 & 10 & 10 & 10 & 10 & 2.65 \\
1 & 1.44 & 1.75 & 1.42 & 1.42 & 10 & 10 & 10 & 10 & 10 & 2.66 \\
1 & 1.43 & 1.74 & 1.42 & 1.42 & 10 & 10 & 10 & 10 & 10 & 2.65 \\
2.65 & 3.76 & 4.6 & 3.75 & 3.75 & 2.65 & 2.65 & 2.65 & 2.66 & 2.65 & 16
\end{bmatrix}$$

Maximal eigenvalues of matrix Z of the diet industrial cluster are 7.0436, 8.9354 and 6.1825, and their topological indexes are 3.5218, 4.4677 and 3.0912.

4.3.2.5 Iron and steel industrial cluster

Matrixes G and Z corresponding to the iron and steel industrial cluster in China are as follows:

$$G_1 = \begin{bmatrix}
1 & 0.03 & 0 & 0 & 0 & 0 & 1 & 0 & 0 & 0 & 0 & 0 & 0 & 0 & 0 & 0 & 0 \\
1 & 0.03 & 0 & 0 & 0 & 0 & 1 & 0 & 0 & 0 & 0 & 0 & 0 & 0 & 0 & 0 & 0 \\
1.41 & 0.04 & 0 & 0 & 0 & 0 & 1 & 0 & 1 & 0 & 0 & 0 & 0 & 0 & 0 & 0 & 0 \\
1.41 & 0.15 & 0 & 0 & 0 & 0 & 1 & 1 & 0 & 0 & 0 & 0 & 0 & 0 & 0 & 0 & 0 \\
2 & 0.11 & 1 & 1 & 1 & 1 & 0 & 0 & 0 & 0 & 0 & 0 & 0 & 0 & 0 & 0 & 0 \\
1 & 0.04 & 0 & 0 & 0 & 1 & 0 & 0 & 0 & 0 & 0 & 0 & 0 & 0 & 0 & 0 & 0 \\
1.41 & 0.08 & 0 & 0 & 1 & 0 & 0 & 0 & 0 & 1 & 0 & 0 & 0 & 0 & 0 & 0 & 0 \\
1.41 & 0.09 & 0 & 0 & 0 & 0 & 0 & 0 & 1 & 0 & 0 & 0 & 0 & 0 & 0 & 0 & 1 \\
1 & 0.05 & 0 & 0 & 0 & 0 & 0 & 0 & 0 & 0 & 0 & 0 & 0 & 0 & 0 & 0 & 1 \\
1 & 0.05 & 0 & 0 & 0 & 0 & 0 & 0 & 0 & 0 & 0 & 0 & 0 & 0 & 0 & 0 & 1 \\
1 & 0.18 & 0 & 0 & 0 & 0 & 0 & 0 & 0 & 0 & 0 & 0 & 0 & 0 & 0 & 0 & 1 \\
1.41 & 0.23 & 0 & 0 & 0 & 0 & 0 & 0 & 0 & 0 & 0 & 0 & 0 & 0 & 0 & 0 & 1 \\
1.41 & 0.11 & 0 & 0 & 0 & 0 & 0 & 0 & 0 & 0 & 0 & 0 & 0 & 0 & 0 & 0 & 1 \\
1 & 0.04 & 0 & 0 & 0 & 0 & 0 & 0 & 0 & 0 & 0 & 0 & 0 & 0 & 0 & 0 & 1 \\
2.65 & 0.17 & 0 & 0 & 0 & 0 & 0 & 0 & 0 & 1 & 1 & 1 & 1 & 1 & 1 & 1 & 0
\end{bmatrix}$$

$$G_2 = \begin{bmatrix}
1 & 0.03 & 0 & 2 & 2 & 2 & 0 & 0 & 0 & 0 & 0 & 0 & 0 & 0 & 0 & 0 & 0 \\
1 & 0.03 & 2 & 0 & 2 & 2 & 0 & 0 & 0 & 0 & 0 & 0 & 0 & 0 & 0 & 0 & 0 \\
1.41 & 0.04 & 2 & 2 & 0 & 2 & 0 & 0 & 0 & 2 & 0 & 0 & 0 & 0 & 0 & 0 & 0 \\
1.41 & 0.15 & 2 & 2 & 2 & 0 & 0 & 0 & 0 & 0 & 0 & 0 & 0 & 0 & 0 & 0 & 0 \\
2 & 0.11 & 0 & 0 & 0 & 0 & 0 & 2 & 2 & 0 & 0 & 0 & 0 & 0 & 0 & 0 & 0 \\
1 & 0.04 & 0 & 0 & 0 & 0 & 2 & 0 & 0 & 0 & 0 & 0 & 0 & 0 & 0 & 0 & 0 \\
1.41 & 0.08 & 0 & 0 & 0 & 0 & 2 & 0 & 0 & 0 & 0 & 0 & 0 & 0 & 0 & 0 & 2 \\
1.41 & 0.09 & 0 & 0 & 2 & 0 & 0 & 0 & 0 & 0 & 2 & 2 & 2 & 2 & 2 & 2 & 0 \\
1 & 0.05 & 0 & 0 & 0 & 0 & 0 & 0 & 0 & 2 & 0 & 2 & 2 & 2 & 2 & 2 & 0 \\
1 & 0.05 & 0 & 0 & 0 & 0 & 0 & 0 & 0 & 2 & 2 & 0 & 2 & 2 & 2 & 2 & 0 \\
1 & 0.18 & 0 & 0 & 0 & 0 & 0 & 0 & 0 & 2 & 2 & 2 & 0 & 2 & 2 & 2 & 0 \\
1.41 & 0.23 & 0 & 0 & 0 & 0 & 0 & 0 & 0 & 2 & 2 & 2 & 2 & 0 & 2 & 2 & 0 \\
1.41 & 0.11 & 0 & 0 & 0 & 0 & 0 & 0 & 0 & 2 & 2 & 2 & 2 & 2 & 0 & 2 & 0 \\
1 & 0.04 & 0 & 0 & 0 & 0 & 0 & 0 & 0 & 2 & 2 & 2 & 2 & 2 & 2 & 0 & 0 \\
2.65 & 0.17 & 0 & 0 & 0 & 0 & 0 & 0 & 2 & 0 & 0 & 0 & 0 & 0 & 0 & 0 & 0
\end{bmatrix}$$

$$G_3 = \begin{bmatrix}
1 & 0.03 & 0 & 0 & 0 & 0 & 0 & 3 & 3 & 0 & 0 & 0 & 0 & 0 & 0 & 0 & 0 \\
1 & 0.03 & 0 & 0 & 0 & 0 & 0 & 3 & 3 & 0 & 0 & 0 & 0 & 0 & 0 & 0 & 0 \\
1.41 & 0.04 & 0 & 0 & 0 & 0 & 0 & 3 & 0 & 0 & 0 & 0 & 0 & 0 & 0 & 0 & 3 \\
1.41 & 0.15 & 0 & 0 & 0 & 0 & 0 & 0 & 3 & 0 & 0 & 0 & 0 & 0 & 0 & 0 & 0 \\
2 & 0.11 & 0 & 0 & 0 & 0 & 0 & 0 & 0 & 3 & 0 & 0 & 0 & 0 & 0 & 0 & 0 \\
1 & 0.04 & 3 & 3 & 3 & 0 & 0 & 0 & 0 & 0 & 0 & 0 & 0 & 0 & 0 & 0 & 0 \\
1.41 & 0.08 & 3 & 3 & 0 & 3 & 0 & 0 & 0 & 0 & 3 & 3 & 3 & 3 & 3 & 3 & 0 \\
1.41 & 0.09 & 0 & 0 & 0 & 0 & 3 & 0 & 0 & 0 & 0 & 0 & 0 & 0 & 0 & 0 & 0 \\
1 & 0.05 & 0 & 0 & 0 & 0 & 0 & 0 & 3 & 0 & 0 & 0 & 0 & 0 & 0 & 0 & 0 \\
1 & 0.05 & 0 & 0 & 0 & 0 & 0 & 0 & 3 & 0 & 0 & 0 & 0 & 0 & 0 & 0 & 0 \\
1 & 0.18 & 0 & 0 & 0 & 0 & 0 & 0 & 3 & 0 & 0 & 0 & 0 & 0 & 0 & 0 & 0 \\
1.41 & 0.23 & 0 & 0 & 0 & 0 & 0 & 0 & 3 & 0 & 0 & 0 & 0 & 0 & 0 & 0 & 0 \\
1.41 & 0.11 & 0 & 0 & 0 & 0 & 0 & 0 & 3 & 0 & 0 & 0 & 0 & 0 & 0 & 0 & 0 \\
1 & 0.04 & 0 & 0 & 0 & 0 & 0 & 0 & 3 & 0 & 0 & 0 & 0 & 0 & 0 & 0 & 0 \\
2.65 & 0.17 & 0 & 0 & 3 & 0 & 0 & 0 & 0 & 0 & 0 & 0 & 0 & 0 & 0 & 0 & 0
\end{bmatrix}$$

$$Z_1 = \begin{bmatrix}
2 & 2 & 2.42 & 2.42 & 2 & 1 & 1.42 & 1.42 & 1 & 1 & 1 & 1.42 & 1.42 & 1 & 2.65 \\
2 & 2 & 2.42 & 2.42 & 2 & 1 & 1.42 & 1.42 & 1 & 1 & 1.01 & 1.42 & 1.42 & 1 & 2.65 \\
2.42 & 2.42 & 4 & 3.01 & 2.83 & 1.42 & 2 & 3 & 1.42 & 1.42 & 1.42 & 2.01 & 2 & 1.42 & 3.75 \\
2.42 & 2.42 & 3.01 & 4.02 & 2.85 & 1.42 & 2.01 & 2.01 & 1.42 & 1.42 & 1.44 & 2.03 & 2.02 & 1.42 & 3.77 \\
2 & 2 & 2.83 & 2.85 & 8.01 & 3 & 3.84 & 2.84 & 2.01 & 2.01 & 2.02 & 2.85 & 2.84 & 2 & 5.31 \\
1 & 1 & 1.42 & 1.42 & 3 & 2 & 1.42 & 1.42 & 1 & 1 & 1.01 & 1.42 & 1.42 & 1 & 2.65 \\
1.42 & 1.42 & 2 & 2.01 & 3.84 & 1.42 & 4.01 & 2.01 & 1.42 & 1.42 & 1.43 & 2.02 & 2.01 & 1.42 & 4.76 \\
1.42 & 1.42 & 3 & 2.01 & 2.84 & 1.42 & 2.01 & 4.01 & 2.42 & 2.42 & 2.43 & 3.02 & 3.01 & 2.42 & 3.76 \\
1 & 1 & 1.42 & 1.42 & 2.01 & 1 & 1.42 & 2.42 & 2 & 2 & 2.01 & 2.43 & 2.42 & 2 & 2.65 \\
1 & 1 & 1.42 & 1.42 & 2.01 & 1 & 1.42 & 2.42 & 2 & 2 & 2.01 & 2.43 & 2.42 & 2 & 2.65 \\
1 & 1.01 & 1.42 & 1.44 & 2.02 & 1.01 & 1.43 & 2.43 & 2.01 & 2.01 & 2.03 & 2.46 & 2.43 & 2.01 & 2.68 \\
1.42 & 1.42 & 2.01 & 2.03 & 2.85 & 1.42 & 2.02 & 3.02 & 2.43 & 2.43 & 2.46 & 3.05 & 3.02 & 2.42 & 3.78 \\
1.42 & 1.42 & 2 & 2.02 & 2.84 & 1.42 & 2.01 & 3.01 & 2.42 & 2.42 & 2.43 & 3.02 & 3.01 & 2.42 & 3.76 \\
1 & 1 & 1.42 & 1.42 & 2 & 1 & 1.42 & 2.42 & 2 & 2 & 2.01 & 2.42 & 2.42 & 2 & 2.65 \\
2.65 & 2.65 & 3.75 & 3.77 & 5.31 & 2.65 & 4.76 & 3.76 & 2.65 & 2.65 & 2.68 & 3.78 & 3.76 & 2.65 & 14
\end{bmatrix}$$

$$Z_2 = \begin{bmatrix}
13 & 9 & 9.42 & 9.42 & 2 & 1 & 1.42 & 5.42 & 1 & 1 & 1 & 1.42 & 1.42 & 1 & 2.65 \\
9 & 13 & 9.42 & 9.42 & 2 & 1 & 1.42 & 5.42 & 1 & 1 & 1.01 & 1.42 & 1.42 & 1 & 2.65 \\
9.42 & 9.42 & 18 & 10 & 2.83 & 1.42 & 2 & 2 & 5.42 & 5.42 & 5.42 & 6.01 & 6 & 5.42 & 3.75 \\
9.42 & 9.42 & 10 & 14 & 2.85 & 1.42 & 2.01 & 6.01 & 1.42 & 1.42 & 1.44 & 2.03 & 2.02 & 1.42 & 3.77 \\
2 & 2 & 2.83 & 2.85 & 12 & 2 & 2.84 & 2.84 & 2.01 & 2.01 & 2.02 & 2.85 & 2.84 & 2 & 9.31 \\
1 & 1 & 1.42 & 1.42 & 2 & 5 & 5.42 & 1.42 & 1 & 1 & 1.01 & 1.42 & 1.42 & 1 & 2.65 \\
1.42 & 1.42 & 2 & 2.01 & 2.84 & 5.42 & 10 & 2.01 & 1.42 & 1.42 & 1.43 & 2.02 & 2.01 & 1.42 & 3.76 \\
5.42 & 5.42 & 2 & 6.01 & 2.84 & 1.42 & 2.01 & 30 & 21.4 & 21.4 & 21.4 & 22 & 22 & 21.4 & 3.76 \\
1 & 1 & 5.42 & 1.42 & 2.01 & 1 & 1.42 & 21.4 & 25 & 21 & 21 & 21.4 & 21.4 & 21 & 2.65 \\
1 & 1 & 5.42 & 1.42 & 2.01 & 1 & 1.42 & 21.4 & 21 & 25 & 21 & 21.4 & 21.4 & 21 & 2.65 \\
1 & 1.01 & 5.42 & 1.44 & 2.02 & 1.01 & 1.43 & 21.4 & 21 & 21 & 25 & 21.5 & 21.4 & 21 & 2.68 \\
1.42 & 1.42 & 6.01 & 2.03 & 2.85 & 1.42 & 2.02 & 22 & 21.4 & 21.4 & 21.5 & 26.1 & 22 & 21.4 & 3.78 \\
1.42 & 1.42 & 6 & 2.02 & 2.84 & 1.42 & 2.01 & 22 & 21.4 & 21.4 & 21.4 & 22 & 26 & 21.4 & 3.76 \\
1 & 1 & 5.42 & 1.42 & 2 & 1 & 1.42 & 21.4 & 21 & 21 & 21 & 21.4 & 21.4 & 25 & 2.65 \\
2.65 & 2.65 & 3.75 & 3.77 & 9.31 & 2.65 & 3.76 & 3.76 & 2.65 & 2.65 & 2.68 & 3.78 & 3.76 & 2.65 & 11
\end{bmatrix}$$

$$
Z_3 = \begin{bmatrix}
19 & 19 & 10.4 & 10.4 & 2 & 1 & 1.42 & 1.42 & 10 & 10 & 10 & 10.4 & 10.4 & 10 & 2.65 \\
19 & 19 & 10.4 & 10.4 & 2 & 1 & 1.42 & 1.42 & 10 & 10 & 10 & 10.4 & 10.4 & 10 & 2.65 \\
10.4 & 10.4 & 20 & 2.01 & 2.83 & 1.42 & 2 & 2 & 1.42 & 1.42 & 1.42 & 2.01 & 2 & 1.42 & 3.75 \\
10.4 & 10.4 & 2.01 & 11 & 2.85 & 1.42 & 2.01 & 2.01 & 10.4 & 10.4 & 10.4 & 11 & 11 & 10.4 & 3.77 \\
2 & 2 & 2.83 & 2.85 & 13 & 2 & 2.84 & 2.84 & 2.01 & 2.01 & 2.02 & 2.85 & 2.84 & 2 & 5.31 \\
1 & 1 & 1.42 & 1.42 & 2 & 28 & 19.4 & 1.42 & 1 & 1 & 1.01 & 1.42 & 1.42 & 1 & 11.7 \\
1.42 & 1.42 & 2 & 2.01 & 2.84 & 19.4 & 83 & 2.01 & 1.42 & 1.42 & 1.43 & 2.02 & 2.01 & 1.42 & 3.76 \\
1.42 & 1.42 & 2 & 2.01 & 2.84 & 1.42 & 2.01 & 11 & 1.42 & 1.42 & 1.43 & 2.02 & 2.01 & 1.42 & 3.76 \\
10 & 10 & 1.42 & 10.4 & 2.01 & 1 & 1.42 & 1.42 & 10 & 10 & 10 & 10.4 & 10.4 & 10 & 2.65 \\
10 & 10 & 1.42 & 10.4 & 2.01 & 1 & 1.42 & 1.42 & 10 & 10 & 10 & 10.4 & 10.4 & 10 & 2.65 \\
10 & 10 & 1.42 & 10.4 & 2.02 & 1.01 & 1.43 & 1.43 & 10 & 10 & 10 & 10.5 & 10.4 & 10 & 2.68 \\
10.4 & 10.4 & 2.01 & 11 & 2.85 & 1.42 & 2.02 & 2.02 & 10.4 & 10.4 & 10.5 & 11.1 & 11 & 10.4 & 3.78 \\
10.4 & 10.4 & 2 & 11 & 2.84 & 1.42 & 2.01 & 2.01 & 10.4 & 10.4 & 10.4 & 11 & 11 & 10.4 & 3.76 \\
10 & 10 & 1.42 & 10.4 & 2 & 1 & 1.42 & 1.42 & 10 & 10 & 10 & 10.4 & 10.4 & 10 & 2.65
\end{bmatrix}
$$

Maximal eigenvalues of matrix Z of the iron and steel industrial cluster are 8.0565, 8.4781 and 9.6692, and their topological indexes are 4.0283, 4.2390 and 4.8346.

4.3.2.6   Petroleum industrial cluster

Matrixes G and Z corresponding to the petroleum industrial cluster in China are as follows:

$$
G_1 = \begin{bmatrix}
1 & 0.04 & 0 & 1 & 0 & 0 & 0 & 0 & 0 & 0 & 0 & 0 & 0 & 0 & 0 \\
1.73 & 0.07 & 1 & 0 & 1 & 0 & 0 & 0 & 0 & 0 & 0 & 0 & 0 & 0 & 1 \\
1.73 & 0.01 & 0 & 1 & 0 & 1 & 1 & 0 & 0 & 0 & 0 & 0 & 0 & 0 & 0 \\
1 & 0.04 & 0 & 0 & 1 & 0 & 0 & 0 & 0 & 0 & 0 & 0 & 0 & 0 & 0 \\
1.41 & 0.03 & 0 & 0 & 1 & 0 & 0 & 0 & 0 & 0 & 0 & 0 & 0 & 0 & 1 \\
1 & 0.09 & 0 & 0 & 0 & 0 & 0 & 0 & 0 & 0 & 0 & 0 & 0 & 0 & 1 \\
1 & 0.15 & 0 & 0 & 0 & 0 & 0 & 0 & 0 & 1 & 0 & 0 & 0 & 0 & 0 \\
1.41 & 0.01 & 0 & 0 & 0 & 0 & 0 & 0 & 1 & 0 & 0 & 0 & 0 & 0 & 1 \\
1 & 0.03 & 0 & 0 & 0 & 0 & 0 & 0 & 0 & 0 & 0 & 0 & 0 & 0 & 1 \\
1 & 0.03 & 0 & 0 & 0 & 0 & 0 & 0 & 0 & 0 & 0 & 0 & 0 & 0 & 1 \\
1 & 0.09 & 0 & 0 & 0 & 0 & 0 & 0 & 0 & 0 & 0 & 0 & 0 & 0 & 1 \\
1 & 0.06 & 0 & 0 & 0 & 0 & 0 & 0 & 0 & 0 & 0 & 0 & 0 & 0 & 1 \\
2.83 & 0.14 & 0 & 1 & 0 & 0 & 1 & 1 & 0 & 1 & 1 & 1 & 1 & 1 & 0
\end{bmatrix}
$$

$$
G_2 = \begin{bmatrix}
1 & 0.04 & 0 & 0 & 2 & 0 & 0 & 0 & 0 & 0 & 0 & 0 & 0 & 0 & 2 \\
1.73 & 0.07 & 0 & 0 & 0 & 2 & 2 & 2 & 0 & 2 & 2 & 2 & 2 & 2 & 0 \\
1.73 & 0.01 & 2 & 0 & 0 & 0 & 0 & 0 & 0 & 0 & 0 & 0 & 0 & 0 & 2 \\
1 & 0.04 & 0 & 2 & 0 & 0 & 2 & 0 & 0 & 0 & 0 & 0 & 0 & 0 & 0 \\
1.41 & 0.03 & 0 & 2 & 0 & 2 & 0 & 2 & 0 & 2 & 2 & 2 & 2 & 2 & 0 \\
1 & 0.09 & 0 & 2 & 0 & 0 & 2 & 0 & 0 & 2 & 2 & 2 & 2 & 2 & 0 \\
1 & 0.15 & 0 & 0 & 0 & 0 & 0 & 0 & 0 & 0 & 0 & 0 & 0 & 0 & 2 \\
1.41 & 0.01 & 0 & 2 & 0 & 0 & 2 & 2 & 0 & 0 & 2 & 2 & 2 & 2 & 0 \\
1 & 0.03 & 0 & 2 & 0 & 0 & 2 & 2 & 0 & 2 & 0 & 2 & 2 & 2 & 0 \\
1 & 0.03 & 0 & 2 & 0 & 0 & 2 & 2 & 0 & 2 & 2 & 0 & 2 & 2 & 0 \\
1 & 0.09 & 0 & 2 & 0 & 0 & 2 & 2 & 0 & 2 & 2 & 2 & 0 & 2 & 0 \\
1 & 0.06 & 0 & 2 & 0 & 0 & 2 & 2 & 0 & 2 & 2 & 2 & 2 & 0 & 0 \\
2.83 & 0.14 & 2 & 0 & 2 & 0 & 0 & 0 & 2 & 0 & 0 & 0 & 0 & 0 & 0
\end{bmatrix}
$$

$$\boldsymbol{G_3} = \begin{bmatrix}
1 & 0.04 & 0 & 0 & 0 & 3 & 3 & 3 & 0 & 3 & 3 & 3 & 3 & 3 & 0 \\
1.73 & 0.07 & 0 & 0 & 0 & 0 & 0 & 0 & 3 & 0 & 0 & 0 & 0 & 0 & 3 \\
1.73 & 0.01 & 0 & 0 & 0 & 0 & 0 & 3 & 0 & 3 & 3 & 3 & 3 & 3 & 0 \\
1 & 0.04 & 3 & 0 & 0 & 0 & 0 & 0 & 0 & 0 & 0 & 0 & 0 & 0 & 3 \\
1.41 & 0.03 & 3 & 0 & 0 & 0 & 0 & 0 & 3 & 0 & 0 & 0 & 0 & 0 & 0 \\
1 & 0.09 & 3 & 0 & 3 & 0 & 0 & 0 & 3 & 0 & 0 & 0 & 0 & 0 & 0 \\
1 & 0.15 & 0 & 3 & 0 & 0 & 3 & 3 & 0 & 0 & 3 & 3 & 3 & 3 & 0 \\
1.41 & 0.01 & 3 & 0 & 3 & 0 & 0 & 0 & 0 & 0 & 0 & 0 & 0 & 0 & 0 \\
1 & 0.03 & 3 & 0 & 3 & 0 & 0 & 0 & 3 & 0 & 0 & 0 & 0 & 0 & 0 \\
1 & 0.03 & 3 & 0 & 3 & 0 & 0 & 0 & 3 & 0 & 0 & 0 & 0 & 0 & 0 \\
1 & 0.09 & 3 & 0 & 3 & 0 & 0 & 0 & 3 & 0 & 0 & 0 & 0 & 0 & 0 \\
1 & 0.06 & 3 & 0 & 3 & 0 & 0 & 0 & 3 & 0 & 0 & 0 & 0 & 0 & 0 \\
2.83 & 0.14 & 0 & 3 & 0 & 3 & 0 & 0 & 0 & 0 & 0 & 0 & 0 & 0 & 0
\end{bmatrix}$$

$$\boldsymbol{Z_1} = \begin{bmatrix}
2 & 1.73 & 2.73 & 1 & 1.42 & 1 & 1.01 & 1.41 & 1 & 1 & 1 & 1 & 3.83 \\
1.73 & 6 & 3 & 2.73 & 4.45 & 2.74 & 1.74 & 3.45 & 2.73 & 2.73 & 2.74 & 2.74 & 4.91 \\
2.73 & 3 & 6 & 1.73 & 2.45 & 1.73 & 1.73 & 2.45 & 1.73 & 1.73 & 1.73 & 1.73 & 6.9 \\
1 & 2.73 & 1.73 & 2 & 2.42 & 1 & 1.01 & 1.41 & 1 & 1 & 1 & 1 & 2.83 \\
1.42 & 4.45 & 2.45 & 2.42 & 4 & 2.42 & 1.42 & 3 & 2.42 & 2.42 & 2.42 & 2.42 & 4 \\
1 & 2.74 & 1.73 & 1 & 2.42 & 2.01 & 1.01 & 2.42 & 2 & 2 & 2.01 & 2.01 & 2.84 \\
1.01 & 1.74 & 1.73 & 1.01 & 1.42 & 1.01 & 2.02 & 1.42 & 1.01 & 1 & 1.01 & 1.01 & 3.85 \\
1.41 & 3.45 & 2.45 & 1.41 & 3 & 2.42 & 1.42 & 4 & 2.41 & 2.41 & 2.42 & 2.42 & 4 \\
1 & 2.73 & 1.73 & 1 & 2.42 & 2 & 1.01 & 2.41 & 2 & 2 & 2 & 2 & 2.83 \\
1 & 2.73 & 1.73 & 1 & 2.42 & 2 & 1 & 2.41 & 2 & 2 & 2 & 2 & 2.83 \\
1 & 2.74 & 1.73 & 1 & 2.42 & 2.01 & 1.01 & 2.42 & 2 & 2 & 2.01 & 2.01 & 2.84 \\
1 & 2.74 & 1.73 & 1 & 2.42 & 2.01 & 1.01 & 2.42 & 2 & 2 & 2.01 & 2 & 2.84 \\
3.83 & 4.91 & 6.9 & 2.83 & 4 & 2.84 & 3.85 & 4 & 2.83 & 2.83 & 2.84 & 2.84 & 16
\end{bmatrix}$$

$$\boldsymbol{Z_2} = \begin{bmatrix}
9 & 1.73 & 5.73 & 1 & 1.42 & 1 & 5.01 & 1.41 & 1 & 1 & 1 & 1 & 6.83 \\
1.73 & 35 & 3 & 5.73 & 30.5 & 25.7 & 1.74 & 26.5 & 25.7 & 25.7 & 25.7 & 25.7 & 4.91 \\
5.73 & 3 & 11 & 1.73 & 2.45 & 1.73 & 5.73 & 2.45 & 1.73 & 1.73 & 1.73 & 1.73 & 8.9 \\
1 & 5.73 & 1.73 & 9 & 5.42 & 9 & 1.01 & 9.41 & 9 & 9 & 9 & 9 & 2.83 \\
1.42 & 30.5 & 2.45 & 5.42 & 34 & 25.4 & 1.42 & 26 & 25.4 & 25.4 & 25.4 & 25.4 & 4 \\
1 & 25.7 & 1.73 & 9 & 25.4 & 29 & 1.01 & 25.4 & 25 & 25 & 25 & 25 & 2.84 \\
5.01 & 1.74 & 5.73 & 1.01 & 1.42 & 1.01 & 5.02 & 1.42 & 1.01 & 1 & 1.01 & 1.01 & 2.85 \\
1.41 & 26.5 & 2.45 & 9.41 & 26 & 25.4 & 1.42 & 30 & 25.4 & 25.4 & 25.4 & 25.4 & 4 \\
1 & 25.7 & 1.73 & 9 & 25.4 & 25 & 1.01 & 25.4 & 29 & 25 & 25 & 25 & 2.83 \\
1 & 25.7 & 1.73 & 9 & 25.4 & 25 & 1 & 25.4 & 25 & 29 & 25 & 25 & 2.83 \\
1 & 25.7 & 1.73 & 9 & 25.4 & 25 & 1.01 & 25.4 & 25 & 25 & 29 & 25 & 2.84 \\
1 & 25.7 & 1.73 & 9 & 25.4 & 25 & 1.01 & 25.4 & 25 & 25 & 25 & 29 & 2.84 \\
6.83 & 4.91 & 8.9 & 2.83 & 4 & 2.84 & 2.85 & 4 & 2.83 & 2.83 & 2.84 & 2.84 & 20
\end{bmatrix}$$

$$\boldsymbol{Z_3} = \begin{bmatrix}
73 & 1.73 & 55.7 & 1 & 1.42 & 1 & 55 & 1.41 & 1 & 1 & 1 & 1 & 11.8 \\
1.73 & 21 & 3 & 10.7 & 11.5 & 10.7 & 1.74 & 2.45 & 10.7 & 10.7 & 10.7 & 10.7 & 4.91 \\
55.7 & 3 & 57 & 1.73 & 2.45 & 1.73 & 46.7 & 2.45 & 1.73 & 1.73 & 1.73 & 1.73 & 4.9 \\
1 & 10.7 & 1.73 & 19 & 10.4 & 10 & 1.01 & 10.4 & 10 & 10 & 10 & 10 & 2.83 \\
1.42 & 11.5 & 2.45 & 10.4 & 20 & 19.4 & 1.42 & 11 & 19.4 & 19.4 & 19.4 & 19.4 & 4 \\
1 & 10.7 & 1.73 & 10 & 19.4 & 28 & 1.01 & 19.4 & 28 & 28 & 28 & 28 & 2.84 \\
55 & 1.74 & 46.7 & 1.01 & 1.42 & 1.01 & 64 & 1.42 & 1.01 & 1 & 1.01 & 1.01 & 11.8 \\
1.41 & 2.45 & 2.45 & 10.4 & 11 & 19.4 & 1.42 & 20 & 19.4 & 19.4 & 19.4 & 19.4 & 4 \\
1 & 10.7 & 1.73 & 10 & 19.4 & 28 & 1.01 & 19.4 & 28 & 28 & 28 & 28 & 2.83 \\
1 & 10.7 & 1.73 & 10 & 19.4 & 28 & 1 & 19.4 & 28 & 28 & 28 & 28 & 2.83 \\
1 & 10.7 & 1.73 & 10 & 19.4 & 28 & 1.01 & 19.4 & 28 & 28 & 28 & 28 & 2.84 \\
1 & 10.7 & 1.73 & 10 & 19.4 & 28 & 1.01 & 19.4 & 28 & 28 & 28 & 28 & 2.84 \\
11.8 & 4.91 & 4.9 & 2.83 & 4 & 2.84 & 11.8 & 4 & 2.83 & 2.83 & 2.84 & 2.84 & 26
\end{bmatrix}$$

Maximal eigenvalues of matrix Z of the petroleum cluster are respectively 9.1459, 11.1417 and 10.1042, and their topological indexes are 4.5730, 5.5709 and 5.0521.

4.3.2.7 Nonferrous metals industrial cluster

Matrixes G and Z corresponding to the nonferrous metals industrial cluster in China are as follows:

$$G_1 = \begin{bmatrix} 1 & 0.08 & 0 & 0 & 1 & 0 & 0 & 0 \\ 1 & 0.05 & 0 & 0 & 1 & 0 & 0 & 0 \\ 1.73 & 0.11 & 1 & 1 & 0 & 1 & 0 & 0 \\ 1.41 & 0.09 & 0 & 0 & 1 & 0 & 1 & 0 \\ 1.41 & 0.15 & 0 & 0 & 0 & 1 & 0 & 1 \\ 1 & 0.09 & 0 & 0 & 0 & 0 & 1 & 0 \end{bmatrix}$$

$$G_2 = \begin{bmatrix} 1 & 0.08 & 0 & 2 & 0 & 2 & 0 & 0 \\ 1 & 0.05 & 2 & 0 & 0 & 2 & 0 & 0 \\ 1.73 & 0.11 & 0 & 0 & 0 & 0 & 2 & 0 \\ 1.41 & 0.09 & 2 & 2 & 0 & 0 & 0 & 2 \\ 1.41 & 0.15 & 0 & 0 & 2 & 0 & 0 & 0 \\ 1 & 0.09 & 0 & 0 & 0 & 2 & 0 & 0 \end{bmatrix}$$

$$G_3 = \begin{bmatrix} 1 & 0.08 & 0 & 0 & 0 & 0 & 3 & 0 \\ 1 & 0.05 & 0 & 0 & 0 & 0 & 3 & 0 \\ 1.73 & 0.11 & 0 & 0 & 0 & 0 & 0 & 3 \\ 1.41 & 0.09 & 0 & 0 & 0 & 0 & 0 & 0 \\ 1.41 & 0.15 & 3 & 3 & 0 & 0 & 0 & 0 \\ 1 & 0.09 & 0 & 0 & 3 & 0 & 0 & 0 \end{bmatrix}$$

$$Z_1 = \begin{bmatrix} 2.01 & 2 & 1.74 & 2.42 & 1.43 & 1.01 \\ 2 & 2 & 1.74 & 2.42 & 1.42 & 1 \\ 1.74 & 1.74 & 6.01 & 2.46 & 3.47 & 1.74 \\ 2.42 & 2.42 & 2.46 & 4.01 & 2.01 & 2.42 \\ 1.43 & 1.42 & 3.47 & 2.01 & 4.02 & 1.43 \\ 1.01 & 1 & 1.74 & 2.42 & 1.43 & 2.01 \end{bmatrix}$$

$$Z_2 = \begin{bmatrix} 9.01 & 5 & 1.74 & 5.42 & 1.43 & 5.01 \\ 5 & 9 & 1.74 & 5.42 & 1.42 & 5 \\ 1.74 & 1.74 & 7.01 & 2.46 & 2.47 & 1.74 \\ 5.42 & 5.42 & 2.46 & 14 & 2.01 & 1.42 \\ 1.43 & 1.42 & 2.47 & 2.01 & 6.02 & 1.43 \\ 5.01 & 5 & 1.74 & 1.42 & 1.43 & 5.01 \end{bmatrix}$$

$$Z_3 = \begin{bmatrix} 1.01 & 1 & 7.74 & 1.42 & 1.43 & 1.01 \\ 1 & 1 & 7.74 & 1.42 & 1.42 & 1 \\ 1.74 & 1.74 & 3.01 & 8.46 & 2.47 & 1.74 \\ 1.42 & 1.42 & 2.46 & 2.01 & 2.01 & 1.42 \\ 7.43 & 7.42 & 2.47 & 14 & 2.02 & 1.43 \\ 1.01 & 1 & 1.74 & 1.42 & 7.43 & 1.01 \end{bmatrix}$$

Maximal eigenvalues of matrix Z of the nonferrous metals cluster are 2.7826, 3.4184 and 3.4554, and their topological indexes are 1.3913, 1.7092 and 1.7277.

4.3.2.8   Construction industrial cluster

Matrixes G and Z corresponding to the construction industrial cluster in China are as follows:

$$
G_1=\begin{bmatrix}
1 & 0.08 & 0 & 1 & 0 & 0 & 0 & 0 & 0 & 0 & 0 & 0 & 0 & 0 & 0 & 0 & 0 & 0 & 0 & 0\\
1.41 & 0.08 & 1 & 0 & 0 & 0 & 0 & 0 & 0 & 0 & 0 & 0 & 0 & 0 & 0 & 0 & 0 & 0 & 0 & 1\\
1 & 0.18 & 0 & 0 & 0 & 0 & 0 & 0 & 0 & 0 & 0 & 0 & 0 & 0 & 0 & 0 & 0 & 0 & 0 & 1\\
1.41 & 0.09 & 0 & 0 & 0 & 0 & 0 & 0 & 0 & 0 & 0 & 0 & 0 & 0 & 0 & 0 & 0 & 0 & 1 & 1\\
1 & 0.05 & 0 & 0 & 0 & 0 & 0 & 0 & 0 & 0 & 0 & 0 & 0 & 0 & 0 & 0 & 0 & 0 & 0 & 1\\
1.41 & 0.15 & 0 & 0 & 0 & 0 & 0 & 0 & 0 & 1 & 0 & 0 & 0 & 0 & 0 & 0 & 0 & 0 & 0 & 1\\
1 & 0.02 & 0 & 0 & 0 & 0 & 0 & 0 & 0 & 0 & 0 & 0 & 0 & 0 & 0 & 0 & 0 & 0 & 0 & 1\\
1.41 & 0.09 & 0 & 0 & 0 & 0 & 0 & 1 & 0 & 0 & 0 & 0 & 0 & 0 & 0 & 0 & 0 & 0 & 0 & 1\\
1 & 0.17 & 0 & 0 & 0 & 0 & 0 & 0 & 0 & 0 & 0 & 0 & 0 & 0 & 0 & 0 & 0 & 0 & 0 & 1\\
1 & 0.06 & 0 & 0 & 0 & 0 & 0 & 0 & 0 & 0 & 0 & 0 & 0 & 0 & 0 & 0 & 0 & 0 & 0 & 1\\
1.41 & 0.09 & 0 & 0 & 0 & 0 & 0 & 0 & 0 & 0 & 0 & 0 & 0 & 0 & 0 & 1 & 0 & 0 & 0 & 1\\
1 & 0.12 & 0 & 0 & 0 & 0 & 0 & 0 & 0 & 0 & 0 & 0 & 0 & 0 & 0 & 0 & 0 & 0 & 0 & 1\\
1 & 0.06 & 0 & 0 & 0 & 0 & 0 & 0 & 0 & 0 & 0 & 0 & 0 & 0 & 0 & 0 & 0 & 0 & 0 & 1\\
1 & 0.04 & 0 & 0 & 0 & 0 & 0 & 0 & 0 & 0 & 0 & 0 & 0 & 0 & 0 & 0 & 0 & 0 & 0 & 1\\
1.41 & 0.12 & 0 & 0 & 0 & 0 & 0 & 0 & 0 & 0 & 0 & 0 & 0 & 0 & 0 & 0 & 0 & 0 & 0 & 1\\
1.41 & 0.06 & 0 & 0 & 0 & 1 & 0 & 0 & 0 & 0 & 0 & 0 & 0 & 0 & 0 & 0 & 0 & 0 & 0 & 1\\
3.87 & 0.23 & 0 & 1 & 1 & 1 & 1 & 1 & 1 & 1 & 1 & 1 & 1 & 1 & 1 & 1 & 1 & 1 & 1 & 0
\end{bmatrix}
$$

$$
G_2=\begin{bmatrix}
1 & 0.08 & 0 & 0 & 0 & 0 & 0 & 0 & 0 & 0 & 0 & 0 & 0 & 0 & 0 & 0 & 0 & 0 & 0 & 2\\
1.41 & 0.08 & 0 & 0 & 2 & 2 & 2 & 2 & 2 & 2 & 2 & 2 & 2 & 2 & 2 & 2 & 2 & 2 & 2 & 0\\
1 & 0.18 & 0 & 2 & 0 & 2 & 2 & 2 & 2 & 2 & 2 & 2 & 2 & 2 & 2 & 2 & 2 & 2 & 2 & 0\\
1.41 & 0.09 & 0 & 2 & 2 & 0 & 2 & 2 & 2 & 2 & 2 & 2 & 2 & 2 & 2 & 2 & 2 & 2 & 2 & 2\\
1 & 0.05 & 0 & 2 & 2 & 2 & 0 & 2 & 2 & 2 & 2 & 2 & 2 & 2 & 2 & 2 & 2 & 2 & 2 & 0\\
1.41 & 0.15 & 0 & 2 & 2 & 2 & 2 & 0 & 2 & 2 & 2 & 2 & 2 & 2 & 2 & 2 & 2 & 2 & 2 & 2\\
1 & 0.02 & 0 & 2 & 2 & 2 & 2 & 2 & 0 & 2 & 2 & 2 & 2 & 2 & 2 & 2 & 2 & 2 & 2 & 0\\
1.41 & 0.09 & 0 & 2 & 2 & 2 & 2 & 2 & 2 & 0 & 2 & 2 & 2 & 2 & 2 & 2 & 2 & 2 & 2 & 2\\
1 & 0.17 & 0 & 2 & 2 & 2 & 2 & 2 & 2 & 2 & 0 & 2 & 2 & 2 & 2 & 2 & 2 & 2 & 2 & 0\\
1 & 0.06 & 0 & 2 & 2 & 2 & 2 & 2 & 2 & 2 & 2 & 0 & 2 & 2 & 2 & 2 & 2 & 2 & 2 & 0\\
1.41 & 0.09 & 0 & 2 & 2 & 2 & 2 & 2 & 2 & 2 & 2 & 2 & 0 & 2 & 2 & 2 & 2 & 2 & 2 & 2\\
1 & 0.12 & 0 & 2 & 2 & 2 & 2 & 2 & 2 & 2 & 2 & 2 & 2 & 0 & 2 & 2 & 2 & 2 & 2 & 0\\
1 & 0.06 & 0 & 2 & 2 & 2 & 2 & 2 & 2 & 2 & 2 & 2 & 2 & 2 & 0 & 2 & 2 & 2 & 2 & 0\\
1 & 0.04 & 0 & 2 & 2 & 2 & 2 & 2 & 2 & 2 & 2 & 2 & 2 & 2 & 2 & 0 & 2 & 2 & 2 & 0\\
1.41 & 0.12 & 0 & 2 & 2 & 2 & 2 & 2 & 2 & 2 & 2 & 2 & 2 & 2 & 2 & 2 & 0 & 2 & 2 & 2\\
1.41 & 0.06 & 0 & 2 & 2 & 2 & 2 & 2 & 2 & 2 & 2 & 2 & 2 & 2 & 2 & 2 & 2 & 0 & 2 & 2\\
3.87 & 0.23 & 2 & 0 & 0 & 2 & 0 & 2 & 0 & 2 & 0 & 0 & 2 & 0 & 0 & 0 & 2 & 2 & 0 & 0
\end{bmatrix}
$$

$$
G_3=\begin{bmatrix}
1 & 0.08 & 0 & 0 & 3 & 3 & 3 & 3 & 3 & 3 & 3 & 3 & 3 & 3 & 3 & 3 & 3 & 3 & 3 & 0\\
1.41 & 0.08 & 0 & 0 & 0 & 3 & 0 & 3 & 0 & 3 & 0 & 0 & 3 & 0 & 0 & 0 & 3 & 3 & 0 & 0\\
1 & 0.18 & 3 & 0 & 0 & 3 & 0 & 3 & 0 & 3 & 0 & 0 & 3 & 0 & 0 & 0 & 3 & 3 & 0 & 0\\
1.41 & 0.09 & 3 & 3 & 3 & 0 & 3 & 3 & 3 & 3 & 3 & 3 & 3 & 3 & 3 & 3 & 3 & 3 & 3 & 0\\
1 & 0.05 & 3 & 0 & 0 & 3 & 0 & 3 & 0 & 3 & 0 & 0 & 3 & 0 & 0 & 0 & 3 & 3 & 0 & 0\\
1.41 & 0.15 & 3 & 3 & 3 & 3 & 3 & 0 & 3 & 0 & 3 & 3 & 3 & 3 & 3 & 3 & 3 & 3 & 3 & 0\\
1 & 0.02 & 3 & 0 & 0 & 3 & 0 & 3 & 0 & 3 & 0 & 0 & 3 & 0 & 0 & 0 & 3 & 3 & 0 & 0\\
1.41 & 0.09 & 3 & 3 & 3 & 3 & 3 & 0 & 3 & 0 & 3 & 3 & 3 & 3 & 3 & 3 & 3 & 3 & 3 & 0\\
1 & 0.17 & 3 & 0 & 0 & 3 & 0 & 3 & 0 & 3 & 0 & 0 & 3 & 0 & 0 & 0 & 3 & 3 & 0 & 0\\
1 & 0.06 & 3 & 0 & 0 & 3 & 0 & 3 & 0 & 3 & 0 & 0 & 3 & 0 & 0 & 0 & 3 & 3 & 0 & 0\\
1.41 & 0.09 & 3 & 3 & 3 & 3 & 3 & 3 & 3 & 3 & 3 & 3 & 0 & 3 & 0 & 0 & 0 & 3 & 3 & 0\\
1 & 0.12 & 3 & 0 & 0 & 3 & 0 & 3 & 0 & 3 & 0 & 0 & 3 & 0 & 0 & 0 & 3 & 3 & 0 & 0\\
1 & 0.06 & 3 & 0 & 0 & 3 & 0 & 3 & 0 & 3 & 0 & 0 & 0 & 0 & 0 & 0 & 3 & 3 & 0 & 0\\
1 & 0.04 & 3 & 0 & 0 & 3 & 0 & 3 & 0 & 3 & 0 & 0 & 0 & 0 & 0 & 0 & 3 & 3 & 0 & 0\\
1.41 & 0.12 & 3 & 3 & 3 & 3 & 3 & 3 & 3 & 3 & 3 & 3 & 0 & 3 & 3 & 3 & 0 & 3 & 3 & 0\\
1.41 & 0.06 & 3 & 3 & 3 & 3 & 3 & 3 & 3 & 3 & 3 & 3 & 3 & 3 & 3 & 3 & 0 & 0 & 0 & 0\\
3.87 & 0.23 & 0 & 0 & 0 & 0 & 0 & 0 & 0 & 0 & 0 & 0 & 0 & 0 & 0 & 0 & 0 & 0 & 0 & 0
\end{bmatrix}
$$

$$
z_1=\begin{bmatrix}
2.01 & 1.42 & 1.01 & 1.42 & 1 & 1.43 & 1 & 1.42 & 1.01 & 1.01 & 1.42 & 1.01 & 1 & 1 & 1.42 & 1.42 & 4.89\\
1.42 & 4.01 & 2.43 & 3.01 & 2.42 & 3.01 & 2.42 & 3.01 & 2.43 & 2.42 & 3.01 & 2.42 & 2.42 & 2.42 & 3.01 & 3 & 5.5\\
1.01 & 2.43 & 2.03 & 2.43 & 2.01 & 2.44 & 2 & 2.43 & 2.03 & 2.01 & 2.43 & 2.02 & 2.01 & 2.01 & 2.44 & 2.43 & 3.91\\
1.42 & 3.01 & 2.43 & 4.01 & 2.42 & 3.01 & 2.42 & 3.01 & 2.43 & 2.42 & 3.01 & 2.43 & 2.42 & 2.42 & 3.01 & 3.01 & 6.5\\
1 & 2.42 & 2.01 & 2.42 & 2 & 2.42 & 2 & 2.42 & 2.01 & 2 & 2.42 & 2.01 & 2 & 2 & 2.42 & 2.42 & 3.88\\
1.43 & 3.01 & 2.44 & 3.01 & 2.42 & 4.02 & 2.42 & 3.01 & 2.44 & 2.42 & 3.01 & 2.43 & 2.42 & 2.42 & 3.02 & 3.01 & 6.51\\
1 & 2.42 & 2 & 2.42 & 2 & 2.42 & 2 & 2.42 & 2 & 2 & 2.42 & 2 & 2 & 2 & 2.42 & 2.42 & 3.88\\
1.42 & 3.01 & 2.43 & 3.01 & 2.42 & 4.01 & 2.42 & 3.01 & 2.43 & 2.42 & 3.01 & 2.43 & 2.42 & 2.42 & 3.01 & 3.01 & 6.5\\
1.01 & 2.43 & 2.03 & 2.43 & 2.01 & 2.44 & 2 & 2.43 & 2.03 & 2.01 & 2.43 & 2.02 & 2.01 & 2.01 & 2.44 & 2.43 & 3.91\\
1.01 & 2.42 & 2.01 & 2.42 & 2 & 2.42 & 2 & 2.42 & 2.01 & 2 & 2.42 & 2.01 & 2 & 2 & 2.42 & 2.42 & 3.89\\
1.42 & 3.01 & 2.43 & 3.01 & 2.42 & 3.01 & 2.42 & 4.01 & 2.43 & 2.42 & 3.01 & 2.43 & 2.42 & 2.42 & 3.01 & 3.01 & 6.5\\
1.01 & 2.42 & 2.02 & 2.43 & 2.01 & 2.43 & 2 & 2.43 & 2.02 & 2.01 & 2.42 & 2.02 & 2.01 & 2.01 & 2.43 & 2.42 & 3.9\\
1 & 2.42 & 2.01 & 2.42 & 2 & 2.42 & 2 & 2.42 & 2.01 & 2 & 2.42 & 2.01 & 2 & 2 & 2.42 & 2.42 & 3.89\\
1 & 2.42 & 2.01 & 2.42 & 2 & 2.42 & 2 & 2.42 & 2.01 & 2 & 2.42 & 2.01 & 2 & 2 & 2.42 & 2.42 & 3.88\\
1.42 & 3.01 & 2.44 & 3.01 & 2.42 & 3.02 & 2.42 & 3.01 & 2.44 & 2.42 & 3.01 & 2.43 & 2.42 & 2.42 & 3.02 & 3.01 & 5.51\\
1.42 & 3 & 2.43 & 3.01 & 2.42 & 3.01 & 2.42 & 3.01 & 2.43 & 2.42 & 3.01 & 2.42 & 2.42 & 2.42 & 3.01 & 4 & 6.49\\
4.89 & 5.5 & 3.91 & 6.5 & 3.88 & 6.51 & 3.88 & 6.5 & 3.91 & 3.89 & 6.5 & 3.9 & 3.89 & 3.88 & 5.51 & 6.49 & 30.1
\end{bmatrix}
$$

$$
Z_2 = \begin{bmatrix}
5.01 & 1.42 & 1.01 & 5.42 & 1 & 5.43 & 1 & 5.42 & 1.01 & 1.01 & 5.42 & 1.01 & 1 & 1 & 5.42 & 5.42 & 3.89 \\
1.42 & 58 & 53.4 & 54 & 53.4 & 54 & 53.4 & 54 & 53.4 & 53.4 & 54 & 53.4 & 53.4 & 53.4 & 54 & 54 & 29.5 \\
1.01 & 53.4 & 57 & 53.4 & 53 & 53.4 & 53 & 53.4 & 53 & 53 & 53.4 & 53 & 53 & 53 & 53.4 & 53.4 & 27.9 \\
5.42 & 54 & 53.4 & 62 & 53.4 & 58 & 53.4 & 58 & 53.4 & 53.4 & 58 & 53.4 & 53.4 & 53.4 & 58 & 58 & 25.5 \\
1 & 53.4 & 53 & 53.4 & 57 & 53.4 & 53 & 53.4 & 53 & 53 & 53.4 & 53 & 53 & 53 & 53.4 & 53.4 & 27.9 \\
5.43 & 54 & 53.4 & 58 & 53.4 & 62 & 53.4 & 58 & 53.4 & 53.4 & 58 & 53.4 & 53.4 & 53.4 & 58 & 58 & 25.5 \\
1 & 53.4 & 53 & 53.4 & 53 & 53.4 & 57 & 53.4 & 53 & 53 & 53.4 & 53 & 53 & 53 & 53.4 & 53.4 & 27.9 \\
5.42 & 54 & 53.4 & 58 & 53.4 & 58 & 53.4 & 62 & 53.4 & 53.4 & 58 & 53.4 & 53.4 & 53.4 & 58 & 58 & 25.5 \\
1.01 & 53.4 & 53 & 53.4 & 53 & 53.4 & 53 & 53.4 & 57 & 53 & 53.4 & 53 & 53 & 53 & 53.4 & 53.4 & 27.9 \\
1.01 & 53.4 & 53 & 53.4 & 53 & 53.4 & 53 & 53.4 & 53 & 57 & 53.4 & 53 & 53 & 53 & 53.4 & 53.4 & 27.9 \\
5.42 & 54 & 53.4 & 58 & 53.4 & 58 & 53.4 & 58 & 53.4 & 53.4 & 62 & 53.4 & 53.4 & 53.4 & 58 & 58 & 25.5 \\
1.01 & 53.4 & 53 & 53.4 & 53 & 53.4 & 53 & 53.4 & 53 & 53 & 53.4 & 57 & 53 & 53 & 53.4 & 53.4 & 27.9 \\
1 & 53.4 & 53 & 53.4 & 53 & 53.4 & 53 & 53.4 & 53 & 53 & 53.4 & 53 & 57 & 53 & 53.4 & 53.4 & 27.9 \\
1 & 53.4 & 53 & 53.4 & 53 & 53.4 & 53 & 53.4 & 53 & 53 & 53.4 & 53 & 53 & 57 & 53.4 & 53.4 & 27.9 \\
5.42 & 54 & 53.4 & 58 & 53.4 & 58 & 53.4 & 58 & 53.4 & 53.4 & 58 & 53.4 & 53.4 & 53.4 & 62 & 58 & 25.5 \\
5.42 & 54 & 53.4 & 58 & 53.4 & 58 & 53.4 & 58 & 53.4 & 53.4 & 58 & 53.4 & 53.4 & 53.4 & 58 & 62 & 25.5 \\
3.89 & 29.5 & 27.9 & 25.5 & 27.9 & 25.5 & 27.9 & 25.5 & 27.9 & 27.9 & 25.5 & 27.9 & 27.9 & 27.9 & 25.5 & 25.5 & 43.1
\end{bmatrix}
$$

$$
Z_3 = \begin{bmatrix}
127 & 55.4 & 55 & 118 & 55 & 109 & 55 & 109 & 55 & 55 & 91.4 & 55 & 46 & 46 & 109 & 118 & 3.89 \\
55.4 & 56 & 55.4 & 47 & 55.4 & 38 & 55.4 & 54 & 55.4 & 55.4 & 38 & 55.4 & 46.4 & 46.4 & 38 & 47 & 5.5 \\
55 & 55.4 & 64 & 55.4 & 64 & 46.4 & 64 & 46.4 & 64 & 64 & 46.4 & 64 & 55 & 55 & 46.4 & 55.4 & 3.91 \\
118 & 47 & 55.4 & 137 & 55.4 & 119 & 55.4 & 119 & 55.4 & 55.4 & 101 & 55.4 & 46.4 & 46.4 & 119 & 128 & 5.5 \\
55 & 55.4 & 64 & 55.4 & 64 & 46.4 & 64 & 46.4 & 64 & 64 & 46.4 & 64 & 55 & 55 & 46.4 & 55.4 & 3.88 \\
109 & 38 & 46.4 & 119 & 46.4 & 128 & 46.4 & 128 & 46.4 & 46.4 & 92 & 46.4 & 37.4 & 37.4 & 110 & 119 & 5.51 \\
55 & 55.4 & 64 & 55.4 & 64 & 46.4 & 64 & 46.4 & 64 & 64 & 46.4 & 64 & 55 & 55 & 46.4 & 55.4 & 3.88 \\
109 & 38 & 46.4 & 119 & 46.4 & 128 & 46.4 & 128 & 46.4 & 46.4 & 92 & 46.4 & 37.4 & 37.4 & 110 & 119 & 5.5 \\
55 & 55.4 & 64 & 55.4 & 64 & 46.4 & 64 & 46.4 & 64 & 64 & 46.4 & 64 & 55 & 55 & 46.4 & 55.4 & 3.91 \\
55 & 55.4 & 64 & 55.4 & 64 & 46.4 & 64 & 46.4 & 64 & 64 & 46.4 & 64 & 55 & 55 & 46.4 & 55.4 & 3.89 \\
91.4 & 38 & 46.4 & 101 & 46.4 & 92 & 46.4 & 92 & 46.4 & 46.4 & 110 & 46.4 & 46.4 & 46.4 & 110 & 101 & 5.5 \\
55 & 55.4 & 64 & 55.4 & 64 & 46.4 & 64 & 46.4 & 64 & 64 & 46.4 & 64 & 55 & 55 & 46.4 & 55.4 & 3.91 \\
46 & 46.4 & 55 & 46.4 & 55 & 37.4 & 55 & 37.4 & 55 & 55 & 46.4 & 55 & 55 & 55 & 46.4 & 46.4 & 3.89 \\
46 & 46.4 & 55 & 46.4 & 55 & 37.4 & 55 & 37.4 & 55 & 55 & 46.4 & 55 & 55 & 55 & 46.4 & 46.4 & 3.89 \\
109 & 38 & 46.4 & 119 & 46.4 & 110 & 46.4 & 110 & 46.4 & 46.4 & 110 & 46.4 & 46.4 & 46.4 & 128 & 119 & 5.51 \\
118 & 47 & 55.4 & 128 & 55.4 & 119 & 55.4 & 119 & 55.4 & 55.4 & 101 & 55.4 & 46.4 & 46.4 & 119 & 137 & 5.49 \\
3.89 & 5.5 & 3.91 & 5.5 & 3.88 & 5.51 & 3.88 & 5.5 & 3.91 & 3.89 & 5.5 & 3.9 & 3.89 & 3.88 & 5.51 & 5.49 & 15.1
\end{bmatrix}
$$

Maximal eigenvalues of matrix Z of the construction cluster are 15.9278, 15.0453 and 10.0093, and their topological indexes are 7.9639, 7.5226 and 5.0046.

4.3.2.9   Textiles industrial cluster

Matrixes G and Z corresponding to the textiles industrial cluster in China are as follows:

$$
G_1 = \begin{bmatrix}
1 & 0.09 & 0 & 0 & 0 & 1 & 0 & 0 & 0 & 0 & 0 & 0 \\
1 & 0.1 & 0 & 0 & 0 & 1 & 0 & 0 & 0 & 0 & 0 & 0 \\
1 & 0.06 & 0 & 0 & 0 & 1 & 0 & 0 & 0 & 0 & 0 & 0 \\
2.24 & 0.16 & 1 & 1 & 1 & 0 & 1 & 0 & 0 & 1 & 0 & 0 \\
1.41 & 0.09 & 0 & 0 & 0 & 1 & 0 & 1 & 0 & 0 & 0 & 0 \\
1.41 & 0.22 & 0 & 0 & 0 & 0 & 1 & 0 & 1 & 0 & 0 & 0 \\
1.41 & 0.03 & 0 & 0 & 0 & 0 & 0 & 1 & 0 & 1 & 0 & 0 \\
2 & 0.11 & 0 & 0 & 0 & 1 & 0 & 0 & 1 & 0 & 1 & 1 \\
1 & 0.09 & 0 & 0 & 0 & 0 & 0 & 0 & 0 & 1 & 0 & 0 \\
1 & 0.1 & 0 & 0 & 0 & 0 & 0 & 0 & 0 & 1 & 0 & 0
\end{bmatrix}
$$

$$
G_2 = \begin{bmatrix}
1 & 0.09 & 0 & 2 & 2 & 0 & 2 & 0 & 0 & 2 & 0 & 0 \\
1 & 0.1 & 2 & 0 & 2 & 0 & 2 & 0 & 0 & 2 & 0 & 0 \\
1 & 0.06 & 2 & 2 & 0 & 0 & 2 & 0 & 0 & 2 & 0 & 0 \\
2.24 & 0.16 & 0 & 0 & 0 & 0 & 0 & 2 & 2 & 0 & 2 & 2 \\
1.41 & 0.09 & 2 & 2 & 2 & 0 & 0 & 0 & 2 & 2 & 0 & 0 \\
1.41 & 0.22 & 0 & 0 & 0 & 2 & 0 & 0 & 0 & 2 & 0 & 0 \\
1.41 & 0.03 & 0 & 0 & 0 & 2 & 2 & 0 & 0 & 0 & 2 & 2 \\
2 & 0.11 & 2 & 2 & 2 & 0 & 2 & 2 & 0 & 0 & 0 & 0 \\
1 & 0.09 & 0 & 0 & 0 & 2 & 0 & 0 & 2 & 0 & 0 & 0 \\
1 & 0.1 & 0 & 0 & 0 & 2 & 0 & 0 & 2 & 0 & 0 & 0
\end{bmatrix}
$$

$$
G_3 = \begin{bmatrix}
1 & 0.09 & 0 & 0 & 0 & 0 & 0 & 3 & 3 & 0 & 3 & 3 \\
1 & 0.1 & 0 & 0 & 0 & 0 & 0 & 3 & 3 & 0 & 3 & 3 \\
1 & 0.06 & 0 & 0 & 0 & 0 & 0 & 3 & 3 & 0 & 3 & 3 \\
2.24 & 0.16 & 0 & 0 & 0 & 0 & 0 & 0 & 3 & 0 & 0 & 0 \\
1.41 & 0.09 & 0 & 0 & 0 & 0 & 0 & 0 & 0 & 3 & 0 & 0 \\
1.41 & 0.22 & 3 & 3 & 3 & 0 & 0 & 0 & 0 & 0 & 3 & 3 \\
1.41 & 0.03 & 3 & 3 & 3 & 3 & 0 & 0 & 0 & 0 & 0 & 0 \\
2 & 0.11 & 0 & 0 & 0 & 0 & 3 & 0 & 0 & 0 & 0 & 0 \\
1 & 0.09 & 3 & 3 & 3 & 0 & 0 & 3 & 0 & 0 & 0 & 0 \\
1 & 0.1 & 3 & 3 & 3 & 0 & 0 & 3 & 0 & 0 & 0 & 0
\end{bmatrix}
$$

$$Z_1 = \begin{bmatrix} 2.01 & 2.01 & 2.01 & 2.25 & 2.42 & 1.43 & 1.42 & 3.01 & 1.01 & 1.01 \\ 2.01 & 2.01 & 2.01 & 2.25 & 2.42 & 1.44 & 1.42 & 3.01 & 1.01 & 1.01 \\ 2.01 & 2.01 & 2 & 2.25 & 2.42 & 1.43 & 1.42 & 3.01 & 1.01 & 1.01 \\ 2.25 & 2.25 & 2.25 & 10 & 3.18 & 4.2 & 4.17 & 4.49 & 3.25 & 3.25 \\ 2.42 & 2.42 & 2.42 & 3.18 & 4.01 & 2.02 & 3 & 3.84 & 1.42 & 1.42 \\ 1.43 & 1.44 & 1.43 & 4.2 & 2.02 & 4.05 & 2.01 & 3.85 & 1.43 & 1.44 \\ 1.42 & 1.42 & 1.42 & 4.17 & 3 & 2.01 & 4 & 2.83 & 2.42 & 2.42 \\ 3.01 & 3.01 & 3.01 & 4.49 & 3.84 & 3.85 & 2.83 & 8.01 & 2.01 & 2.01 \\ 1.01 & 1.01 & 1.01 & 3.25 & 1.42 & 1.43 & 2.42 & 2.01 & 2.01 & 2.01 \\ 1.01 & 1.01 & 1.01 & 3.25 & 1.42 & 1.44 & 2.42 & 2.01 & 2.01 & 2.01 \end{bmatrix}$$

$$Z_2 = \begin{bmatrix} 17 & 13 & 13 & 2.25 & 13.4 & 5.43 & 5.42 & 14 & 1.01 & 1.01 \\ 13 & 17 & 13 & 2.25 & 13.4 & 5.44 & 5.42 & 14 & 1.01 & 1.01 \\ 13 & 13 & 17 & 2.25 & 13.4 & 5.43 & 5.42 & 14 & 1.01 & 1.01 \\ 2.25 & 2.25 & 2.25 & 21 & 7.18 & 3.2 & 11.2 & 8.49 & 6.25 & 6.25 \\ 13.4 & 13.4 & 13.4 & 7.18 & 22 & 6.02 & 2 & 14.8 & 5.42 & 5.42 \\ 5.43 & 5.44 & 5.43 & 3.2 & 6.02 & 10.1 & 6.01 & 2.85 & 5.43 & 5.44 \\ 5.42 & 5.42 & 5.42 & 11.2 & 2 & 6.01 & 18 & 6.83 & 5.42 & 5.42 \\ 14 & 14 & 14 & 8.49 & 14.8 & 2.85 & 6.83 & 24 & 2.01 & 2.01 \\ 1.01 & 1.01 & 1.01 & 6.25 & 5.42 & 5.43 & 5.42 & 2.01 & 9.01 & 9.01 \\ 1.01 & 1.01 & 1.01 & 6.25 & 5.42 & 5.44 & 5.42 & 2.01 & 9.01 & 9.01 \end{bmatrix}$$

$$Z_3 = \begin{bmatrix} 37 & 37 & 37 & 11.3 & 1.42 & 19.4 & 1.42 & 2.01 & 10 & 10 \\ 37 & 37 & 37 & 11.3 & 1.42 & 19.4 & 1.42 & 2.01 & 10 & 10 \\ 37 & 37 & 37 & 11.2 & 1.42 & 19.4 & 1.42 & 2.01 & 10 & 10 \\ 11.3 & 11.3 & 11.2 & 14 & 3.18 & 3.2 & 3.17 & 4.49 & 2.25 & 2.25 \\ 1.42 & 1.42 & 1.42 & 3.18 & 11 & 2.02 & 2 & 2.84 & 1.42 & 1.42 \\ 19.4 & 19.4 & 19.4 & 3.2 & 2.02 & 47.1 & 29 & 2.85 & 28.4 & 28.4 \\ 1.42 & 1.42 & 1.42 & 3.17 & 2 & 29 & 38 & 2.83 & 28.4 & 28.4 \\ 2.01 & 2.01 & 2.01 & 4.49 & 2.84 & 2.85 & 2.83 & 13 & 2.01 & 2.01 \\ 10 & 10 & 10 & 2.25 & 1.42 & 28.4 & 28.4 & 2.01 & 37 & 37 \\ 10 & 10 & 10 & 2.25 & 1.42 & 28.4 & 28.4 & 2.01 & 37 & 37 \end{bmatrix}$$

Maximal eigenvalues of matrix Z of the textiles cluster are 5.3420, 6.4033 and 4.6490, and their topological indexes are 2.6710, 3.2017 and 2.3245.

4.3.2.10   Chemistry industrial cluster

Matrixes G and Z corresponding to the chemistry industrial cluster in China are as follows:

$$G_1 = \begin{bmatrix} 1 & 0.14 & 0 & 0 & 0 & 1 \\ 1 & 0.09 & 0 & 0 & 0 & 1 \\ 1 & 0.1 & 0 & 0 & 0 & 1 \\ 1.73 & 0.13 & 1 & 1 & 1 & 0 \end{bmatrix}$$

$$G_2 = \begin{bmatrix} 1 & 0.14 & 0 & 2 & 2 & 0 \\ 1 & 0.09 & 2 & 0 & 2 & 0 \\ 1 & 0.1 & 2 & 2 & 0 & 0 \\ 1.73 & 0.13 & 0 & 0 & 0 & 0 \end{bmatrix}$$

$$G_3 = \begin{bmatrix} 1 & 0.14 & 0 & 0 & 0 & 0 \\ 1 & 0.09 & 0 & 0 & 0 & 0 \\ 1 & 0.1 & 0 & 0 & 0 & 0 \\ 1.73 & 0.13 & 0 & 0 & 0 & 0 \end{bmatrix}$$

$$Z_1 = \begin{bmatrix} 2.02 & 2.01 & 2.01 & 1.75 \\ 2.01 & 2.01 & 2.01 & 1.74 \\ 2.01 & 2.01 & 2.01 & 1.74 \\ 1.75 & 1.74 & 1.74 & 6.02 \end{bmatrix}$$

$$Z_2 = \begin{bmatrix} 9.02 & 5.01 & 5.01 & 1.75 \\ 5.01 & 9.01 & 5.01 & 1.74 \\ 5.01 & 5.01 & 9.01 & 1.74 \\ 1.75 & 1.74 & 1.74 & 3.02 \end{bmatrix}$$

$$Z_3 = \begin{bmatrix} 1.02 & 1.01 & 1.01 & 1.75 \\ 1.01 & 1.01 & 1.01 & 1.74 \\ 1.01 & 1.01 & 1.01 & 1.74 \\ 1.75 & 1.74 & 1.74 & 3.02 \end{bmatrix}$$

Maximal eigenvalues of matrix Z of the chemistry industrial cluster are 3.9997, 2.7935 and 4.0000, and their topological indexes are 1.9999, 1.3968 and 2.0000.

4.3.2.11   Administration industrial cluster

Matrixes G and Z corresponding to the administration industrial cluster in China are as follows:

$$G_1 = \begin{bmatrix} 1 & 0.02 & 0 & 1 & 0 & 0 & 0 & 0 & 0 & 0 \\ 1.73 & 0.05 & 1 & 0 & 1 & 0 & 1 & 0 & 0 & 0 \\ 1 & 0.04 & 0 & 1 & 0 & 0 & 0 & 0 & 0 & 0 \\ 1 & 0.06 & 0 & 0 & 0 & 0 & 1 & 0 & 0 & 0 \\ 2 & 0.1 & 0 & 1 & 0 & 1 & 0 & 1 & 1 & 0 \\ 1 & 0.03 & 0 & 0 & 0 & 0 & 1 & 0 & 0 & 0 \\ 1.41 & 0.04 & 0 & 0 & 0 & 0 & 1 & 0 & 0 & 1 \\ 1 & 0.01 & 0 & 0 & 0 & 0 & 0 & 0 & 1 & 0 \end{bmatrix}$$

$$G_2 = \begin{bmatrix} 1 & 0.02 & 0 & 0 & 2 & 0 & 2 & 0 & 0 & 0 \\ 1.73 & 0.05 & 0 & 0 & 0 & 2 & 0 & 2 & 2 & 0 \\ 1 & 0.04 & 2 & 0 & 0 & 0 & 2 & 0 & 0 & 0 \\ 1 & 0.06 & 0 & 2 & 0 & 0 & 0 & 2 & 2 & 0 \\ 2 & 0.1 & 2 & 0 & 2 & 0 & 0 & 0 & 0 & 2 \\ 1 & 0.03 & 0 & 2 & 0 & 2 & 0 & 0 & 2 & 0 \\ 1.41 & 0.04 & 0 & 2 & 0 & 2 & 0 & 2 & 0 & 0 \\ 1 & 0.01 & 0 & 0 & 0 & 0 & 2 & 0 & 0 & 0 \end{bmatrix}$$

$$G_3 = \begin{bmatrix} 1 & 0.02 & 0 & 0 & 0 & 3 & 0 & 3 & 3 & 0 \\ 1.73 & 0.05 & 0 & 0 & 0 & 0 & 0 & 0 & 0 & 3 \\ 1 & 0.04 & 0 & 0 & 0 & 3 & 0 & 3 & 3 & 0 \\ 1 & 0.06 & 3 & 0 & 3 & 0 & 0 & 0 & 0 & 3 \\ 2 & 0.1 & 0 & 0 & 0 & 0 & 0 & 0 & 0 & 0 \\ 1 & 0.03 & 3 & 0 & 3 & 0 & 0 & 0 & 0 & 3 \\ 1.41 & 0.04 & 3 & 0 & 3 & 0 & 0 & 0 & 0 & 0 \\ 1 & 0.01 & 0 & 3 & 0 & 3 & 0 & 3 & 0 & 0 \end{bmatrix}$$

$$Z_1 = \begin{bmatrix} 2 & 1.73 & 2 & 1 & 3 & 1 & 1.42 & 1 \\ 1.73 & 6 & 1.73 & 2.73 & 3.47 & 2.73 & 3.45 & 1.73 \\ 2 & 1.73 & 2 & 1 & 3 & 1 & 1.42 & 1 \\ 1 & 2.73 & 1 & 2 & 2.01 & 2 & 2.42 & 1 \\ 3 & 3.47 & 3 & 2.01 & 8.01 & 2 & 2.83 & 3 \\ 1 & 2.73 & 1 & 2 & 2 & 2 & 2.42 & 1 \\ 1.42 & 3.45 & 1.42 & 2.42 & 2.83 & 2.42 & 4 & 1.41 \\ 1 & 1.73 & 1 & 1 & 3 & 1 & 1.41 & 2 \end{bmatrix}$$

$$Z_2 = \begin{bmatrix} 9 & 1.73 & 5 & 1 & 6 & 1 & 1.42 & 5 \\ 1.73 & 15 & 1.73 & 9.73 & 3.47 & 9.73 & 10.5 & 1.73 \\ 5 & 1.73 & 9 & 1 & 6 & 1 & 1.42 & 5 \\ 1 & 9.73 & 1 & 13 & 2.01 & 9 & 9.42 & 1 \\ 6 & 3.47 & 6 & 2.01 & 16 & 2 & 2.83 & 2 \\ 1 & 9.73 & 1 & 9 & 2 & 13 & 9.42 & 1 \\ 1.42 & 10.5 & 1.42 & 9.42 & 2.83 & 9.42 & 14 & 1.41 \\ 5 & 1.73 & 5 & 1 & 2 & 1 & 1.41 & 5 \end{bmatrix}$$

$$Z_3 = \begin{bmatrix} 28 & 1.73 & 28 & 1 & 2 & 1 & 1.42 & 19 \\ 1.73 & 12 & 1.73 & 10.7 & 3.47 & 10.7 & 2.45 & 1.73 \\ 28 & 1.73 & 28 & 1 & 2 & 1 & 1.42 & 19 \\ 1 & 10.7 & 1 & 28 & 2.01 & 28 & 19.4 & 1 \\ 2 & 3.47 & 2 & 2.01 & 4.01 & 2 & 2.83 & 2 \\ 1 & 10.7 & 1 & 28 & 2 & 28 & 19.4 & 1 \\ 1.42 & 2.45 & 1.42 & 19.4 & 2.83 & 19.4 & 20 & 1.41 \\ 19 & 1.73 & 19 & 1 & 2 & 1 & 1.41 & 28 \end{bmatrix}$$

Maximal eigenvalues of matrix Z of the administration industrial cluster are 4.0737, 6.3949 and 5.5444, and their topological indexes are 2.0368, 3.1974 and 2.7722.

4.3.3 Study on relations between structures and outputs of industrial clusters in China

Correlation analyses between topological indexes and total outputs of each industrial cluster is conducted, and the results are shown in Table 4.23.

Table 4.23   Correlation analyses between topological indexes and total outputs

Topological index	R	S	N
Am_1	0.7515	46876097	11
Am_2	0.7004	50719033	11
$Am3$	0.6020	56736052	11
Am_1 and Am_2	0.7566	49282848	11
Am_1 and Am_3	0.7533	49567685	11
Am_2 and Am_3	0.7004	53793380	11
Am_1, Am_2 and Am_3	0.7570	52647780	11

From Table 4.23, we can see the best result is single index Am_1, whose correlation coefficient is 0.7515 and standard deviation is 46876097. When Am_1 combines with Am_2, we can also get a more perfect result ($R = 0.7566$, $S = 49282848$). Integrated consideration correlation coefficients and standard deviations make clear that the best correlation between the topological index and the whole outputs of each industrial cluster is Am_1.

Regression analyses are carried through using the correlation between the topological index and the total outputs of each cluster, and the regression equation is obtained:

$$Y = 55892952.89 + 22298218.11 \ Am_1$$

Then we calculate the total outputs of the 11 clusters using the regression equation, and the results are shown in Table 4.24.

Table 4.24 Am_1, Am_2, Am_3 and the real and calculated total outputs of industrial clusters in China

Industrial clusters	Am_1	Am_2	Am_3	Y_1	Y_2
Industrial cluster 1	8.0715	7.6381	5.4610	206648140.0	235872781.1
Industrial cluster 2	2.5111	3.3294	2.7806	134938207.6	111885891.5
Industrial cluster 3	4.4302	4.4535	5.2496	154369367.7	154679591.7
Industrial cluster 4	3.5218	4.4677	3.0912	155830385.5	134423236.3
Industrial cluster 5	4.0283	4.2390	4.8346	202035452.6	145715772.5
Industrial cluster 6	4.5730	5.5709	5.0521	118353031.0	157861663.7
Industrial cluster 7	1.3913	1.7092	1.7277	84006711.8	86916729.1
Industrial cluster 8	7.9639	7.5226	5.0046	246217339.4	233473663.2
Industrial cluster 9	2.6710	3.2017	2.3245	185036277.5	115451976.4
Industrial cluster 10	1.9999	1.3968	2.0000	76604613.2	100486146.1
Industrial cluster 11	2.0368	3.1974	2.7722	14038327.6	101310597.3

Note: Y_1 and Y_2 are respectively real total outputs and calculated total outputs of each industrial cluster.

CHAPTER 5

Function Analysis of Industrial Clusters in China

An industrial cluster is an industrial system (called a system for short) which is made up of different industrial sectors. Functions of industrial clusters are the effects possessed by the industrial system in its internal integral part. The industrial clusters system of China is a huge open industrial system composed of 11 industrial clusters and has two functions, that is, external functions and internal functions. External functions are the links that one industrial cluster has with other clusters, and the contacts with external groups (including other industrial clusters or nonindustrial clusters) according to the industrial clusters' internal demands in order to facilitate developing and growing to maturity. Internal functions are the links among internal industries of industrial clusters. Functions of industrial clusters display all kinds of flow input, transition and output among industries in and out of the system. The study of the functions of industrial clusters is actually the study of these flows. In order to maintain stable and ordered development of industrial clusters in China, and to realize adjustment and to optimize growth of industrial structures in China, we must macro-control all kinds of flows among industries, make them harmonious and unimpeded, make the system develop towards a more ordered advanced phase, and offer decision-making bases for government decision-making sectors.

5.1 ANALYSIS ON BASIC FUNCTIONS OF INDUSTRIAL CLUSTERS IN CHINA

5.1.1 Indexes of measuring basic functions of industrial cluster in China

Figure 5.1 shows the basic format of China's realigned input-output table. The member industries of an industrial cluster are arranged in m columns and m rows, of which m is the number of all industries of industrial clusters in China and n is the number of industries of interindustry transaction in China's input-output table, $n = 124$. The final demand is $n + 1, \ldots, l$ and the primary input is $n + 1, \ldots, h$. The main indexes of measuring the size of the distributions of China's industrial clusters make up of 5 sections in the input-output table: A_1, A_2, A_3, A_4, and A_5.

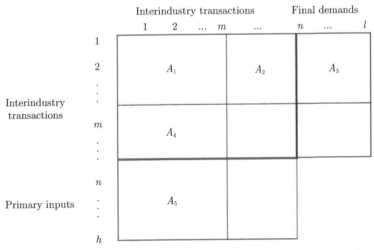

Figure 5.1   The basic format of the realigned input-output table of China

5.1.1.1   Total output of an industrial cluster

Total output of an industrial cluster is defined as:

$$\sum_{i=1}^{m}\sum_{j=1}^{l} x_{ij}^{t} = \sum_{i=1}^{h}\sum_{j=1}^{m} x_{ij}^{t}$$

where x_{ij}^{t} is an interindustry transaction at time t. Total output of the industrial cluster equals area $(A_1 + A_2 + A_3)$, and also equals area $(A_1 + A_4 + A_5)$ in Figure 5.1.

5.1.1.2   Interindustry input of an industrial cluster

Interindustry input of an industrial cluster is defined as:

$$\sum_{i=1}^{n}\sum_{j=1}^{m} x_{ij}^{t}$$

which is equal to area $(A_1 + A_4)$ in Figure 5.1.

5.1.1.3   Interindustry output of an industrial cluster

Interindustry output of an industrial cluster is defined as:

$$\sum_{i=1}^{m}\sum_{j=1}^{n} x_{ij}^{t}$$

which equals area $(A_1 + A_2)$ in Figure 5.1.

5.1.1.4   Intracluster transaction of an industrial cluster

Intracluster transaction of an industrial cluster is defined as:

$$\sum_{i=1}^{m}\sum_{j=1}^{m} x_{ij}^{t}$$

which equals areas (A_1) in Figure 5.1. The index will be used most widely to measure the size of industrial clusters.

5.1.1.5   Ratio of interindustry input to total output of an industrial cluster

The ratio is defined as:

$$\sum_{i=1}^{n}\sum_{j=1}^{m} x_{ij}^{t} \Big/ \sum_{i=1}^{m}\sum_{j=1}^{l} x_{ij}^{t} = \frac{A_1 + A_4}{A_1 + A_4 + A_5} \times 100\% = \frac{A_1 + A_4}{A_1 + A_2 + A_3} \times 100\%$$

If the ratio is larger, then interindustry economic links of an industrial cluster mainly display interindustry-input orientation; otherwise, they display primary-input oriented.

5.1.1.6   Ratio of interindustry output to total output of an industrial cluster

The ratio is defined as:

$$\sum_{i=1}^{m}\sum_{j=1}^{n} x_{ij}^{t} \Big/ \sum_{i=1}^{m}\sum_{j=1}^{l} x_{ij}^{t} \times 100\% = \frac{A_1 + A_2}{A_1 + A_4 + A_5} \times 100\% = \frac{A_1 + A_2}{A_1 + A_2 + A_3} \times 100\%$$

If the ratio is larger, then interindustry economic links of an industrial cluster mainly display interindustry-output orientation; otherwise, they display final-demand oriented.

5.1.1.7 Ratio of intracluster transaction to interindustry input of an industrial cluster

The ratio is defined as:

$$\sum_{i=1}^{m}\sum_{j=1}^{m}x_{ij}^{t}\bigg/\sum_{i=1}^{n}\sum_{j=1}^{m}x_{ij}^{t}\times 100\% = \frac{A_1}{A_1+A_4}\times 100\%$$

If the ratio is larger, then interindustry economic links of an industrial cluster are relatively independent of industries outside the industrial cluster, thereby becoming more specialized among themselves relative to industries outside the industrial cluster; otherwise, the industrial cluster will depend more on industries outside the industrial cluster.

5.1.1.8 The ratio of intracluster transaction to interindustry output of an industrial cluster

The ratio is defined as:

$$\sum_{i=1}^{m}\sum_{j=1}^{m}x_{ij}^{t}\bigg/\sum_{i=1}^{m}\sum_{j=1}^{n}x_{ij}^{t}\times 100\% = \frac{A_1}{A_1+A_2}\times 100\%$$

If the ratio is larger, then member industries of the industrial cluster absorb more interindustry output than industries outside the industrial cluster; otherwise, member industries of the industrial cluster will absorb fewer interindustry output than industries outside the industrial cluster.

5.1.2 The size distributions of industrial clusters in China

Size distributions of a number of industries, total output, interindustry input, interindustry output, intracluster transaction of 11 industrial clusters in China in 1997 are shown in Table 5.1.

Table 5.1 The size distributions of industrial clusters in China

Types of industrial clusters	Number of industries	Total output of an industrial cluster/hundred million yuan	Interindustry input of an industrial cluster/hundred million yuan	Interindustry output of an industrial cluster/hundred million yuan	Intracluster transaction of an industrial cluster/hundred million yuan
Planting industrial cluster	18	39044.3	21477.7	20650.8	15277.5
Electronics industrial cluster	9	18958.6	11262.2	13480.8	4513.2
Commerce industrial cluster	15	23994.4	12748.2	15284.6	4550.8
Diet industrial cluster	11	31196.7	16734.4	15575.9	9086.7
Iron and steel industrial cluster	15	37819.4	26409.5	19243.2	9694.8
Petroleum industrial cluster	13	13363.7	7129.9	11821.7	3248.8
Nonferrous metal industrial cluster	6	8854.5	5886.7	8209.0	2768.5
Construction industrial cluster	17	43739.6	30833.8	24095.3	15319.3
Textile industrial cluster	10	28507.1	15556.3	18440.1	8745.3
Chemistry industrial cluster	4	7169.7	5303.2	7301.9	2281.7
Administration industrial cluster	8	6753.5	3635.2	1403.8	700.3

The planting industrial cluster is the largest and is made up of 18 industries; the construction industrial cluster is the second largest and is composed of 17 industries. The cluster whose industrial number is the lowest is the chemistry industrial cluster, which is composed of 4 industries. Closely following the chemistry industrial cluster is the nonferrous metal industrial cluster, which is composed of 6 industries. Other industrial clusters' numbers are between 8 (administration industrial cluster) and 15 (commerce industrial cluster and iron and steel industrial cluster).

The size distribution of the total output of an industrial cluster reflects the relative status of each industrial cluster in China's economy. The largest industrial cluster in China in 1997 is the construction industrial cluster and its total output size is 43739.6 hundred million yuan. The planting industrial cluster and the iron and steel industrial cluster are the second and the third largest with the total output of each industrial cluster exceeding 37800 hundred million yuan. Except for the iron and steel industrial cluster, the total outputs of heavy manufacturing industrial clusters such as the petroleum industrial cluster, chemistry industrial cluster and nonferrous metal industrial cluster are lower than those of light manufacturing industrial clusters, and this point is unexpected for us. Except for the administration industrial cluster, the total outputs of the nonferrous metal industrial cluster and the chemistry industrial cluster are the lowest in the heavy manufacturing industrial clusters.

The size distributions of interindustry input, interindustry output and intracluster transaction of an industrial cluster are similar to the size distribution of total output of an industrial cluster as shown in Table 5.2. What is different is that seating arrangements among the planting industrial cluster, the iron and steel industrial cluster, and the diet industrial cluster have some changes. In general, the size distribution of total output of an industrial cluster can basically explain the size distribution of the industrial cluster in China.

Table 5.2   Seating arrangements about indexes of size distributions of industrial clusters in China

Seating arrangement	Total outputs of an industrial cluster	Interindustry input of an industrial cluster	Interindustry outputs of an industrial cluster	Intracluster transaction of an industrial cluster
1	Construction industrial cluster	Construction industrial cluster	Construction industrial cluster	Construction industrial cluster
2	Planting industrial cluster	Iron and steel industrial cluster	Planting industrial cluster	Planting industrial cluster
3	Iron and steel industrial cluster	Planting industrial cluster	Iron and steel industrial cluster	Iron and steel industrial
4	Diet industrial cluster	Diet industrial cluster	Diet industrial cluster	Diet industrial cluster
5	Textile industrial cluster	Textile industrial cluster	Textile industrial cluster	Textile industrial cluster
6	Commerce industrial cluster	Commerce industrial cluster	Commerce industrial cluster	Commerce industrial cluster
7	Electronics industrial cluster	Electronics industrial cluster	Electronics industrial cluster	Electronics industrial cluster
8	Petroleum industrial cluster	Petroleum industrial cluster	Petroleum industrial cluster	Petroleum industrial cluster
9	Nonferrous metal industrial cluster	Nonferrous metal industrial cluster	Nonferrous metal industrial cluster	Nonferrous metal industrial cluster
10	Chemistry industrial cluster	Chemistry industrial cluster	Chemistry industrial cluster	Chemistry industrial cluster
11	Administration industrial cluster	Administration industrial cluster	Administration industrial cluster	Administration industrial cluster

5.1.3   Functional types of industrial clusters in China

The partition of functional types of industrial clusters in China is according to the ratio of interindustry input to total output of an industrial cluster, the ratio of interindustry output to total output of an industrial cluster, the ratio of intracluster transaction to interindustry input of an industrial cluster and the ratio of intracluster transaction to interindustry output of industrial clusters (Table 5.3). Basic six types can thereby be divided: primary input−oriented industrial clusters, interindustry input−oriented industrial clusters, final demand−oriented industrial clusters, interindustry output−oriented industrial clusters, intracluster demand oriented industrial clusters and intracluster supply−oriented industrial clusters.

Table 5.3　Functional type indexes of industrial clusters in China (Units: %)

Types of industrial clusters	Interindustry input/total output of an industrial cluster	Interindustry output/total output of an industrial cluster	Intracluster transaction/ interindustry input of an industrial cluster	Intracluster transaction/ interindustry output of an industrial cluster
Planting industrial cluster	55.01	52.89	71.13	73.98
Electronics industrial cluster	59.40	71.11	40.07	33.48
Commerce industrial cluster	53.13	63.70	35.70	29.77
Diet industrial cluster	53.64	49.93	54.30	58.34
Iron and steel industrial cluster	69.83	50.88	36.71	50.38
Petroleum industrial cluster	53.35	88.46	45.57	27.48
Nonferrous metal industrial cluster	66.48	92.71	47.03	33.73
Construction industrial cluster	70.49	55.09	49.68	63.58
Textile industrial cluster	54.57	64.69	56.22	47.43
Chemistry industrial cluster	73.97	101.84	43.02	31.25
Administration industrial cluster	53.83	20.79	19.26	49.88

5.1.3.1　Primary input−oriented and interindustry input−oriented industrial clusters

The chemistry industrial cluster, construction industrial cluster, iron and steel industrial cluster and nonferrous metal industrial cluster have higher ratios of interindustry input to total output of an industrial cluster and belong to the interindustry input−oriented industrial clusters. The electronics industrial cluster, planting industrial cluster, textile industrial cluster, administration industrial cluster, diet industrial cluster, petroleum industrial cluster and commerce industrial cluster have lower ratios of interindustry input to total output of an industrial cluster and belong to the primary input−oriented industrial cluster. In general, the heavy manufacturing industrial cluster and construction industrial cluster belong to interindustry input−oriented industrial clusters, while the light manufacturing industrial cluster and service industrial cluster belong to primary input−oriented industrial clusters.

5.1.3.2　Final demand−oriented and interindustry output−oriented industrial clusters

Different from the ratio of interindustry input to total output of an industrial cluster, the difference in the ratio of interindustry output to total output of an industrial cluster is very large among industrial clusters. The chemistry industrial cluster, nonferrous metal industrial cluster, petroleum industrial cluster, electronics industrial cluster, textile industrial cluster and commerce industrial cluster have higher ratios of interindustry output to total output of an industrial cluster and belong to the interindustry output−oriented industrial clusters. The construction industrial cluster, planting industrial cluster, iron and steel industrial cluster, diet industrial cluster and administration industrial cluster have lower ratios of interindustry output to total output of an industrial cluster and belong to the final demand−oriented industrial cluster.

5.1.3.3　Intracluster demand−oriented and intracluster supply−oriented industrial clusters

According to relative size about the ratio of intracluster transaction to interindustry input of an industrial cluster and the ratio of intracluster transaction to interindustry output of an industrial cluster, intracluster demand−oriented industrial clusters and intracluster supply−oriented industrial clusters are divided. For the former, the ratio of intracluster transaction to interindustry input of an industrial cluster is bigger than that of intracluster transaction to interindustry output of an industrial cluster; the latter is just the reverse. Industrial clusters belonging to intracluster demand—oriented industrial clusters are petroleum industrial cluster, nonferrous metal industrial cluster, textile industrial cluster, electronics industrial cluster, chemistry industrial cluster and commerce industrial cluster. Industrial clusters belonging to intracluster supply—oriented industrial clusters are planting industrial cluster, construction industrial cluster, diet industrial cluster, iron and steel industrial cluster, administration industrial cluster.

Considering relationships among the above six types of industrial clusters, 11 industrial clusters in China in 1997 can be respectively placed within different functional types (Table 5.4).

Table 5.4   Functional types of industrial clusters in China and corresponding industrial clusters

Types of industrial clusters	Primary input oriented—industrial clusters	Interindustry input—oriented industrial clusters	Final demand—oriented industrial clusters	Interindustry output—oriented industrial clusters	Intracluster demand—oriented industrial clusters	Intracluster supply—oriented industrial clusters
Primary input—oriented industrial clusters	Electronics industrial cluster, planting industrial cluster, textile industrial cluster, administration industrial cluster, diet industrial cluster, petroleum industrial cluster, commerce industrial cluster		Planting industrial cluster, diet industrial cluster	Petroleum industrial cluster, textile industrial cluster, commerce industrial cluster, electronics industrial cluster	Electronics industrial cluster, commerce industrial cluster	Planting industrial cluster, diet industrial cluster, administration industrial cluster, textile industrial cluster, petroleum industrial cluster
Interindustry input—oriented industrial clusters		Chemistry industrial cluster, construction industrial cluster, iron and steel industrial cluster, nonferrous metal industrial cluster	Construction industrial cluster, iron and steel industrial cluster	Chemistry industrial cluster, nonferrous metal industrial cluster	Nonferrous metal industrial cluster, chemistry industrial cluster	Construction industrial cluster, iron and steel industrial cluster
Final demand—oriented industrial clusters	Planting industrial cluster, diet industrial cluster	Construction industrial cluster, iron and steel industrial cluster	Construction industrial cluster, planting industrial cluster, iron and steel industrial cluster, diet industrial cluster, administration industrial cluster			Planting industrial cluster, construction industrial cluster, diet industrial cluster, iron and steel industrial cluster, administration industrial cluster

Continued

Types of industrial clusters	Primary input–oriented industrial clusters	Interindustry input–oriented industrial clusters	Final demand–oriented industrial clusters	Interindustry output–oriented industrial clusters	Intracluster demand oriented industrial clusters	Intracluster supply–oriented industrial clusters
Interindustry output–oriented industrial clusters	Petroleum industrial cluster, textile industrial cluster, commerce industrial cluster, electronics industrial cluster	Chemistry industrial cluster, nonferrous metal industrial cluster		Chemistry industrial cluster, nonferrous metal industrial cluster, petroleum industrial cluster, electronics industrial cluster, textile industrial cluster, commerce industrial cluster	Nonferrous metal industrial cluster, electronics industrial cluster, chemistry industrial cluster, commerce industrial cluster	Petroleum industrial cluster, textile industrial cluster
Intracluster demand–oriented industrial clusters	Electronics industrial cluster, commerce industrial cluster	Nonferrous metal industrial cluster, chemistry industrial cluster		Nonferrous metal industrial cluster, electronics industrial cluster, chemistry industrial cluster, commerce industrial cluster	Petroleum industrial cluster, nonferrous metal industrial cluster, textile industrial cluster, electronics industrial cluster, chemistry industrial cluster, commerce industrial cluster	
Intracluster supply–oriented industrial clusters	Planting industrial cluster, diet industrial cluster, administration industrial cluster, textile industrial cluster, petroleum industrial cluster	Construction industrial cluster, iron and steel industrial cluster	Planting industrial cluster, construction industrial cluster, diet industrial cluster, iron and steel industrial cluster, administration industrial cluster	Textile industrial cluster, petroleum industrial cluster		Planting industrial cluster, construction industrial cluster, diet industrial cluster, iron and steel industrial cluster, administration industrial cluster

5.1.4 Central industries and node industries of industrial clusters in China

5.1.4.1 Central industries of industrial clusters in China

Complicated economic links exist among industries within industrial clusters. Every industrial cluster has one or several industries which have close links ($e_{ij}^t \geqslant 0.2$) with other sectors of this industrial cluster. These industries are called central industries. Different from Czamanski's (1974) operable definition, we define central industries as those industries which have 3 or more close links ($e_{ij}^t \geqslant 0.2$) with other members of this industrial cluster at the time of our research. There are 21 central clusters in 11 industrial clusters in China in 1997. The number of close links for each central industry, that is, connecting degrees of central industries is shown in Table 5.5.

The industrial cluster with the maximum number of central industries is the planting industrial cluster. There are 4 central industries in the industrial cluster in 1997. The next is the petroleum industrial cluster which has 3 central industries. The electronics industrial cluster, chemistry industrial cluster, nonferrous metal industrial cluster and construction industrial cluster each have only 1 central industry.

Table 5.5   Central industries and their connecting degrees of industrial clusters in China

Types of industrial clusters	Central industries	Connecting degrees
Planting industrial cluster	Crop cultivation	15
	Grain mill products, vegetable oil and forage	3
	Livestock and livestock products	3
	Sugar refining	3
Electronics industrial cluster	Electronic element and device	5
Commerce industrial cluster	Wholesale and retail trade	8
	Paper and products	4
Diet industrial cluster	Eating and drinking places	7
	Grain mill products, vegetable oil and forage	3
Iron and steel industrial cluster	Steel processing	7
	Coal mining and processing	4
Petroleum industrial cluster	Petroleum refining	8
	Railway freight transport	3
	Social welfare	3
Nonferrous metal industrial cluster	Nonferrous metal smelting	3
Construction industrial cluster	Construction	15
Textile industrial cluster	Cotton textiles	5
	Wearing apparel	4
Chemistry industrial cluster	Other chemical products	3
Administration industrial cluster	Public administration and other sectors	4
	Air passenger transport	3

There are great differences for the number of close links or connection degrees of central industries in different central industries. The connection degrees of the planting industrial cluster and construction industrial cluster are the highest at 15. The commerce industrial cluster and petroleum industrial cluster take second place with both connection degrees at 8. The connection degrees of the rest are between 3 and 7.

As far as industrial clusters in China are concerned, central industries are normally maximal industrial sectors or larger industrial sectors. They profoundly influence structures, sizes and functions of industrial clusters.

In every industrial cluster, the number of close economic links among industries, the number of close economic links with central industries and their average intensities measured by e_{ij}^t coefficient are shown in Table 5.6.

Table 5.6   The status of interindustry economic links in industrial clusters

Types of industrial clusters	The number of interindustry links	The number of links with central industries	Average intensities of interindustry links	Average intensities of links with central industries
Planting industrial cluster	21	20	0.4584	0.4624
Electronics industrial cluster	8	5	0.3735	0.3352
Commerce industrial cluster	14	12	0.3011	0.2979
Diet industrial cluster	13	12	0.3866	0.3702
Iron and steel industrial cluster	14	11	0.3443	0.3133
Petroleum industrial cluster	13	12	0.3761	0.3771
Nonferrous metal industrial cluster	5	3	0.3970	0.4311
Construction industrial cluster	19	15	0.3753	0.4005
Textile industrial cluster	10	8	0.3497	0.3124
Chemistry industrial cluster	3	3	0.3249	0.3249
Administration industrial cluster	7	6	0.2106	0.3177
Dispersion coefficients			0.1773	0.1515

Table 5.7   The nodal industries of industrial clusters in China

Types of nodal industries	Nodal industries	Industrial clusters belonged to
Industries belonging to two industrial clusters	Logging and transport of timber and bamboo	Commerce industrial cluster and construction industrial cluster
	Grain mill products, vegetable oil and forage	Planting industrial cluster and diet industrial cluster
	Slaughtering, meat processing, eggs and dairy products	Planting industrial cluster and diet industrial cluster
	Other food products	Planting industrial cluster and diet industrial cluster
	Wines, spirits and liquors	Planting industrial cluster and diet industrial cluster
	Hemp textiles	Planting industrial cluster and textile industrial cluster
	Other textiles	Planting industrial cluster and textile industrial cluster
	Chemical pesticides	Planting industrial cluster and commerce industrial cluster
	Chemical fibers	Textile industrial cluster and chemistry industrial cluster
	Rubber products	Commerce industrial cluster and textile industrial cluster
	Steel processing	Iron and steel industrial cluster and construction industrial cluster
	Metal products	Iron and steel industrial cluster and construction industrial cluster
	Other electric machinery and equipment	Nonferrous metal industrial cluster and construction industrial cluster
	Electronic appliances	Electronics industrial cluster and commerce industrial cluster
	Electricity production and supply	Iron and steel industrial cluster petroleum industrial cluster
	Construction	Iron and steel industrial cluster and construction industrial cluster
	Telecommunication	Nonferrous metal industrial cluster and construction industrial cluster
	Wholesale and retail trade	Electronics industrial cluster and commerce industrial cluster
	Railway passenger transport	Petroleum industrial cluster and administration industrial cluster
	Social welfare	Petroleum industrial cluster and administration industrial cluster
	Technical services for agriculture, forestry, livestock and fishing	Planting industrial cluster and diet industrial cluster
Industries belonging to three industrial clusters	Crop cultivation	Planting industrial cluster, diet industrial cluster and textile industrial cluster
	Nonalcoholic beverage	Planting industrial cluster, commerce industrial cluster and diet industrial cluster
	Tobacco products	Planting industrial cluster, commerce industrial cluster and diet industrial cluster
	Storage and warehousing	Planting industrial cluster, commerce industrial cluster and construction industrial cluster

We can see from Table 5.6 that all industrial clusters have industries which do not link with central industries except for the chemistry industrial cluster. The number of industries that do not link closely with central industries is most in the construction industrial cluster, they is up to 4 industries. The industries whose average intensities of links with central industries are higher than average the intensities of all links in industrial clusters are the nonferrous metal industrial cluster, the planting industrial cluster, the administration industrial cluster, the construction industrial cluster and the petroleum industrial cluster. Other industrial clusters are just the reverse; their average intensities of links with central industries are lower than the average intensities of all links in industrial clusters. This indicates that economic link intensities among noncentral industries are not low for most industrial clusters in China.

The differences in economic link intensities among industries in industrial clusters in China are not large because the dispersion coefficient of economic link intensities among all industries in all industrial clusters in China is 0.1773. The maximum of average link intensities is 0.4584, while the minimum of average link intensities is 0.2106. The difference of economic link intensities between industrial clusters and central industries in China is also not large since the dispersion coefficient of average intensities of links with central industries in all industrial clusters in China is only 0.1515. The maximal average link intensity is 0.4624, while the minimal average link intensity is 0.2979.

5.1.4.2   The nodal industries of industrial clusters in China

Each industrial cluster is a part of the industrial system, so there are definite economic links among industrial clusters. It is due to the economic links among different industrial clusters that industrial clusters in China form a gigantic industrial cluster system. We call the industries belonging to two or more industrial clusters nodal industries. Fifty-five of 177 industries (32.2%) are divided as nodal industries in the study of American economy in 1963 (Czamanski, 1974). There are 25 nodal industries in 11 industrial clusters, of which 21 belong to two industrial clusters and 4 belong to three industrial clusters in the study of the Chinese economy in 1997 (Table 5.7).

Those industries which are both central industries and nodal industries have especially important roles in industrial development, adjustment and optimization of industrial structures. In industrial clusters in China in 1997, such industries are crop cultivation, grain mill products, vegetable oil and forage, wholesale and retail trade, steel processing, social welfare and construction.

5.2   ANALYSIS OF TECHNOLOGICAL ECONOMY STRUCTURE RELATIONSHIPS OF INDUSTRIAL CLUSTERS IN CHINA

The technological economy structures of industrial clusters are the quantitative proportion relationships reflecting economic links and link ways among industries of industrial clusters; that is, quantizing proportion relationships of inputs and outputs among industries of industrial clusters, they reflect industrial structure relationships of industrial clusters from quantitative aspects.

The abilities of assimilation and absorbency for technologies tend to have great differences in different industries of different industrial clusters, and even in the same industries of different industrial clusters. There exist great differences in input structures and output structures among industries of different industrial clusters. As far as industrial clusters in China in 1997 are concerned, total outputs of 11 industrial clusters are up to 166035.91 hundred million yuan which accounts for 83.1% of total outputs of entire industries. The study of technological economy structure status of industrial cluster in China is the foundation of industrial structure adjustment and economy growth not only for industrial clusters, but also for China.

5.2.1   Analysis on production technology links of industrial clusters

Technological economy structures of industrial clusters are interindustry interdependent structural relationships formed on the bases of interindustry production technology links of industrial clusters. So technological economy structure levels first depend on interindustry production technology link levels of industrial clusters.

The foundations of technological economy structures are interindustry production technology links of industrial clusters, and these links are realized by movements of intermediate products, namely, they take place by producing and using intermediate products. Therefore, direct consumption coefficients (input coefficients) of intermediate products of industrial clusters reflect production technology links among industrial sectors. The higher input coefficients of unit product indicate the greater material consumptions and the lower technological levels and vice versa. In technological economic structural relationships, the matrix composed of all input coefficients reflects overall production technology links among industries in the region, so we call it technological matrix A, and it can be used as a measuring index of interindustry technological links.

Because interindustry technological links can be denoted by technological matrixes, overall levels of interindustry technological links can be reflected by levels of technological matrixes. Technological matrixes as the sum of input coefficients are materially technological structures in each industry of a society. Generally speaking, in industrial technological structure, production technological process of each industry selected is that which most microcosmic enterprises adopt in this industry, so it has the meaning of average production technological levels. Overall levels of technological matrixes are the sum of specifically technological levels of each industry, and they can be given by weighted average values of technological levels of industrial sectors:

$$G_A = \sum G_j F_j$$

where G_A is technological matrix level, G_j is the average technological matrix level of sector j and F_j is the weighted number of the average technological level of sector j (it is output composings of industrial sectors commonly). G_A reflects changes of material consumption levels. The greater its value is, the higher the material consumption level is and the lower the technological matrix level is.

Input coefficients and technological matrix levels of industrial clusters in China are shown in Table 5.8. The input coefficient of the chemistry industrial cluster is the highest, while that of commerce industrial cluster is the lowest, which indicates that in the industrial system made up of all industrial clusters, the technological level of chemistry industrial cluster is the lowest and that of commerce industrial cluster is the highest. The industrial clusters whose technological level is highest are the construction industrial cluster, the iron and steel industrial cluster, the nonferrous metal industrial cluster, the electronics industrial cluster, the planting industrial cluster, the administration industrial cluster, the diet industrial cluster, the textile industrial cluster and the petroleum industrial cluster.

We can see from Table 5.8 that the technological matrix level of the construction industrial cluster is the highest, while the administration industrial cluster is the lowest. This indicates that the material consumption level among industrial sectors of the construction industrial cluster is the highest and the overall level of technological links among industrial sectors of the cluster is the lowest, the material consumption level among industrial sectors of administration industrial cluster is the lowest and the overall level of technological links among industrial sectors of the cluster is the lowest. Industrial clusters whose technological link level among industrial sectors within industrial clusters is highest are the iron and steel industrial cluster, the planting industrial cluster, the diet industrial cluster, the textile industrial cluster, commerce industrial cluster, the electronics industrial cluster, the petroleum industrial cluster, the nonferrous metal industrial cluster and the chemistry industrial cluster.

Table 5.8   Input coefficients and technological matrix levels of industrial clusters in China

Types of industrial clusters	Input coefficients	Technological matrix levels
Planting industrial cluster	0.5501	0.1075
Electronics industrial cluster	0.5940	0.0564
Commerce industrial cluster	0.5313	0.0638
Diet industrial cluster	0.5364	0.0837
Iron and steel industrial cluster	0.6983	0.1322
Petroleum industrial cluster	0.5335	0.0357
Nonferrous metal industrial cluster	0.6648	0.0295
Construction industrial cluster	0.7049	0.1543
Textile industrial cluster	0.5359	0.0764
Chemistry industrial cluster	0.7397	0.0265
Administration industrial cluster	0.5383	0.0182

What deserves our attention is that as far as overall technological levels of industrial clusters reflected by input coefficients and levels of technological links among industries within industrial clusters reflected by the technological matrix levels are concerned, some clusters are consistent and some clusters are not. This is because the technological matrix levels of industrial clusters are not only concerned with the technological level of each industry but also with the output of each industry. The technological level and output of each industry are shown in Table 5.9. Because ordering among them is not consistent, it leads to the incompletely consistent phenomena of overall technological levels of industrial clusters and levels of technological links among industries within industrial clusters.

Table 5.9   The ordering status of technological level and output of industrial clusters

Technological level	Ordering(from low to high)	Output	Ordering(from low to high)
Chemistry industrial cluster	1	Construction industrial cluster	1
Construction industrial cluster	2	Planting industrial cluster	2
Iron and steel industrial cluster	3	Iron and steel industrial cluster	3
Nonferrous metal industrial cluster	4	Diet industrial cluster	4
Electronics industrial cluster	5	Textile industrial cluster	5
Planting industrial cluster	6	Commerce industrial cluster	6
Administration industrial cluster	7	Electronics industrial cluster	7
Diet industrial cluster	8	Petroleum industrial cluster	8
Textile industrial cluster	9	Nonferrous metal industrial cluster	9
Petroleum industrial cluster	10	Chemistry industrial cluster	10
Commerce industrial cluster	11	Administration industrial cluster	11

5.2.2   Industry structure relationships of industrial clusters

5.2.2.1   Measuring indexes of industry structure relationships of industrial clusters

The following two indexes can be used to analyze the status of industry structure relationships of industrial clusters. The first one is the ratio of inputs of the inner industrial cluster of one industry to the interindustry inputs of the industry. The ratio is defined as

$$\sum_{i=1}^{m} x_{ij}^{t} \bigg/ \sum_{i=1}^{n} x_{ij}^{t} \times 100 = A/B \times 100$$

If the ratio of one industry is higher, that industry will depend more on other industries in industrial cluster.

The second one is the ratio of outputs of the inner industrial cluster of one industry to the interindustry outputs of the industry. The ratio is defined as

$$\sum_{j=1}^{m} x_{ij}^t \bigg/ \sum_{j=1}^{n} x_{ij}^t \times 100 = C/D \times 100$$

If the ratio of one industry is higher, outputs of that industry will be absorbed more by other industries of the industrial cluster.

If both ratios of one industry are higher, this indicates that levels of the structural relationships of the industry within industrial clusters are higher; otherwise, those of structural relationships of the industry within industrial clusters are lower.

5.2.2.2 Analysis on industry structure relationships of industrial clusters in China

Tables 5.10 through Table 5.20 are the indexes of 11 industrial clusters in China. Each shows the inputs of the inner industrial cluster of one industry, interindustry inputs of industry, the ratio of inputs of the inner industrial cluster of that industry to the interindustry inputs of the industry, the outputs of the inner industrial cluster of that industry, interindustry outputs of industry, and the ratio of the outputs in the inner industrial cluster of that industry to interindustry outputs of the industry.

Table 5.10 **Status of industry structure relationships of planting industrial cluster (Units: hundred million yuan)**

Industrial composing	Intracluster inputs(A)	Interindustry inputs(B)	A/B	Intracluster outputs(C)	Interindustry outputs(D)	C/D
Crop cultivation	3388.77	4490.08	0.7547	7470.39	8176.04	0.9137
Livestock and livestock products	3366.05	3841.99	0.8761	1794.66	2871.19	0.6251
Other agriculture products	256.78	482.57	0.5321	285.38	656.52	0.4347
Natural gas products	3.03	49.89	0.0607	34.43	76.48	0.4501
Grain mill products, vegetable oil and forage	3012.91	3549.03	0.8489	2280.80	2792.33	0.8168
Sugar refining	198.74	250.14	0.7945	192.87	214.70	0.8983
Slaughtering, meat processing, eggs and dairy products	1062.69	1235.95	0.8598	99.84	547.53	0.1823
Other food products	1323.80	2001.62	0.6614	164.14	289.53	0.5669
Wines, spirits and liquors	688.88	1098.27	0.6272	165.74	513.49	0.3228
Nonalcoholic beverages	364.09	609.44	0.5974	42.07	221.24	0.1902
Tobacco products	399.70	657.50	0.6079	174.96	364.71	0.4797
Hemp textiles	86.10	118.54	0.7264	37.35	129.10	0.2893
Other textiles	571.14	749.20	0.7623	146.88	1067.67	0.1376
Chemical fertilizers	181.61	1047.30	0.1734	1525.24	1598.04	0.9544
Chemical pesticides	79.87	324.83	0.2459	309.70	410.34	0.7548
Agriculture, forestry, animal husbandry and fishing machinery	157.12	742.68	0.2116	308.47	368.32	0.8375
Storage and warehousing	28.80	50.75	0.5676	15.84	76.34	0.2074
Technical services for agriculture, forestry, livestock and fishing	107.47	177.89	0.6041	228.78	277.27	0.8251

(1) Planting industrial cluster.

We can see from Table 5.10 that the levels of structural relationships within the industrial cluster of industries such as crop cultivation, grain mill products, vegetable oil and forage and sugar refining are higher, while the levels of structural relationships within the industrial cluster of industries such as natural gas products, non-alcoholic beverage and storage and warehousing are lower.

(2) Electronics industrial cluster.

We can see from Table 5.11 that the levels of structural relationships within the industrial cluster of industries such as electronic computer, electronic appliances and other electronic and communication equipment are higher, while the levels of structural relationships within the industrial cluster of industries such as wholesale and retail trade, glass and glass products and other social services are lower.

(3) Commerce industrial cluster.

We can see from Table 5.12 that the levels of structural relationships within the industrial cluster of industries such as tobacco products and printing and record medium reproduction are higher, while the levels of structural relationships within the industrial cluster of industries such as logging and transport of timber and bamboo, cultural goods, storage and warehousing and real estate are lower.

Table 5.11   Status of industry structure relationships of electronics industrial cluster (Units: hundred million yuan)

Industrial composing	Intracluster inputs(A)	Interindustry inputs(B)	A/B	Intracluster outputs(C)	Interindustry outputs(D)	C/D
Glass and glass products industry	89.89	465.38	0.1931	223.10	613.75	0.3635
Electronic computer	606.15	776.30	0.7808	503.71	640.87	0.7860
Electronic appliances	583.42	958.65	0.6086	211.63	313.93	0.6741
Electronic element and device	535.47	1014.13	0.5280	1459.03	1937.54	0.7530
Other electronic and communication equipment	608.70	905.60	0.6722	409.46	583.74	0.7014
Cultural and office equipment	77.21	202.79	0.3808	54.36	166.53	0.3264
Wholesale and retail trade	1426.40	5413.64	0.2635	1240.71	7299.54	0.1700
Other social services	528.47	1355.76	0.3898	378.14	1855.47	0.2038
Scientific research	57.49	169.97	0.3382	33.06	69.47	0.4759

Table 5.12   Status of industry structure relationships of commerce industrial cluster (Units: hundred million yuan)

Industrial composing	Intracluster inputs(A)	Interindustry inputs(B)	A/B	Intracluster outputs(C)	Interindustry outputs(D)	C/D
Forestry	81.83	220.49	0.3711	327.76	688.78	0.4758
Logging and transport of timber and bamboo	57.43	113.75	0.5049	24.63	359.28	0.0686
Nonalcoholic beverage	104.59	609.44	0.1716	110.81	221.24	0.5009
Tobacco products	311.55	657.50	0.4738	279.58	364.71	0.7666
Paper and products	764.23	1721.60	0.4439	1189.17	2529.69	0.4701
Printing and record medium reproduction	439.55	613.83	0.7161	448.50	990.84	0.4526
Cultural goods	53.65	129.98	0.4128	17.40	197.39	0.0881
Chemical pesticides	94.59	324.83	0.2912	161.06	410.34	0.3925
Rubber products	361.93	980.08	0.3693	130.66	994.19	0.1314
Electronic appliances	111.45	958.65	0.1163	209.17	313.93	0.6663
Storage and warehousing	5.60	50.75	0.1103	26.92	76.34	0.3526
Wholesale and retail trade	1857.08	5413.64	0.3430	1363.21	7299.54	0.1868
Real estate	43.57	447.02	0.0975	176.18	561.29	0.3139
Recreational services	49.36	140.94	0.3502	27.24	102.53	0.2657
Culture and arts, radio, film and television services	214.36	365.74	0.5861	58.46	174.53	0.3350

(4) Diet industrial cluster.

We can see from Table 5.13 that the levels of structural relationships within the industrial cluster of industries such as prepared fish and seafood, fishery and tobacco products are higher, while the levels of structural relationships within the industrial cluster of industries such as slaughtering, meat processing, eggs and dairy products and eating and drinking places are lower.

Table 5.13   Status of industry structure relationships of diet industrial cluster (Units: hundred million yuan)

Industrial composing	Intracluster inputs(A)	Interindustry inputs(B)	A/B	Intracluster outputs(C)	Interindustry outputs(D)	C/D
Crop cultivation	1481.98	4490.08	0.3301	5294.00	8176.04	0.6475
Fishery	517.55	900.67	0.5746	937.95	1019.68	0.9198
Grain mill products, vegetable oil and forage	3073.10	3549.03	0.8659	1331.52	2792.33	0.4768
Slaughtering, meat processing, eggs and dairy products	66.24	1235.95	0.0536	253.07	547.53	0.4622
Prepared fish and seafood	481.70	564.91	0.8527	153.18	183.34	0.8355
Other food products	1006.99	2001.62	0.5031	222.69	289.53	0.7691
Wines, spirits and liquors	684.67	1098.27	0.6234	298.67	513.49	0.5816
Nonalcoholic beverage	300.24	609.44	0.4927	105.90	221.24	0.4787
Tobacco products	400.70	657.50	0.6094	248.17	364.71	0.6804
Eating and drinking places	962.94	1449.05	0.6645	44.81	1190.76	0.0376
Technical services for agriculture, forestry, livestock and fishing	110.58	177.89	0.6216	196.73	277.27	0.7095

(5) Iron and steel industrial cluster.

We can see from Table 5.14 that the levels of structural relationships within the industrial cluster of industries such as ferrous ore mining, coking, steel processing and bicycle are higher, while the levels of structural relationships within the industrial cluster of industries such as electricity production and supply, steam and hot water production and supply, and construction are lower.

Table 5.14   Status of industry structure relationships of iron and steel industrial cluster (Units: hundred million yuan)

Industrial composing	Intracluster inputs(A)	Interindustry inputs(B)	A/B	Intracluster outputs(C)	Interindustry outputs(D)	C/D
Coal mining and processing	370.18	1083.50	0.3416	1086.01	2138.97	0.5077
Ferrous ore mining	118.04	220.69	0.5349	436.83	487.84	0.8954
Coking	140.73	232.94	0.6042	176.11	248.12	0.7098
Iron-smelting	240.22	451.69	0.5318	359.20	494.00	0.7271
Steel-smelting	280.87	536.37	0.5236	475.88	742.81	0.6406
Steel processing	1903.51	2986.77	0.6373	3224.58	4511.95	0.7147
Alloy iron smelting	122.66	270.84	0.4529	199.13	289.62	0.6876
Metal products	2175.33	3820.23	0.5694	1985.31	3966.62	0.5005
Other special industrial equipment	718.77	1625.49	0.4422	519.29	1255.65	0.4136
Bicycle	194.51	278.41	0.6987	96.69	103.94	0.9302
Electricity production and supply	853.97	2127.99	0.4013	973.92	3501.49	0.2781
Steam and hot water production and supply	44.67	92.61	0.4824	21.18	125.66	0.1686
Gas production and supply	51.82	102.74	0.5044	18.97	60.54	0.3133
Water production and supply	112.76	191.21	0.5897	76.78	287.44	0.2671
Construction	2366.80	12388.02	0.1911	44.98	1028.51	0.0437

(6) Petroleum industrial cluster.

We can see from Table 5.15 that the overall levels of structural relationships within the industrial cluster are lower than 1, with levels of structural relationships within the industrial cluster of industries such as electricity production and supply, railway freight transport, highway freight transport, railway passenger transport, highway passenger transport and water passenger transport all lower than 1.

Table 5.15   Status of industry structure relationships of petroleum industrial cluster (Units: hundred million yuan)

Industrial composing	Intracluster inputs(A)	Interindustry inputs(B)	A/B	Intracluster outputs(C)	Interindustry outputs(D)	C/D
Crude petroleum products	85.97	377.27	0.2279	1626.48	1875.31	0.8673
Petroleum refining	1763.59	2181.78	0.8083	954.59	3089.94	0.3089
Railroad transport equipment	49.15	196.75	0.2498	108.70	117.61	0.9242
Electricity production and supply	482.65	2127.99	0.2268	292.28	3501.49	0.0835
Railway freight transport	152.65	312.49	0.4885	78.52	844.28	0.0930
Highway freight transport	263.44	742.37	0.3549	61.44	1425.57	0.0431
Pipeline transport	11.28	19.13	0.5898	20.24	33.06	0.6120
Water freight transport	54.95	138.16	0.3977	20.65	133.64	0.1545
Railway passenger transport	54.94	126.05	0.4359	6.29	250.37	0.0251
Highway passenger transport	61.02	152.82	0.3993	10.87	128.47	0.0846
Water passenger transport	59.33	146.28	0.4056	6.27	141.82	0.0442
Public services	205.24	560.28	0.3663	58.11	275.80	0.2107
Social welfare	4.57	48.57	0.0940	4.36	4.36	1.0000

(7) Nonferrous metal cluster.

We can see from Table 5.16 that the level of structural relationships within the industrial cluster of nonferrous metal smelting is higher, while the levels of structural relationships within the industrial cluster of these industries such as scrap and waste and telecommunication are lower.

Table 5.16   Status of industry structure relationships of nonferrous metal industrial cluster (Units: hundred million yuan)

Industrial composing	Intracluster inputs(A)	Interindustry inputs(B)	A/B	Intracluster outputs(C)	Interindustry outputs(D)	C/D
Nonferrous metal mining	178.76	550.73	0.3246	663.89	876.14	0.7577
Nonferrous metal smelting	717.14	1139.28	0.6295	980.52	1520.00	0.6451
Nonferrous metal processing	633.71	806.05	0.7862	507.51	1032.14	0.4917
Other electric machinery and equipment	1055.69	2674.58	0.3947	450.79	2785.94	0.1618
Scrap and waste	0	0	0	113.12	514.62	0.2198
Telecommunication	183.23	716.06	0.2559	52.70	1480.17	0.0356

(8) Construction industrial cluster.

We can see from Table 5.17 that the levels of structural relationships within the industrial cluster of industries such as sawmills and fiberboard and cement and asbestos products are higher, while the levels of structural relationships within the industrial cluster of industries such as logging and transport of timber and bamboo, construction and storage and warehousing are lower.

(9) Textile industrial cluster.

We can see from Table 5.18 that the levels of structural relationships within the industrial cluster of industries such as silk textiles, cotton textiles, other textiles and hemp textiles are higher, while the levels of structural relationships within the industrial cluster of industries such as crop cultivation and rubber products are lower.

(10) Chemistry industrial cluster.

We can see from Table 5.19 that there are no industries with levels of structural relationships higher than 1 within the industrial cluster in this industrial cluster, while overall levels of structural relationships within the industrial cluster of this industrial cluster are lower.

(11) Administration industrial cluster.

We can see from Table 5.20 that the level of structural relationships within the industrial cluster of tourism is higher, but overall the levels of structural relationships within the industrial cluster of this industrial cluster are lower.

Table 5.17   Status of industry structure relationships of construction industrial cluster (Units: hundred million yuan)

Industrial composing	Intracluster inputs(A)	Interindustry inputs(B)	A/B	Intracluster outputs(C)	Interindustry outputs(D)	C/D
Nonmetal minerals and other ore mining	204.64	824.46	0.2482	996.90	1293.20	0.7709
Logging and transport of timber and bamboo	12.89	113.75	0.1134	204.16	359.28	0.5683
Sawmills and fiberboard	267.89	605.43	0.4425	745.42	903.04	0.8255
Furniture and products of wood, bamboo, cane, palm, straw, etc.	505.49	1009.61	0.5007	254.45	601.89	0.4228
Cement	441.87	1714.14	0.2578	1846.25	2076.03	0.8893
Cement and asbestos products	521.34	869.44	0.5996	985.00	1073.21	0.9178
Bricks, tiles, lime and lightweight building materials	516.40	1586.87	0.3254	1858.94	2132.60	0.8717
Pottery, china and earthenware	182.07	472.77	0.3851	423.66	566.84	0.7474
Fireproof products	120.07	392.80	0.3057	385.44	515.67	0.7475
Steel processing	974.54	2986.77	0.3263	3005.53	4511.95	0.6661
Metal products	1878.88	3820.23	0.4918	2256.20	3966.62	0.5688
Other electric machinery and equipment	570.04	2674.58	0.2131	1256.52	2785.94	0.4510
Instruments, meters and other measuring equipment	134.13	367.72	0.3648	235.21	487.11	0.4829
Construction	8601.50	12388.02	0.6943	100.96	1028.51	0.0982
Storage and warehousing	6.70	50.75	0.1321	23.73	76.34	0.3109
Telecommunication	280.80	716.06	0.3921	570.52	1480.17	0.3854
Geological prospecting and water conservancy	100.00	240.36	0.4160	170.37	236.93	0.7191

Table 5.18   Status of industry structure relationships of textile industrial cluster (Units: hundred million yuan)

Industrial composing	Intracluster inputs(A)	Interindustry inputs(B)	A/B	Intracluster outputs(C)	Interindustry outputs(D)	C/D
Crop cultivation	1178.01	4490.08	0.2624	1865.26	8176.04	0.2281
Cotton textiles	2226.56	2925.26	0.7611	2498.60	3516.97	0.7104
Wool textiles	475.62	872.23	0.5453	899.22	1145.40	0.7851
Hemp textiles	98.72	118.54	0.8328	84.58	129.10	0.6551
Silk textiles	851.85	1116.22	0.7632	1095.60	1294.30	0.8465
Knitted mills	654.24	881.96	0.7418	76.46	147.25	0.5192
Other textiles	564.17	749.20	0.7530	940.35	1067.67	0.8807
Wearing apparel	2008.51	2470.31	0.8131	81.95	596.72	0.1373
Chemical fibers	315.62	952.39	0.3314	1083.01	1372.43	0.7891
Rubber products	372.00	980.08	0.3796	120.27	994.19	0.1210

Table 5.19   Status of industry structure relationships of chemistry industrial cluster (Units: hundred million yuan)

Industrial composing	Intracluster inputs(A)	Interindustry inputs(B)	A/B	Intracluster outputs(C)	Interindustry outputs(D)	C/D
Other chemical products	233.85	1435.88	0.1629	1289.69	2473.03	0.5215
Chemical fibers	510.87	952.39	0.5364	277.56	1372.43	0.2022
Plastic products	1334.75	2193.87	0.6084	606.10	2536.60	0.2389
Other manufacturing products	202.21	721.06	0.2804	108.32	919.81	0.1178

Table 5.20   Status of industry structure relationships of administration industrial cluster (Units: hundred million yuan)

Industrial composing	Intracluster inputs(A)	Interindustry inputs(B)	A/B	Intracluster outputs(C)	Interindustry outputs(D)	C/D
Aircraft	20.12	62.55	0.3216	68.06	90.75	0.7500
Post services	0.10	117.11	0.0009	30.66	93.87	0.3266
Railway passenger transport	3.34	126.05	0.0265	108.16	250.37	0.4320
Air passenger transport	55.93	231.72	0.2414	174.08	310.52	0.5606
Hotels	17.17	303.00	0.0567	262.65	591.12	0.4443
Tourism	186.74	312.26	0.5980	55.76	62.84	0.8872
Social welfare	2.62	48.57	0.0540	0.92	4.36	0.2106
Public administration and other sectors	414.26	2433.93	0.1702	0	0	0

5.2.3   Technological economy structure relationship degrees of industrial clusters

5.2.3.1   Indexes of technological economy structure relationship degrees of industrial clusters

Technological economy structure relationships of industrial clusters not only are the basis of definite production technology levels, but also display corresponding industrial relationships. These relationships display transaction sizes of intermediate products among industrial sectors. Therefore, we can use two indexes about influence force coefficient and induction degree coefficient to measure total relationship degrees of industrial clusters.

As far as economic meanings are concerned, the sum of column J in the inverse matrix coefficient of input and output $(I - A)^{-1}$, which reflects the absolute levels of production demand and pull produced by industry sector J to all industrial sectors (namely, productive outputs required to each sector by direct and indirect relationships when we add unit final demand to the sector j), is called influence degree. Its expression is

$$f_j = \sum_{i=1}^{n} b_{ij}$$

The sum of the column for each sector divided by the average value of the sum of each column is the influence force coefficient and its expression is

$$F_j = \sum_{i=1}^{n} b_{ij} \bigg/ \frac{1}{n} \sum_{i=1}^{n} \sum_{j=1}^{n} b_{ij}$$

The coefficient reflects that when we add one unit of final demand to one sector, it will bring production demand affection degree to each sector in the national economy. If the influence force coefficient is larger than 1, this denotes that production affection degrees which productions of this sector give birth to productions of other sectors exceed the average influence force levels of a society. The greater influence force coefficient denotes the greater pull functions of this sector to other sectors and the deeper backward total relationship degrees of the sector with other sectors. The greater influence force coefficients also show greater intermediate consumptions of industries.

The sum of row i in the inverse matrix coefficient of input and output $(I - A)^{-1}$, which reflects the absolute levels of demand induction of sector i received from all sectors of national economy (namely, productive outputs provided by the sector for production of other sectors when we add unit final demand in all sectors of national economy), is called induction degree. Its expression is

$$e_i = \sum_{j=1}^{n} b_{ij}$$

The sum of the row for each sector divided by the average value of the sum of each row is the induction degree coefficient and its expression is

$$E_i = \sum_{j=1}^{n} b_{ij} \left/ \frac{1}{n} \sum_{i=1}^{n} \sum_{j=1}^{n} b_{ij} \right.$$

The coefficient reflects that when one unit final demand is added to each sector in a national economy, one sector gets the demand induction degree. If the induction degree coefficient is larger than 1, this denotes that the induction degree of the sector is higher than the average induction degree of a society. If the induction degree is smaller than 1, this denotes that the induction degree of this sector is under the average induction degree of a society. The greater induction degree coefficient denotes the greater demand of this sector from other sectors (the greater pull function of this sector) and the deeper total forward relationship degree of the sector with other sectors. The greater induction degree coefficient also accounts for the greater middle demand of industries.

Generally speaking, if the influence force coefficient and the induction degree coefficient of one industry are both larger than 1, this indicates that the total relationship degree of the industry with other industries is very high. Industry (cluster) structure adjustment should emphatically support the industries (clusters) in which force coefficient and induction degree coefficient of one industry are both larger than 1.

5.2.3.2   Analysis on general technological economy structure relationship degrees of industrial clusters

The influence force coefficient and the induction degree coefficient of industrial clusters in China in 1997 are shown in Table 5.21. Industrial clusters with an influence force coefficient larger than 1 are, in order, iron and steel industrial cluster (influence degree is 3.5783, that is, when we add consumption of every 1 yuan to iron and steel industrial cluster, it will drive the total production value of 3.5783 yuan; influence force coefficient is 1.1237, that is, its affection degree is 1.1237 times as large as the average level of all industries, noted similarly hereinafter), chemistry industrial cluster (3.5777, 1.1235), construction industrial cluster (3.5772, 1.1234), nonferrous metal industrial cluster (3.3592, 1.0549), and planting industrial cluster (3.2399, 1.0175).

Industrial clusters with an induction degree coefficient larger than 1 are, in order, construction industrial cluster (the induction degree is 13.3545, that is, when we add final consumption of 1 yuan to all sectors, the industrial cluster needs to produce 13.3545 yuan; induction degree coefficient is 4.1939, that is, its induction degree is 4.1939 times as large as the average level of all industries, noted similarly hereinafter), iron and steel industrial cluster (11.1725, 3.5086), commerce industrial cluster (8.5087, 2.6721), electronics industrial cluster (7.8206, 2.4560), petroleum industrial cluster (7.3269, 2.3010), planting industrial

cluster (7.1545, 2.2468), textile industrial cluster (6.7705, 2.1262), nonferrous metal industrial cluster (5.9876, 1.8804), diet industrial cluster (5.5482, 1.7424), chemistry industrial
cluster (4.9472, 1.5536), administration industrial cluster (1.7688, 0.5555).

Table 5.21   Influence force and induction degree of industrial clusters in China

Types of industrial clusters	Influence force	Influence force coefficient	Induction degree	Induction degree coefficient
Planting industrial cluster	3.2399	1.0175	7.1545	2.2468
Electronics industrial cluster	3.0746	0.9655	7.8206	2.4560
Commerce industrial cluster	2.9362	0.9221	8.5087	2.6721
Diet industrial cluster	3.1785	0.9982	5.5482	1.7424
Iron and steel industrial cluster	3.5783	1.1237	11.1725	3.5086
Petroleum industrial cluster	2.7881	0.8756	7.3269	2.3010
Nonferrous metal industrial cluster	3.3592	1.0549	5.9876	1.8804
Construction industrial cluster	3.5772	1.1234	13.3545	4.1939
Textile industrial cluster	3.0308	0.9518	6.7705	2.1262
Chemistry industrial cluster	3.5777	1.1235	4.9472	1.5536
Administration industrial cluster	2.8600	0.8982	1.7688	0.5555

The industrial clusters whose influence force coefficient and induction degree coefficient
are both larger than 1 are iron and steel industrial cluster, chemistry industrial cluster,
construction industrial cluster, nonferrous metal industrial cluster and planting industrial
cluster. Total relationship degree of these industrial clusters with other industrial sectors
in national economies is very high. These industrial clusters and their industrial sectors are
the keystone of industry structure adjustment.

5.2.3.3   Aanlyses on technological economy structure relationship degree of each industrial
          cluster

(1) Planting industrial cluster.

We can see from Table 5.22 that industries with greater influence force and influence
force coefficient of the industrial cluster are in order slaughtering, meat processing, eggs
and dairy products (2.5550, 1.3800), grain mill products, vegetable oil and forage (2.2342,
1.2066), sugar refining (2.2088, 1.1930), other textiles (2.1324, 1.1517), other food products
(2.1021, 1.1353), hemp textiles (2.0850, 1.1261), non-alcoholic beverage (2.0229, 1.0926),
storage and warehousing (2.0083, 1.0847) and wines, spirits and liquors (1.9472, 1.0517).

Industries with larger induction degree and induction degree coefficient of the industrial
cluster are in order crop cultivation (7.813, 4.220), chemical fertilizers (2.786, 1.505), grain
mill products, vegetable oil and forage (2.444, 1.320), livestock and livestock products (2.302,
1.243), agriculture, forestry, animal husbandry and fishing machinery (1.938, 1.047) and
chemical pesticides (1.922, 1.038).

The industry whose influence force coefficient and induction degree coefficient are both
larger than 1 in the cluster is only grain mill products, vegetable oil and forage. It is the
sector that is emphatically supported in the structure adjustment of the industrial cluster.

(2) Electronics industrial cluster.

We can see from Table 5.23 that industries with greater influence force and influence force
coefficient of the industrial cluster are, in order, wholesale and retail trade (3.6485, 1.6426),
electronic element and device (3.4344, 1.5462), cultural and office equipment (2.7528, 1.2393)
and electronic appliances (2.5685, 1.1564).

The industries of the cluster with the larger induction degree and induction degree
coefficient are, in order, electronic element and device (4.7542, 2.1404), wholesale and retail
trade (4.1694, 1.8771) and glass and glass products (2.2532, 1.0144).

In this cluster, industries whose influence force coefficient and induction degree coefficient are both larger than 1 are wholesale and retail trade and electronic element and device. They are sectors which are emphatically supported in the structure adjustment of the industrial cluster.

Table 5.22   Influence force and induction degree of planting industrial cluster

Industries composing	Influence force	Influence force coefficient	Induction degree	Induction degree coefficient
Crop cultivation	1.4592	0.7881	7.8130	4.2197
Livestock and livestock	1.8464	0.9972	2.3017	1.2431
Other agriculture products products	1.4344	0.7747	1.2269	0.6627
Natural gas products	1.0517	0.5680	1.2406	0.6700
Grain mill products, vegetable oil and forage	2.2342	1.2066	2.4438	1.3199
Sugar refining	2.2088	1.1930	1.2810	0.6918
Slaughtering, meat processing, eggs and dairy products	2.5550	1.3800	1.0682	0.5769
Other food products	2.1021	1.1353	1.1827	0.6388
Wines, spirits and liquors	1.9472	1.0517	1.1942	0.6450
Nonalcoholic beverage	2.0229	1.0926	1.0607	0.5729
Tobacco products	1.4781	0.7983	1.1665	0.6300
Hemp textiles	2.0850	1.1261	1.0906	0.5890
Other textiles	2.1324	1.1517	1.3682	0.7390
Chemical fertilizers	1.5514	0.8379	2.7863	1.5048
Chemical pesticides	1.8430	0.9954	1.9225	1.0383
Agriculture, forestry, animal husbandry and fishing machinery	1.7588	0.9499	1.9384	1.0469
Storage and warehousing	2.0083	1.0847	1.0759	0.5811
Technical services for agriculture, forestry, livestock and fishing	1.6087	0.8689	1.1666	0.6301

Table 5.23   Influence force and induction degree of electronics industrial cluster

Industries composing	Influence force	Influence force coefficient	Induction degree	Induction degree coefficient
Glass and glass products	1.0227	0.4604	2.2532	1.0144
Electronic computer	1.3803	0.6214	1.6166	0.7278
Electronic appliances	2.5685	1.1564	1.4833	0.6678
Electronic element and device	3.4344	1.5462	4.7542	2.1404
Other electronic and communication equipment	2.2103	0.9951	1.6347	0.7360
Cultural and office equipment	2.7528	1.2393	1.1574	0.5211
Wholesale and retail trade	3.6485	1.6426	4.1694	1.8771
Other social services	1.7266	0.7773	1.8690	0.8414
Scientific research	1.2464	0.5612	1.0529	0.4740

(3) Commerce industrial cluster.

We can see from Table 5.24 that industries with greater influence force and influence force coefficient of the industrial cluster are, in order, wholesale and retail trade (4.0817, 2.0482), tobacco products (4.0204, 2.0174), printing and record medium reproduction (3.3063, 1.6591), culture and arts, radio, film and television services (2.8965, 1.4534) and paper and products (2.2167, 1.1123).

The industries of the cluster with greater induction degree and induction degree coefficient are, in order, paper and products (5.9320, 2.9766), wholesale and retail trade (5.0716, 2.5449), printing and record medium reproduction (2.5722, 1.2907) and tobacco products (2.4337, 1.2212).

Table 5.24   **Influence force and induction degree of commerce industrial cluster**

Industries composing	Influence force	Influence force coefficient	Induction degree	Induction degree coefficient
Forestry	1.0145	0.5091	1.3954	0.7002
Logging and transport of timber and bamboo	1.0216	0.5126	1.0724	0.5381
Nonalcoholic beverage	1.3711	0.6880	1.2724	0.6385
Tobacco products	4.0204	2.0174	2.4337	1.2212
Paper and products	2.2167	1.1123	5.9320	2.9766
Printing and record medium reproduction	3.3063	1.6591	2.5722	1.2907
Cultural goods	1.5058	0.7556	1.0681	0.5360
Chemical pesticides	1.1657	0.5850	1.3145	0.6596
Rubber products	1.6791	0.8426	1.2474	0.6259
Electronic appliances	1.8575	0.9321	1.6343	0.8201
Storage and warehousing	1.0121	0.5079	1.0682	0.5360
Wholesale and retail trade	4.0817	2.0482	5.0716	2.5449
Real estate	1.3698	0.6874	1.5053	0.7553
Recreational services	1.3739	0.6894	1.0771	0.5405
Culture and arts, radio, film and television services	2.8965	1.4534	1.2284	0.6164

In this cluster, industries whose influence force coefficient and induction degree coefficient are both larger than 1 are paper and products, wholesale and retail trade, printing and record medium reproduction and tobacco products. They are sectors which are emphatically supported in the structure adjustment of the industrial cluster.

(4) Diet industrial cluster.

We can see from Table 5.25 that industries with greater influence force and influence force coefficient of the industrial cluster are, in order, other food products (2.6028, 1.4820), eating and drinking places (2.3558, 1.3414), wines, spirits and liquors (2.1709, 1.2361), slaughtering, meat processing, eggs and dairy products (2.1410, 1.2191) and grain mill products, vegetable oil and forage (2.0916, 1.1910).

Table 5.25   **Influence force and induction degree of diet industrial cluster**

Industries composing	Influence force	Influence force coefficient	Induction degree	Induction degree coefficient
Crop cultivation	1.1947	0.6803	5.0191	2.8579
Fishery	1.1914	0.6784	1.7587	1.0014
Grain mill products, vegetable oil and forage	2.0916	1.1910	2.1431	1.2203
Slaughtering, meat processing, eggs and dairy products	2.1410	1.2191	1.9193	1.0928
Prepared fish and seafood	1.4552	0.8286	1.1205	0.6380
Other food products	2.6028	1.4820	1.3895	0.7912
Wines, spirits and liquors	2.1709	1.2361	1.3499	0.7686
Nonalcoholic beverage	1.3622	0.7756	1.1402	0.6492
Tobacco products	1.5879	0.9042	1.3073	0.7444
Eating and drinking places	2.3558	1.3414	1.0677	0.6079
Technical services for agriculture, forestry, livestock and fishing	1.1650	0.6634	1.1034	0.6283

The industries of the cluster with greater induction degree and induction degree co-efficient are, in order, crop cultivation (5.0191, 2.8579), grain mill products, vegetable oil and forage (2.1431, 1.2203), slaughtering, meat processing, eggs and dairy products (1.9193, 1.0928) and fishery(1.7587, 1.0014).

In this cluster, the industries whose influence force coefficient and induction degree coefficient are both larger than 1 are grain mill products, vegetable oil and forage and slaughtering, meat processing, eggs and dairy products. They are the sectors which are emphatically supported in the structure adjustment of the industrial cluster.

(5) Iron and steel industrial cluster.

We can see from Table 5.26 that industries with greater influence force and influence force coefficient of the industrial cluster are, in order, construction (3.7607, 1.9584), steel processing (3.1111, 1.6201), metal products (2.8939, 1.5070), other special industrial equipment (2.5300, 1.3175), bicycle (2.0845, 1.0855) and iron-smelting (2.0412, 1.0630).

Table 5.26   Influence force and induction degree of iron and steel industrial cluster

Industries composing	Influence force	Influence force coefficient	Induction degree	Induction degree coefficient
Coal mining and processing	1.0942	0.5698	3.2121	1.6727
Ferrous ore mining	1.0562	0.5500	2.1215	1.1048
Coking	1.2282	0.6396	1.6621	0.8655
Iron-smelting	2.0412	1.0630	1.6217	0.8445
Steel-smelting	1.4317	0.7456	1.9351	1.0077
Steel processing	3.1111	1.6201	4.6368	2.4147
Alloy iron-smelting	1.5896	0.8278	1.4518	0.7560
Metal products	2.8939	1.5070	2.4371	1.2691
Other special industrial equipment	2.5300	1.3175	1.5938	0.8300
Bicycle	2.0845	1.0855	1.2725	0.6626
Electricity production and supply	1.6610	0.8650	2.5615	1.3339
Steam and hot water production and supply	1.6274	0.8475	1.0567	0.5503
Gas production and supply	1.2141	0.6323	1.0469	0.5452
Water production and supply	1.4804	0.7709	1.1395	0.5934
Construction	3.7607	1.9584	1.0553	0.5495

Industries of the cluster with greater induction degree and induction degree coefficient are, in order, steel processing (4.6368, 2.4147), coal mining and processing (3.2121, 1.6727), electricity production and supply (2.5615, 1.3339), metal products (2.4371, 1.2691) ferrous ore mining (2.1215, 1.1048) and steel-smelting (1.9351, 1.0077).

In this cluster, industries whose influence force coefficient and induction degree coefficient are both larger than 1 are steel processing and metal products. They are sectors which are emphatically supported in the structure adjustment of the industrial cluster.

(6) Petroleum industrial cluster.

We can see from Table 5.27 that industries with greater influence force and influence force coefficient of the industrial cluster are, in order, electricity production and supply (2.4303, 1.7172), public services (2.3721, 1.6760) and highway freight transport (2.2030, 1.5566).

Table 5.27   Influence force and induction degree of petroleum industrial cluster

Industries composing	Influence force	Influence force coefficient	Induction degree	Induction degree coefficient
Crude petroleum products	1.0193	0.7202	2.2202	1.5687
Petroleum refining	1.3441	0.9497	3.5409	2.5019
Railroad transport equipment	1.1010	0.7779	1.1711	0.8275
Electricity production and supply	2.4303	1.7172	1.6797	1.1868
Railway freight transport	1.2288	0.8682	1.2025	0.8496
Highway freight transport	2.2030	1.5566	1.1680	0.8252
Pipeline transport	1.0974	0.7754	1.0225	0.7225
Water freight transport	1.0769	0.7609	1.0355	0.7317
Railway passenger transport	1.1293	0.7979	1.0245	0.7239
Highway passenger transport	1.2735	0.8998	1.0410	0.7355
Water passenger transport	1.0899	0.7701	1.0183	0.7195
Public services	2.3721	1.6760	1.2648	0.8937
Social welfare	1.0332	0.7300	1.0097	0.7134

Industries of the cluster with greater induction degree and induction degree coefficient are, in order, petroleum refining (3.5409, 2.5019), crude petroleum products (2.2202, 1.5687), and electricity production and supply (1.6797, 1.1868).

In this cluster, the industry whose influence force coefficient and induction degree coefficient are both greater than 1 is electricity production and supply. It is the sector which is emphatically supported in the structure adjustment of the industrial cluster.

(7) Nonferrous metal industrial cluster.

We can see from Table 5.28 that industries with greater influence force and influence force coefficient of the industrial cluster are, in order, telecommunication (3.0227, 1.7304) and other electric machinery and equipment (2.6566, 1.5208).

Table 5.28   Influence force and induction degree of nonferrous metal industrial cluster

Industries composing	Influence force	Influence force coefficient	Induction degree	Induction degree coefficient
Nonferrous metal mining	1.0227	0.5854	1.4107	0.8076
Nonferrous metal smelting	1.1860	0.6789	2.5530	1.4615
Nonferrous metal processing	1.5930	0.9120	2.0413	1.1686
Other electric machinery and equipment	2.6566	1.5208	2.3510	1.3459
Scrap and waste	1.0000	0.5725	1.0678	0.6113
Telecommunication	3.0227	1.7304	1.0573	0.6052

Industries of the cluster with greater induction degree and induction degree coefficient are, in order, nonferrous metal smelting (2.5530, 1.4615), other electric machinery and equipment (2.3510, 1.3459) and nonferrous metal processing (2.0413, 1.1686).

In the cluster, the industry whose influence force coefficient and induction degree coefficient are both larger than 1 is other electric machinery and equipment. It is the sector which is emphatically supported in the structure adjustment of the industrial cluster.

(8) Construction industrial cluster.

We can see from Table 5.29 that industries with greater influence force and influence force coefficient of the industrial cluster are, in order, other electric machinery and equipment (4.3074, 1.7739), steel processing (3.6785, 1.5149), geological prospecting and water conservancy (3.5825, 1.4754), construction (3.5800, 1.4743), metal products (3.3995, 1.4000), telecommunication (3.2191, 1.3257), cement and asbestos products (3.1143, 1.2825) and furniture and products of wood, bamboo, cane, palm, straw, etc. (2.9886, 1.2308).

Table 5.29   Influence force and induction degree of construction industrial cluster

Industries composing	Influence force	Influence force coefficient	Induction degree	Induction degree coefficient
Nonmetal minerals and other ore mining	1.0508	0.4327	2.4772	1.0202
Logging and transport of timber and bamboo	1.0082	0.4152	1.2274	0.5055
Sawmills and fiberboard	1.4671	0.6042	2.2647	0.9327
Furniture and products of wood, bamboo, cane, palm, straw, etc.	2.9886	1.2308	1.4655	0.6035
Cement	1.8081	0.7446	2.2746	0.9367
Cement and asbestos products	3.1143	1.2825	1.3876	0.5715
Bricks, tiles, lime and lightweight building materials	2.3440	0.9653	2.0841	0.8583
Pottery, china and earthenware	1.2350	0.5086	1.3468	0.5546
Fireproof products	1.3069	0.5382	1.9216	0.7914
Steel processing	3.6785	1.5149	9.5033	3.9136
Metal products	3.3995	1.4000	4.4201	1.8203
Other electric machinery and equipment	4.3074	1.7739	3.8808	1.5982
Instruments, meters and other measuring equipment	2.0452	0.8423	1.5694	0.6463
Construction	3.5800	1.4743	1.4396	0.5928
Storage and warehousing	1.1448	0.4715	1.0446	0.4302
Telecommunication	3.2191	1.3257	1.6918	0.6967
Geological prospecting and water conservancy	3.5825	1.4754	1.2811	0.5276

Industries of the cluster with greater induction degree and induction degree coefficient are, in order, steel processing (9.5033, 3.9136), metal products (4.4201, 1.8203), other electric machinery and equipment (3.8808, 1.5982) and nonmetal minerals and other ore mining (2.4772, 1.0202).

In this cluster, industries whose influence force coefficient and induction degree coefficient are both larger than 1 are steel processing, metal products and other electric machinery and equipment. They are sectors which are emphatically supported in the structure adjustment of the industrial cluster.

(9) Textile industrial cluster.

We can see from Table 5.30 that industries with greater influence force and influence force coefficient of the industrial cluster are, in order, knitted mills (2.6099, 1.3811), wearing apparel (2.2216, 1.1756), rubber products (2.0456, 1.0825), other textiles (1.9924, 1.0543), silk textiles (1.9801, 1.0478) and hemp textiles (1.8999, 1.0054).

Table 5.30   Influence force and induction degree of textile industrial cluster

Industries composing	Influence force	Influence force coefficient	Induction degree	Induction degree coefficient
Crop cultivation	1.1431	0.6049	2.9492	1.5607
Cotton textiles	1.6070	0.8504	2.8952	1.5321
Wool textiles	1.7680	0.9356	1.9695	1.0422
Hemp textiles	1.8999	1.0054	1.0833	0.5733
Silk textiles	1.9801	1.0478	2.1422	1.1336
Knitted mills	2.6099	1.3811	1.1008	0.5825
Other textiles	1.9924	1.0543	2.1417	1.1334
Wearing apparel	2.2216	1.1756	1.0503	0.5558
Chemical fibers	1.6295	0.8623	2.3783	1.2585
Rubber products	2.0456	1.0825	1.1866	0.6279

Industries of the cluster with greater induction degree and induction degree coefficient are, in order, crop cultivation (2.9492, 1.5607), cotton textiles (2.8952, 1.5321), chemical fibers (2.3783, 1.2585), silk textiles (2.1422, 1.1336), other textiles (2.1417, 1.1334) and wool textiles (1.9695, 1.0422).

In this cluster, industries whose influence force coefficient and induction degree coefficient are both larger than 1 are other textiles and silk textiles. They are sectors which are emphatically supported in the structure adjustment of the industrial cluster.

(10) Chemistry industrial cluster.

We can see from Table 5.31 that industries with greater influence force and influence force coefficient of the industrial cluster are, in order, other manufacturing products (2.3819, 1.4659) and plastic products (1.9542, 1.2027).

Table 5.31   Influence force and induction degree of chemistry industrial cluster

Industries composing	Influence force	Influence force coefficient	Induction degree	Induction degree coefficient
Other chemical products	1.0324	0.6354	1.9704	1.2127
Chemical fibers	1.1309	0.6960	1.3731	0.8451
Plastic products	1.9542	1.2027	1.6213	0.9978
Other manufacturing products	2.3819	1.4659	1.5345	0.9444

The only industries of the cluster with greater induction degree and induction degree coefficient is other chemical products (1.9704, 1.2127).

In this cluster there are not industries whose influence force coefficient and induction degree coefficient are both bigger than 1.

(11) Administration industrial cluster.

We can see from Table 5.32 that industries with greater influence force and influence force coefficient of the industrial cluster are, in order, tourism (2.1394, 1.7035), air passenger transport (1.5037, 1.1973) and public administration and other sectors (1.3585, 1.0817).

Table 5.32   **Influence force and induction degree of administration industrial cluster**

Industries composing	Influence force	Influence force coefficient	Induction degree	Induction degree coefficient
Aircraft	1.0024	0.7982	1.6243	1.2934
Post services	1.0000	0.7963	1.0262	0.8171
Railway passenger transport	1.0053	0.8004	1.2612	1.0042
Air passenger transport	1.5037	1.1973	1.4809	1.1791
Hotel	1.0241	0.8155	1.3938	1.1098
Tourism	2.1394	1.7035	1.2590	1.0025
Social welfare	1.0137	0.8072	1.0018	0.7977
Public administration and other sectors	1.3585	1.0817	1.0000	0.7962

Industries of the cluster with greater induction degree and induction degree coefficient are, in order, aircraft (1.6243, 1.2934), air passenger transport (1.4809, 1.1791), hotels (1.3938, 1.1098), railway passenger transport (1.2612, 1.0042) and tourism (1.2590, 1.0025).

In this cluster, industries whose influence force coefficient and induction degree coefficient are both larger than 1 are tourism and air passenger transport. They are sectors which are emphatically supported in the structure adjustment of the industrial cluster.

5.2.4   Analysis on aggregation quality of technological economy structure relationship of industrial clusters

5.2.4.1   Indexes of measuring aggregation quality of technological economy structure relationship of industrial clusters

Aggregation quality of technological economy structure relationship of industrial clusters refers to the coupling state among industrial clusters and the resulting whole function of the system. It concerns the interdependent coupling state among industrial clusters from the angle of technological economy structure relationship integrity. There are main indexes for measuring aggregation quality of technological economy structure relationship of industrial clusters.

(1) Proportion unbalancedness.

Amount and proportion relationships of input and output among industries exist in a national economy, and the relationships of supplies and demands among industries, reflect the integrity functions of industry structure relationship. They reflect the relatively balanceable relationships between one sector's output and the demand of other sectors. The index reflecting this kind of relatively balanceable relationships can be denoted by proportion unbalancedness k_i. A greater proportion unbalancedness $|k_i|$ denotes the smaller integrity of the industry structure relationship of industrial clusters. Suppose that the real output of industrial cluster i is X_i and the induction share of final demand (the outputs of satisfying total demand) is X_i^*, then the proportion unbalance degree coefficient of the industrial cluster k_i is

$$k_i = \frac{X_i - X_i^*}{X_i^*}$$

(2) Production induced coefficient.

Production induced coefficient denotes the product share of each sector induced by one unit final demand, and reflects that final demand items (for example, consumption, investment and imports and exports) induce the action force size degree of each industrial sector production. Its computing expression is

$$W_{is} = \sum_{j=1}^{n} r_{ij} f_s \Big/ f_{is}$$

where f_s is item s column vector element of final demand and r_{ij} is the element of row i column j of the Leontief inverse matrix $(\boldsymbol{I\text{-}A})^{-1}$. As far as industrial clusters are concerned, the greater production induced coefficient is, the greater production repercussion effect is, which explains   that   the greater pull function of final demand to industrial cluster growth needs more production to satisfy it.

(3) Dependence degree coefficient of final demand.

Dependence degree of each industrial sector on final demand is the ratio of induced share of one final demand item towards production to the sum of induced share of the each final demand item towards production. It is the index reflecting the dependence degree of production of each industrial sector on final demand item. Its computing expressions is

$$Z_{is} = \sum_{j=1}^{n} r_{ij} f_s \Big/ \sum_{j=1}^{s} \sum_{i=1}^{n} f_s$$

The greater the dependence degree coefficient of final demand is, the greater dependence degree of industrial production on final demand is, and production must to adapt changes of demand. Furthermore, the size of the coefficient also reflects demand contribution of each kind of final demand to the value added of one sector, that is, how much of the realizing value added of one sector is through satisfying consumption demand, and how much is through satisfying investment demand and export demand as well. The stimulating growth of one industrial cluster or sector depends on more effective consumption, investment or export demand stimulating.

5.2.4.2   Analysis of unbalance degree of whole industrial clusters in China

We can see from Table 5.33 that only the administration industrial cluster has an unbalance degree coefficient of final demand of $0 < |k_i| < 0.5$, there are highest integrity function of technological economy structure and the best aggregation quality in this industrial cluster. Clusters whose unbalance degree coefficient of final demand is $0.5 < |k_i| < 1$ are nonferrous metal industrial cluster, petroleum industrial cluster, electronics industrial cluster, textile industrial cluster, commerce industrial cluster, planting industrial cluster, construction industrial cluster, diet industrial cluster, iron and steel industrial cluster. The deviation scope between their real final demands and final demands produced by the technological economy level (induced share of final demand) is 50%~100%. There are higher integrity function of technological economy structure and better aggregation quality in the industrial clusters. There is only chemistry industrial cluster whose unbalance degree coefficient of final demand is $|k_i| \geqslant 1$, and its integrity function induced by final demand is poorest, and aggregation quality lowest.

As far as the unbalance degree coefficient of consumption is concerned, there is only the administration industrial cluster whose unbalance degree coefficient of consumption is $0 < |k_i| < 0.5$, and this cluster has the highest integrity function of technological economy structure and the best aggregation quality. Unbalance degree coefficients of consumption of other industrial clusters are $0.5 < |k_i| < 1$, and the deviation scope between their real final demands and final demands produced by the technological economy level (induced share of final demand) is 50%~100%. There are higher integrity function of technological economy structure and better aggregation quality in the industrial clusters.

As far as unbalance degree coefficient of investment is concerned, there is only the iron and steel industrial cluster whose unbalance degree coefficient of investment is $0 < |k_i| < 0.5$,

and this cluster has the highest integrity function of technological economy structure and
the best aggregation quality. Unbalance degree coefficients of investment of other industrial
clusters are $0.5 < |k_i| < 1$. The deviation scope between their real final demands and final
demands produced by the technological economy level (induced share of final demand) is
50%~100%. There are higher integrity function of technological economy structure and
better aggregation quality in the industrial clusters.

Table 5.33   Unbalance degree coefficient of final demand of industrial clusters in China

Types of industrial clusters	Total consumption	Total investment	Net export	Total demand
Planting industrial cluster	0.6340	0.7527	1.1108	0.6669
Electronics industrial cluster	0.7728	0.8700	0.4693	0.7803
Commerce industrial cluster	0.6780	0.8794	0.4589	0.7224
Diet industrial cluster	0.5855	0.8449	0.7810	0.6271
Iron and steel industrial cluster	0.9046	0.4823	−19.0574	0.6009
Petroleum industrial cluster	0.8070	0.9679	1.5694	0.8907
Nonferrous metal industrial cluster	0.9408	0.9631	0.8131	0.9490
Construction industrial cluster	0.8765	0.5869	0.6273	0.6614
Textile industrial cluster	0.7352	0.8543	0.4559	0.7264
Chemistry industrial cluster	0.9672	0.9297	−4.6954	1.0126
Administration industrial cluster	0.1871	0.5804	8.7304	0.2300

As far as the unbalance degree coefficient of net export is concerned, clusters whose
unbalance degree coefficient of net export is $0 < |k_i| < 0.5$ are electronics industrial cluster,
commerce industrial cluster and textile industrial cluster and these clusters have the highest
integrity function of technological economy structure and best aggregation quality. Clusters
whose unbalance degree coefficient of net export is $0.5 < |k_i| < 1$ are nonferrous metal
industrial cluster, diet industrial cluster and construction industrial cluster, deviation scope
between their real final demands and final demands produced by the technological economy
level (induced share of final demand) is 50%~100%. There are higher integrity function of
technological economy structure and better aggregation quality in the industrial clusters.
Clusters whose unbalance degree coefficient of net export is $|k_i| > 1$ are iron and steel
industrial cluster, administration industrial cluster, chemistry industrial cluster, petroleum
industrial cluster and planting industrial cluster. The integrity function induced by net
export is lowest and their aggregation quality is poorest.

5.2.4.3   Analysis on unbalance degree of each industrial cluster

(1) Planting industrial cluster.

We can see from Table 5.34 that the industries whose unbalance degree coefficient of
final demand in the planting industrial cluster is $0 < |k_i| < 0.1$ are other food products, non-
alcoholic beverage and other textiles, and there are highest integrity function of economic
technology structure and best aggregation quality for these industries in the planting indus-
trial cluster. Industries whose unbalance degree coefficient of final demand is $0.1 \leqslant |k_i| < 0.5$
are respectively hemp textiles, other agriculture products, livestock and livestock products,
tobacco products, wines, spirits and liquors and slaughtering, meat processing, eggs and
dairy products, and there are higher integrity function of economic technology structure
and better aggregation quality for these industries. Industries whose unbalance degree coef-
ficient is $0.5 \leqslant |k_i| < 1$ are respectively chemical pesticides, natural gas products, technical
services for agriculture, forestry, livestock and fishing, sugar refining, grain mill products,
vegetable oil and forage, crop cultivation and agriculture, forestry, animal husbandry and
fishing machinery, and there are lower integrity function of economic technology structure
and poorer aggregation quality for these industries. Industries whose unbalance degree coef-
ficient is $|k_i| \geqslant 1$ are chemical fertilizers and storage and warehousing, and there are lowest

integrity function of economic technology structure and poorest aggregation quality for these industries in the planting industrial cluster.

Table 5.34   Unbalance degree coefficient of planting industrial cluster

Items	Total consumption	Total investment	Net export	Total demand
Crop cultivation	0.5751	0.6969	0.8410	0.5818
Livestock and livestock products	0.2426	0.1326	0.7808	0.2386
Other agriculture products	0.3567	2.5948	0.3246	0.3745
Natural gas products	0.9300	1.0000	41.4098	0.7842
Grain mill products, vegetable oil and forage	0.5644	0.5072	−0.1454	0.5901
Sugar refining	0.6681	0.6893	6.6950	0.7016
Slaughtering, meat processing, eggs and dairy products	0.1036	0.1370	0.0619	0.1008
Other food products	0.0710	0.0782	0.0631	0.0710
Wines, spirits and liquors	0.1376	0.1459	0.2610	0.1390
Nonalcoholic beverage	0.0682	0.0681	0.0590	0.0674
Tobacco products	0.1421	0.1421	0.1420	0.1421
Hemp textiles	0.7258	0.2516	0.0550	0.4610
Other textiles	1.0000	0.1779	0.1234	−0.0438
Chemical fertilizers	1.0000	0.9225	0.3450	1.1309
Chemical pesticides	1.0000	0.9139	0.1307	0.9702
Agriculture, forestry, animal husbandry and fishing machinery	1.0000	0.4420	0.5006	0.5461
Storage and warehousing	1.0000	1.0000	1.0000	1.0000
Technical services for agriculture, forestry, livestock and fishing	0.6860	1.0000	1.0000	0.7034

As far as unbalance degree coefficient of consumption is concerned, industries whose unbalance degree coefficient is $0 < |k_i| < 0.1$ are other food products and nonalcoholic beverage, and there are highest integrity function of economic technology structure and best aggregation quality for these two industries in the planting industrial cluster. Industries whose unbalance degree coefficient is $0.1 \leqslant |k_i| < 0.5$ are respectively other agriculture products, livestock and livestock products, tobacco products, wines, spirits and liquors and slaughtering, meat processing, eggs and dairy products, and there are higher integrity function of economic technology structure and better aggregation quality for these industries. Industries whose unbalance degree coefficient is $0.5 \leqslant |k_i| < 1$ are respectively natural gas products, hemp textiles, technical services for agriculture, forestry, livestock and fishing, sugar refining, crop cultivation, and grain mill products, vegetable oil and forage, and there are lower integrity function of economic technology structure and poorer aggregation quality for these industries. Industries whose unbalance degree coefficient is $|k_i| \geqslant 1$ are other textiles, chemical fertilizers, chemical pesticides, agriculture, forestry, animal husbandry and fishing machinery, and storage and warehousing, and there are lowest integrity function of economic technology structure and poorest aggregation quality for these industries in the planting industrial cluster.

As far as unbalance degree coefficient of investment is concerned, industries whose unbalance degree coefficient is $0 < |k_i| < 0.1$ are other food products and nonalcoholic beverage, and there are highest integrity function of economic technology structure and best aggregation quality for these two industries in the planting industrial cluster. Industries whose unbalance degree coefficient is $0.1 \leqslant |k_i| < 0.5$ are respectively agriculture, forestry, animal husbandry and fishing machinery, hemp textiles, other textiles, wines, spirits and liquors, tobacco products, slaughtering, meat processing, eggs and dairy products and livestock and livestock products, and there are higher integrity function of economic technology structure and better aggregation quality for these industries. Industries whose unbalance degree coefficient is $0.5 \leqslant |k_i| < 1$ are respectively chemical fertilizers, chemical pesticides, crop cultivation, sugar refining and grain mill products, vegetable oil and forage, and there are lower integrity function of economic technology structure and poorer aggregation quality for these industries. Industries whose unbalance degree coefficient is $|k_i| \geqslant 1$ are other agriculture products, natural gas products, storage and warehousing and technical services for agriculture, forestry, livestock and fishing, and there are lowest integrity function of economic technology structure and poorest aggregation quality for these industries in the planting industrial cluster.

As far as unbalance degree coefficient of net export is concerned, industries whose unbalance degree coefficient is $0 < |k_i| < 0.1$ are other food products, slaughtering, meat processing, eggs and dairy products, nonalcoholic beverage and hemp textiles, and there are highest integrity function of economic technology structure and best aggregation quality for these industries in the planting industrial cluster. Industries whose unbalance degree coefficient is $0.1 \leqslant |k_i| < 0.5$ are respectively chemical fertilizers, other agriculture products, wines, spirits and liquors, grain mill products, vegetable oil and forage, tobacco products, chemical pesticides and other textiles, and there are higher integrity function of economic technology structure and better aggregation quality for these industries. Industries whose unbalance degree coefficient is $0.5 \leqslant |k_i| < 1$ are respectively crop cultivation, livestock and livestock products and agriculture, forestry, animal husbandry and fishing machinery, and there are lower integrity function of economic technology structure and poorer aggregation quality for these industries. Industries whose unbalance degree coefficient is $|k_i| \geqslant 1$ are natural gas products, sugar refining, storage and warehousing and technical services for agriculture, forestry, livestock and fishing, and there are lowest integrity function of economic technology structure and poorest aggregation quality for these industries in the planting industrial cluster.

(2) Electronics industrial cluster.

We can see from Table 5.35 that the industries whose unbalance degree coefficient of final demand in the electronics industrial cluster is $0.1 \leqslant |k_i| < 0.5$ are other electronic and communication equipment, and scientific research, and there are higher integrity function of economic technology structure and better aggregation quality for these industries. Industries whose unbalance degree coefficient is $|k_i| \geqslant 1$ are electronic element and device and other social services, and there are lowest integrity function of economic technology structure and poorest aggregation quality for these industries in electronics industrial cluster. Unbalance degree coefficient of final demand of the other industries is $0.5 \leqslant |k_i| < 1$, and there are lower integrity function of economic technology structure and poorer aggregation quality for these industries.

As far as unbalance degree coefficient of consumption is concerned, the only industry whose unbalance degree coefficient is $0.1 \leqslant |k_i| < 0.5$ is scientific research, and there is higher integrity function of economic technology structure and better aggregation quality for this industry. The only industry whose unbalance degree coefficient is $|k_i| \geqslant 1$ is electronic element and device, and there is lowest integrity function of economic technology structure and poorest aggregation quality for the industries in electronics industrial cluster. Unbalance degree coefficient of consumption of the other industries is $0.5 \leqslant |k_i| < 1$, and there are lower integrity function of economic technology structure and poorer aggregation quality for these industries.

As far as unbalance degree coefficient of investment is concerned, the only industry whose unbalance degree coefficient is $0.1 \leqslant |k_i| < 0.5$ is other electronic and communication equipment, and there is higher integrity function of economic technology structure and better aggregation quality for these industries. Industries whose unbalance degree coefficient is $0.5 \leqslant |k_i| < 1$ are respectively glass and glass products, electronic appliances, wholesale and retail trade and electronic computer, and there are lower integrity function of economic technology structure and poorer aggregation quality for these industries. Industries whose unbalance degree coefficient is $|k_i| \geqslant 1$ are electronic element and device, cultural and office equipment, other social services and scientific research, and there are lowest integrity function of economic technology structure and poorest aggregation quality for these industries in the planting industrial cluster.

Table 5.35   Unbalance degree coefficient of electronics industrial cluster

Items	Total consumption	Total investment	Net export	Total demand
Glass and glass products	0.8766	0.8075	1.1410	0.8419
Electronic computer	0.8908	0.5726	0.3229	0.5784
Electronic appliances	0.5005	0.6706	0.5051	0.5235
Electronic element and device	1.0000	1.2783	−0.9298	1.4341
Other electronic and communication equipment	0.6079	0.2384	−12.3773	0.3797
Cultural and office equipment	0.9130	1.0500	0.3943	0.6600
Wholesale and retail trade	0.5973	0.6113	0.5617	0.5891
Other social services	0.9692	1.0000	1.1479	1.0206
Scientific research	0.1027	1.0000	1.0000	0.1639

As far as unbalance degree coefficient of net export is concerned, industries whose unbalance degree coefficient is $0.1 \leqslant |k_i| < 0.5$ are cultural and office requirement and electronic computers and there is higher integrity function of economic technology structure and better aggregation quality for these industries. Industries whose unbalance degree coefficient is $0.5 \leqslant |k_i| < 1$ are respectively electronic element and device, wholesale and retail trade, and electronic appliances and there are lower integrity function of economic technology structure and poorer aggregation quality for these industries. Unbalance degree coefficients of net export of the rest are $|k_i| \geqslant 1$, and there are lowest integrity function of economic technology structure and poorest aggregation quality for these industries in the electronics industrial cluster.

(3) Commerce industrial cluster.

We can see from Table 5.36 that the industries whose unbalance degree coefficient of final demand in the commerce industrial cluster is $0.1 \leqslant |k_i| < 0.5$ are respectively recreational services, real estate, nonalcoholic beverage, rubber products and culture and arts, radio, film and television services, and there are higher integrity function of economic technology structure and better aggregation quality for these industries. Industries whose unbalance degree coefficient is $|k_i| \geqslant 1$ are logging and transport of timber and bamboo, paper and products, printing and record medium reproduction and storage and warehousing, and there are lowest integrity function of economic technology structure and poorest aggregation quality for these industries in the commerce industrial cluster. Unbalance degree coefficient of final demand of the other industries is $0.5 \leqslant |k_i| < 1$, and there are lower integrity function of economic technology structure and poorer aggregation quality for these industries.

As far as an unbalance degree coefficient of consumption is concerned, industries whose unbalance degree coefficient is $0.1 \leqslant |k_i| < 0.5$ are respectively real estate, nonalcoholic beverage and culture and arts, radio, film and television services, and there are higher integrity function of economic technology structure and better aggregation quality for these industries. Industries whose unbalance degree coefficient is $|k_i| \geqslant 1$ are logging and transport of timber and bamboo, chemical pesticides and storage and warehousing, and there are lowest integrity function of economic technology structure and poorest aggregation quality for these industries in the commerce industrial cluster. Unbalance degree coefficient of consumption of the other industries is $0.5 \leqslant |k_i| < 1$, and there are lower integrity function of economic technology structure and poorer aggregation quality for these industries.

As far as unbalance degree coefficient of investment is concerned, industries whose unbalance degree coefficient is $0.1 \leqslant |k_i| < 0.5$ are real estate and forestry, and there are higher integrity function of economic technology structure and better aggregation quality for these industries. Industries whose unbalance degree coefficient is $|k_i| \geqslant 1$ are logging and transport of timber and bamboo, storage and warehousing, recreational services, and culture and arts, radio, film and television services, and there are lowest integrity function of economic technology structure and poorest aggregation quality for these industries in the commerce industrial cluster. Unbalance degree coefficient of investment of the other industries is $0.5 \leqslant |k_i| < 1$, and there are lower integrity function of economic technology structure and poorer aggregation quality for these industries.

Table 5.36   Unbalance degree coefficient of commerce industrial cluster

Items	Total consumption	Total investment	Net export	Total demand
Forestry	0.6624	0.1355	1.6771	0.5536
Logging and transport of timber and bamboo	1.0000	3.3701	−0.1508	157.4373
Nonalcoholic beverage	0.2989	0.6509	0.6583	0.3839
Tobacco products	0.5860	0.7143	0.9131	0.6314
Paper and products	0.9733	0.8610	1.9058	1.0331
Printing and record medium reproduction	0.9930	0.9939	1.0328	1.0018
Cultural goods	0.5651	0.7227	−1.1321	0.7314
Chemical pesticides	1.0000	0.9560	0.9308	0.9735
Rubber products	0.5091	0.5615	0.2168	0.3586
Electronic appliances	0.5415	0.6572	0.5141	0.5468
Storage and warehousing	1.0000	1.0000	1.0000	1.0000
Wholesale and retail trade	0.6499	0.5910	0.5607	0.6168
Real estate	0.3223	0.3959	1.0000	0.4038
Recreational services	0.7562	1.0000	0.1973	0.4452
Culture and arts, radio, film and television services	0.1853	1.0000	0.3361	0.1883

As far as unbalance degree coefficient of net export is concerned, industries whose unbalance degree coefficient is $0.1 \leqslant |k_i| < 0.5$ are respectively culture and arts, radio, film and television services, rubber products, recreational services and logging and transport of timber and bamboo. Industries whose unbalance degree coefficient is $0.5 \leqslant |k_i| < 1$ are respectively chemical pesticides, tobacco products, nonalcoholic beverage, wholesale and retail trade and electronic appliances, and there are lower integrity function of economic technology structure and poorer aggregation quality for these industries. Unbalance degree coefficients of net export of the rest are $|k_i| \geqslant 1$, and there are lowest integrity function of economic technology structure and poorest aggregation quality for these industries in the commerce industrial cluster.

(4) Diet industrial cluster.

We can see from Table 5.37 that the only industry whose unbalance degree coefficient of final demand in the diet industrial cluster is $0 \leqslant |k_i| < 0.1$ is eating and drinking places, and there is highest integrity function of economic technology structure and best aggregation quality for this industry. Industries whose unbalance degree coefficient of final demand in the diet industrial cluster is $0.1 \leqslant |k_i| < 0.5$ are respectively fishery, wines, spirits and liquors, tobacco products, prepared fish and seafood, nonalcoholic beverage and other food products, and there are higher integrity function of economic technology structure and better aggregation quality for these industries. Unbalance degree coefficient of final demand of the other industries is $0.5 \leqslant |k_i| < 1$, and there are lower integrity function of economic technology structure and poorer aggregation quality for these industries.

As far as unbalance degree coefficient of consumption is concerned, the industry whose unbalance degree coefficient of consumption in the diet industrial cluster is $0 \leqslant |k_i| < 0.1$

is only eating and drinking places, and there is highest integrity function of economic technology structure and best aggregation quality for this industry. Industries whose unbalance degree coefficient is $0.1 \leqslant |k_i| < 0.5$ are respectively fishery, prepared fish and seafood, wines, spirits and liquors, tobacco products, non-alcoholic beverage and other food products, and there are higher integrity function of economic technology structure and better aggregation quality for these industries. Unbalance degree coefficient of consumption of the other industries is $0.5 \leqslant |k_i| < 1$, and there are lower integrity function of economic technology structure and poorer aggregation quality for these industries.

As far as unbalance degree coefficient of investment is concerned, industries whose unbalance degree coefficient is $|k_i| \geqslant 1$ are eating and drinking places and technical services for agriculture, forestry, livestock and fishing, and there are lowest integrity function of economic technology structure and poorest aggregation quality for these industries. The only industry whose unbalanced degree coefficient is $0.5 \leqslant |k_i| < 1$ is crop cultivation, and there are lower integrity function of economic technology structure and poorer aggregation quality for this industry. Unbalance degree coefficient of investment of the other industries is $0.1 \leqslant |k_i| < 0.5$, and there are higher integrity function of economic technology structure and better aggregation quality for these industries.

Table 5.37   Unbalance degree coefficient of diet industrial cluster

Items	Total consumption	Total investment	Net export	Total demand
Crop cultivation	0.5583	0.5998	−27.7058	0.5624
Fishery	0.3495	0.3252	0.7748	0.3712
Grain mill products, vegetable oil and forage	0.5389	0.3377	−0.4361	0.5591
Slaughtering, meat processing, eggs and dairy products	0.5344	0.4959	0.4701	0.5273
Prepared fish and seafood	0.2773	0.2473	0.0337	0.2214
Other food products	0.1778	0.1473	0.2071	0.1779
Wines, spirits and liquors	0.2522	0.1771	0.6214	0.2544
Nonalcoholic beverage	0.1918	0.1073	0.1571	0.1854
Tobacco products	0.2350	0.1882	0.3024	0.2336
Eating and drinking places	0.0699	1.0000	0.0541	0.0720
Technical services for agriculture, forestry, livestock and fishing	0.6674	1.0000	1.0000	0.6815

As far as unbalance degree coefficient of net export is concerned, industries whose unbalance degree coefficient of consumption in diet industrial cluster is $0 \leqslant |k_i| < 0.1$ are eating and drinking places and prepared fish and seafood, and there is highest integrity function of economic technology structure and best aggregation quality for the industry. Industries whose unbalance degree coefficient is $0.5 \leqslant |k_i| < 1$ are fishery and wines, spirits and liquors, and there are lower integrity function of economic technology structure and poorer aggregation quality for these industries. Industries whose unbalance degree coefficient is $|k_i| \geqslant 1$ are crop cultivation and technical services for agriculture, forestry, livestock and fishing, and there are lowest integrity function of economic technology structure and poorest aggregation quality for these industries. Unbalance degree coefficient of investment of the other industries is $0.1 \leqslant |k_i| < 0.5$, and there are higher integrity function of economic technology structure and better aggregation quality for these industries.

(5) Iron and steel industrial cluster.

We can see from Table 5.38 that the only industry whose unbalance degree coefficient of final demand in the iron and steel industrial cluster is $0 \leqslant |k_i| < 0.1$ is construction, and there is highest integrity function of economic technology structure and best aggregation quality for this industry. The only industry whose unbalance degree coefficient of final demand in the iron and steel industrial cluster is $0.1 \leqslant |k_i| < 0.5$ is bicycle, and there is higher integrity function of economic technology structure and better aggregation quality for the industry. Industries whose unbalance degree coefficient is $|k_i| \geqslant 1$ are ferrous ore mining, steel processing, and steel-smelting, and there are lowest integrity function of economic technology structure and poorest aggregation quality for these industries. Unbalance degree coefficient of final demand of the other industries is $0.5 \leqslant |k_i| < 1$, and there are lower integrity function of economic technology structure and poorer aggregation quality for these industries.

As far as unbalance degree coefficient of consumption is concerned, the only industry whose unbalance degree coefficient of consumption in the iron and steel industrial cluster is $0 \leqslant |k_i| < 0.1$ is gas production and supply, and there is highest integrity function of economic technology structure and best aggregation quality for the industry. Industries whose unbalance degree coefficient is $0.1 \leqslant |k_i| < 0.5$ are respectively other special industrial equipment, metal products, electricity production and supply, bicycle, steam and hot water production and supply and water production and supply, and there are higher integrity function of economic technology structure and better aggregation quality for these industries. Industries whose unbalance degree coefficient is $0.5 \leqslant |k_i| < 1$ are steel processing and coal mining and processing, and there are lower integrity function of economic technology structure and poorer aggregation quality for these industries. Unbalance degree coefficient of consumption of the other industries is $|k_i| \geqslant 1$, and there are lowest integrity function of economic technology structure and poorest aggregation quality for these industries.

As far as unbalance degree coefficient of investment is concerned, the only industry whose unbalance degree coefficient is $0 \leqslant |k_i| < 0.1$ is construction, and there is highest integrity function of economic technology structure and best aggregation quality for this industry. The only industry whose unbalance degree coefficient is $0.1 \leqslant |k_i| < 0.5$ is only bicycle, and there is higher integrity function of economic technology structure and better aggregation quality for this industry. Industries whose unbalance degree coefficient is $0.5 \leqslant |k_i| < 1$ are respectively coking, coal mining and processing, metal products and other special industrial equipment, and there are lower integrity function of economic technology structure and poorer aggregation quality for these industries. Unbalance degree coefficient of consumption of the other industries is $|k_i| \geqslant 1$, and there are lowest integrity function of economic technology structure and poorest aggregation quality for these industries.

Table 5.38   Unbalance degree coefficient of iron and steel industrial cluster

Items	Total consumption	Total investment	Net export	Total demand
Coal mining and processing	0.7989	0.9903	4.9953	0.9551
Ferrous ore mining	1.0000	1.0024	0.3245	1.0593
Coking	1.0000	0.9947	−0.1427	0.9479
Iron-smelting	1.0000	1.0112	−1.3863	0.9492
Steel-smelting	1.0000	1.0077	0.9015	1.0134
Steel processing	0.9591	1.0134	0.5782	1.0341
Alloy iron smelting	1.0000	1.0024	−27.5167	0.9576
Metal products	0.3180	0.9669	−0.2257	0.9166
Other special industrial equipment	0.4276	0.5676	0.1580	0.6765
Bicycle	0.2141	0.2141	0.2141	0.2141
Electricity production and supply	0.2392	1.0000	1.6295	0.8964
Steam and hot water production and supply	0.1504	1.0000	1.0000	0.8852
Gas production and supply	0.0401	1.0000	1.0000	0.5053
Water production and supply	0.1209	1.0000	1.0000	0.7623
Construction	1.0000	0.0102	0.2142	0.0102

As far as unbalance degree coefficient of net export is concerned, industries whose unbalance degree coefficient is $0.1 \leqslant |k_i| < 0.5$ are ferrous ore mining, metal products, construction, bicycle, other special industrial equipment, coking and there are higher integrity function of economic technology structure and better aggregation quality for these industries. Industries whose unbalance degree coefficient is $0.5 \leqslant |k_i| < 1$ are steel-smelting and steel processing, and there are lower integrity function of economic technology structure and poorer aggregation quality for these industries. Unbalance degree coefficient of consumption of other industries is $|k_i| \geqslant 1$, and there are lowest integrity function of economic technology structure and poorest aggregation quality for these industries.

(6) Petroleum industrial cluster.

We can see from Table 5.39 that the industries whose unbalance degree coefficient of final demand in the petroleum industrial cluster is $0 \leqslant |k_i| < 0.1$ are railway passenger transport, highway passenger transport and social welfare, and there are highest integrity function of economic technology structure and best aggregation quality for these industries. The only industry whose unbalance degree coefficient of final demand in the petroleum industrial cluster is $0.5 \leqslant |k_i| < 1$ is pipeline transport, and there is lower integrity function of economic technology structure and poorer aggregation quality for the industry. Industries whose unbalance degree coefficient is $|k_i| \geqslant 1$ are crude petroleum products and petroleum refining, and there are lowest integrity function of economic technology structure and poorest aggregation quality for these industries. Unbalance degree coefficient of final demand of the other industries is $0.1 \leqslant |k_i| < 0.5$, and there are higher integrity function of economic technology structure and better aggregation quality for these industries.

Table 5.39 Unbalance degree coefficient of petroleum industrial cluster

Items	Total consumption	Total investment	Net export	Total demand
Crude petroleum products	1.0000	−5.9504	0.1038	4.5934
Petroleum refining	0.9465	0.3948	−1.0528	1.1759
Rail road transport equipment	1.0000	0.0574	0.5525	0.1886
Electricity production and supply	0.3899	1.0000	0.4288	0.3991
Railway freight transport	0.6266	0.0620	0.1636	0.4330
Highway freight transport	0.3491	0.0664	0.1121	0.2341
Pipeline transport	0.9348	1.3727	0.0686	0.6839
Water freight transport	0.6246	0.1200	0.0159	0.1435
Railway passenger transport	0.0783	1.0000	0.8103	0.0959
Highway passenger transport	0.0507	1.0000	0.9309	0.0602
Water passenger transport	0.1338	1.0000	0.0663	0.1193
Public services	0.1767	1.0000	0.1869	0.1794
Social welfare	0.0228	1.0000	1.0000	0.0244

As far as unbalance degree coefficient of consumption is concerned, industries whose unbalance degree coefficient of consumption in the petroleum industrial cluster is $0 \leqslant |k_i| < 0.1$ are railway passenger transport, highway passenger transport and social welfare, and there are highest integrity function of economic technology structure and best aggregation quality for these industries. Industries whose unbalance degree coefficient is $0.1 \leqslant |k_i| < 0.5$ are respectively electricity production and supply, road freight transport, public services and water passenger transport, and there are higher integrity function of economic technology structure and better aggregation quality for these industries. Industries whose unbalance degree coefficient is $0.5 \leqslant |k_i| < 1$ are petroleum refining, pipeline transport, railway freight transport and water freight transport, and there are lower integrity function of economic technology structure and poorer aggregation quality for these industries. Unbalance degree coefficient of consumption of the other industries is $|k_i| \geqslant 1$, and there are lowest integrity function of economic technology structure and poorest aggregation quality for these industries.

As far as unbalance degree coefficient of investment is concerned, industries whose unbalance degree coefficient is $0 \leqslant |k_i| < 0.1$ are highway freight transport, railway freight transport and railroad transport equipment, and there are highest integrity function of economic technology structure and best aggregation quality for these industries. Industries whose unbalance degree coefficient is $0.1 \leqslant |k_i| < 0.5$ are petroleum refining and water freight transport, and there are higher integrity function of economic technology structure and better aggregation quality for these industries. Unbalance degree coefficient of consumption of the other industries is $|k_i| \geqslant 1$, and there are lowest integrity function of economic technology structure and poorest aggregation quality for these industries.

As far as unbalance degree coefficient of net export is concerned, industries whose unbalance degree coefficient of consumption in the petroleum industrial cluster is $0 \leqslant |k_i| < 0.1$ are pipeline transport, water passenger transport and water freight transport, and there are highest integrity function of economic technology structure and best aggregation quality for these industries. Industries whose unbalance degree coefficient is $0.1 \leqslant |k_i| < 0.5$ are respectively electricity production and supply, public services, railway freight transport, highway freight transport and crude petroleum products, and there are higher integrity function of economic technology structure and better aggregation quality for these industries. Industries whose unbalance degree coefficient is $0.5 \leqslant |k_i| < 1$ are highway passenger transport, railway passenger transport and railroad transport equipment, and there are lower integrity function of economic technology structure and poorer aggregation quality for these industries. Industries whose unbalance degree coefficient is $|k_i| \geqslant 1$ are only social welfare and petroleum refining, and there are lowest integrity function of economic technology structure and poorest aggregation quality for these industries.

(7) Nonferrous metal industrial cluster.

We can see from Table 5.40 that the only industry whose unbalance degree coefficient of final demand in the nonferrous metal industrial cluster is $0 \leqslant |k_i| < 0.1$ is telecommunication, and there is highest integrity function of economic technology structure and best aggregation quality for this industry. The only industry whose unbalance degree coefficient of final demand in the nonferrous metal industrial cluster is $0.1 \leqslant |k_i| < 0.5$ is other electric machinery and equipment, and there is higher integrity function of economic technology structure and better aggregation quality for this industry. Unbalance degree coefficient of final demand of the other industries is $|k_i| \geqslant 1$, and there are lowest integrity function of economic technology structure and poorest aggregation quality for these industries.

Table 5.40   Unbalance degree coefficient of nonferrous metal industrial cluster

Items	Total consumption	Total investment	Net export	Total demand
Nonferrous metal mining	1.0000	0.5796	−0.3434	1.8335
Nonferrous metal refining	1.0000	1.0048	3.8252	1.3029
Nonferrous metal processing	1.0000	1.2274	1.3321	1.1774
Other electric machinery and equipment	0.8091	0.2446	0.4221	0.4726
Scrap and waste	1.0000	1.0000	1.0000	1.0000
Telecommunication	0.0252	1.0000	0.0649	0.0569

As far as unbalance degree coefficient of consumption is concerned, it is similar to unbalance degree coefficient of final demand.

As far as unbalance degree coefficient of investment is concerned, the only industry whose unbalance degree coefficient is $0.1 \leqslant |k_i| < 0.5$ is other electric machinery and equipment, and there is higher integrity function of economic technology structure and better aggregation quality for this industry. The only industry whose unbalance degree coefficient is $0.5 \leqslant |k_i| < 1$ is nonferrous metal mining, and there is lower integrity function of economic technology structure and poorer aggregation quality for this industry. Unbalance degree coefficient of consumption of the other industries is $|k_i| \geqslant 1$, and there are lowest integrity function of economic technology structure and poorest aggregation quality for these industries.

As far as unbalance degree coefficient of net export is concerned, the industry whose unbalance degree coefficient is $0 \leqslant |k_i| < 0.1$ is only telecommunication, and there is highest integrity function of economic technology structure and best aggregation quality for the industry. Industries whose unbalance degree coefficient is $0.1 \leqslant |k_i| < 0.5$ are other electric machinery and equipment and non-ferrous metal mining, and there are higher integrity function of economic technology structure and better aggregation quality for these industries. Unbalance degree coefficient of consumption of the other industries is $|k_i| \geqslant 1$, and there are lowest integrity function of economic technology structure and poorest aggregation quality for these industries.

(8) Construction industrial cluster.

We can see from Table 5.41 that the industry whose unbalance degree coefficient of final demand in the construction industrial cluster is $0 \leqslant |k_i| < 0.1$ is only construction, and there is highest integrity function of economic technology structure and best aggregation quality for the industry. The industries whose unbalance degree coefficient of final demand in the construction industrial cluster is $0.1 \leqslant |k_i| < 0.5$ is only furniture and products of wood, bamboo, cane, palm, straw, etc., and there is higher integrity function of economic technology structure and better aggregation quality for the industry. Industries whose unbalance degree coefficient is $|k_i| \geqslant 1$ are logging and transport of timber and bamboo, steel processing, sawmills and fiberboard and storage and warehousing, and there are lowest integrity function of economic technology structure and poorest aggregation quality for these industries. Unbalance degree coefficient of final demand of the other industries is $0.5 \leqslant |k_i| < 1$, and there are lower integrity function of economic technology structure and poorer aggregation quality for these industries.

Table 5.41   Unbalance degree coefficient of construction industrial cluster

Items	Total consumption	Total investment	Net export	Total demand
Nonmetal minerals and other ore mining	0.9250	0.9924	0.5568	0.9790
Logging and transport of timber and bamboo	1.0000	1.0185	−0.3186	1.1759
Sawmills and fiberboard	0.8993	1.0062	1.5265	1.0196
Furniture and products of wood, bamboo, cane, palm, straw, etc.	0.2252	0.7771	0.2161	0.4910
Cement	0.7448	0.9852	0.2564	0.9683
Cement and asbestos products	0.5642	0.9799	0.3432	0.9580
Bricks, tiles, lime and lightweight building materials	0.3744	0.9726	0.3816	0.9068
Pottery, china and earthenware	0.4087	0.9727	0.1633	0.8452
Fireproof products	1.0000	0.9863	1.1532	0.9879
Steel processing	0.9885	1.0187	−0.9270	1.0431
Metal products	0.6536	0.9447	0.4463	0.8782
Other electric machinery and equipment	0.8662	0.9382	0.5212	0.9011
Instruments, meters and other measuring equipment	0.6422	0.9279	1.5338	0.9069
Construction	1.0000	0.0179	−1.5197	0.0246
Storage and warehousing	1.0000	1.0000	1.0000	1.0000
Telecommunication	0.3091	1.0000	0.3148	0.8268
Geological prospecting and water conservancy	0.2050	1.0000	1.0000	0.5670

As far as unbalance degree coefficient of consumption is concerned, industries whose unbalance degree coefficient is $0.1 \leqslant |k_i| < 0.5$ are respectively pottery, china and earthenware, bricks, tiles, lime and lightweight building materials, telecommunication, furniture and products of wood, bamboo, cane, palm, straw, etc., and geological prospecting and water conservancy, and there are higher integrity function of economic technology structure and better aggregation quality for these industries. Industries whose unbalance degree coefficient is $|k_i| \geqslant 1$ are logging and transport of timber and bamboo, fireproof products, construction and storage and warehousing, and there are lowest integrity function of economic technology structure and poorest aggregation quality for the industry. Unbalance degree coefficient of consumption of the other industries is $0.5 \leqslant |k_i| < 1$, and there are lower integrity function of economic technology structure and poorer aggregation quality for these industries.

As far as unbalance degree coefficient of investment is concerned, the only industry whose unbalance degree coefficient is $0 \leqslant |k_i| < 0.1$ is construction, and there is highest integrity function of economic technology structure and best aggregation quality for the industry. Industries whose unbalance degree coefficient is $|k_i| \geqslant 1$ are steel processing, logging and transport of timber and bamboo, sawmills and fiberboard, storage and warehousing, telecommunication and geological prospecting and water conservancy, and there are lowest integrity function of economic technology structure and poorest aggregation quality for the industry. Unbalance degree coefficient of consumption of the other industries is $0.5 \leqslant |k_i| < 1$, and there are lower integrity function of economic technology structure and poorer aggregation quality for these industries.

As far as unbalance degree coefficient of net export is concerned, industries whose unbalance degree coefficient is $0.5 \leqslant |k_i| < 1$ are nonmetal minerals and other ore mining, other electric machinery and equipment and steel processing, and there are lower integrity function of economic technology structure and poorer aggregation quality for these industries. Industries whose unbalance degree coefficient is $|k_i| \geqslant 1$ are instruments, meters and other measuring equipment, sawmills and fiberboard, construction, fireproof products, storage and warehousing, and geological prospecting and water conservancy, and there are lowest integrity function of economic technology structure and poorest aggregation quality for the industry. Unbalance degree coefficient of consumption of the other industries is $0.1 \leqslant |k_i| < 0.5$, and there is highest integrity function of economic technology structure and best aggregation quality for these industries.

(9) Textile industrial cluster.

We can see from Table 5.42 that the industries whose unbalance degree coefficient of final demand in the textile industrial cluster is $0 \leqslant |k_i| < 0.1$ are knitted mills and wearing apparel, and there are highest integrity function of economic technology structure and best aggregation quality for these industries. Industries whose unbalance degree coefficient of final demand in the textile industrial cluster is $0.1 \leqslant |k_i| < 0.5$ are rubber products and crop cultivation, and there are higher integrity function of economic technology structure and better aggregation quality for these industries. Industries whose unbalance degree coefficient is $0.5 \leqslant |k_i| < 1$ are wool textiles, silk textiles, cotton textiles and hemp textiles, and there are lower integrity function of economic technology structure and poorer aggregation quality for these industries. Unbalance degree coefficient of final demand of the other industries is $|k_i| \geqslant 1$, and there are lowest integrity function of economic technology structure and poorest aggregation quality for these industries.

As far as unbalance degree coefficient of consumption is concerned, the only industry whose unbalance degree coefficient of consumption in the textile industrial cluster is $0 \leqslant |k_i| < 0.1$ is wearing apparel, and there is highest integrity function of economic technology structure and best aggregation quality for this industry. Industries whose unbalance degree coefficient is $0.1 \leqslant |k_i| < 0.5$ are respectively rubber products, crop cultivation and knitted mills, and there are higher integrity function of economic technology structure and better aggregation quality for these industries. Industries whose unbalance degree coefficient is $0.5 \leqslant |k_i| < 1$ are silk textiles, hemp textiles, wool textiles and cotton textiles, and there are lower integrity function of economic technology structure and poorer aggregation quality for these industries. Unbalance degree coefficient of consumption of the other industries is $|k_i| \geqslant 1$, and there are lowest integrity function of economic technology structure and poorest aggregation quality for these industries.

Table 5.42   Unbalance degree coefficient of textile industrial cluster

Items	Total consumption	Total investment	Net export	Total demand
Crop cultivation	0.1651	0.3240	1.1246	0.1987
Cotton textiles	0.7585	0.4976	0.8932	0.7832
Wool textiles	0.7935	0.5941	1.0510	0.8748
Hemp textiles	0.8251	0.4743	0.6803	0.7221
Silk textiles	0.8425	0.6330	0.9199	0.8446
Knitted mills	0.1197	0.0960	0.0738	0.0842
Other textiles	1.0000	0.7092	1.8239	1.1457
Wearing apparel	0.0219	0.0265	0.0242	0.0232
Chemical fibers	1.0000	0.6552	1.7285	1.1533
Rubber products	0.2563	0.2429	0.1798	0.2079

As far as unbalance degree coefficient of investment is concerned, industries whose unbalance degree coefficient is $0 \leqslant |k_i| < 0.1$ are knitted mills and wearing apparel, and there are highest integrity function of economic technology structure and best aggregation quality for these industries. Industries whose unbalance degree coefficient is $0.1 \leqslant |k_i| < 0.5$ are cotton textiles, hemp textiles, crop cultivation and rubber products, and there are higher integrity function of economic technology structure and better aggregation quality for these industries. Unbalance degree coefficient of consumption of the other industries is $0.5 \leqslant |k_i| < 1$, and there are lower integrity function of economic technology structure and poorer aggregation quality for these industries.

As far as unbalance degree coefficient of net export is concerned, industries whose unbalance degree coefficient is $0 \leqslant |k_i| < 0.1$ are knitted mills and wearing apparel, and there are highest integrity function of economic technology structure and best aggregation quality for these industries. The only industry whose unbalance degree coefficient is $0.1 \leqslant |k_i| < 0.5$ is rubber products, and there is higher integrity function of economic technology structure and better aggregation quality for this industry. Industries whose unbalance degree coefficient is $0.5 \leqslant |k_i| < 1$ are silk textiles, cotton textiles and hemp textiles, and there are lower integrity function of economic technology structure and poorer aggregation quality for these industries. Unbalance degree coefficient of consumption of the other industries is $|k_i| \geqslant 1$, and there are lowest integrity function of economic technology structure and poorest aggregation quality for these industries.

(10) Chemistry industrial cluster.

We can see from Table 5.43 that the only industry whose unbalance degree coefficient of final demand in the chemistry industrial cluster is $0.1 \leqslant |k_i| < 0.5$ are plastic products and other manufacturing products, and there are highest integrity function of economic technology structure and best aggregation quality for this industry. The industries whose unbalanced degree coefficient of final demand is $0.5 \leqslant |k_i| < 1$ are other chemical products and chemical fibers, and there are lower integrity function of economic technology structure and poorer aggregation for these industries.

As far as unbalance degree coefficient of consumption is concerned, industries whose unbalance degree coefficient is $0.1 \leqslant |k_i| < 0.5$ are other manufacturing products and plastic products, and there are higher integrity function of economic technology structure and better aggregation quality for these industries. Unbalance degree coefficient of consumption of the other industries is $|k_i| \geqslant 1$, and these are the lowest integrity function of economic technology structure and the poorest aggregation quality for these industries.

Table 5.43   Unbalance degree coefficient of chemistry industrial cluster

Items	Total consumption	Total investment	Net export	Total demand
Other chemical products	1.0000	0.4183	−0.1359	−0.8103
Chemical fibers	1.0000	0.1299	−0.0978	−0.5628
Plastic products	0.3378	0.2925	0.3272	0.3202
Other manufacturing products	0.3502	0.3811	0.3440	0.3498

As far as unbalance degree coefficient of investment is concerned, unbalance degree coefficients of all industries are $0.1 \leqslant |k_i| < 0.5$, and there are higher integrity function of economic technology structure and better aggregation quality for all industries.

As far as unbalance degree coefficient of net export is concerned, the only industry whose unbalance degree coefficient is $0.1 \leqslant |k_i| < 0.5$ is chemical fibers, and there is higher integrity function of economic technology structure and better aggregation quality for this industry. Unbalance degree coefficient of consumption of the other industries is $0.1 \leqslant |k_i| < 0.5$, and there are higher integrity function of economic technology structure and better aggregation quality for these industries.

(11) Administration industrial cluster.

We can see from Table 5.44 that the industries whose unbalance degree coefficient of final demand in the administration industrial cluster is $0 \leqslant |k_i| < 0.1$ are social welfare and public administration and other sectors, and there are highest integrity function of economic technology structure and best aggregation quality for these two industries. Industries whose unbalance degree coefficient of final demand in administration industrial cluste is $0.1 \leqslant |k_i| < 0.5$ are post services and tourism, and there are higher integrity function of economic technology structure and better aggregation quality for these two industries. Unbalance degree coefficient of final demand of the other industries is $0.5 \leqslant |k_i| < 1$, and there are lower integrity function of economic technology structure and poorer aggregation quality for these industries.

Table 5.44   Unbalance degree coefficient of administration industrial cluster

Items	Total consumption	Total investment	Net export	Total demand
Aircraft	1.0000	0.0024	−0.2351	0.9228
Post services	0.4932	1.0000	−0.0153	0.4809
Rail way passenger transport	0.7312	1.0000	0.9812	0.7442
Air passenger transport	0.9752	1.0000	0.3413	0.8473
Hotels	0.8182	1.0000	0.3529	0.7957
Tourism	0.2646	1.0000	0.2003	0.2455
Social welfare	0.0060	1.0000	1.0000	0.0064
Public administration and other sectors	0.0000		0.0000	0.0000

As far as unbalance degree coefficient of consumption is concerned, industries whose unbalance degree coefficient is $0 \leqslant |k_i| < 0.1$ are social welfare and public administration and other sectors, and there are highest integrity function of economic technology structure and best aggregation quality for these two industries. Industries whose unbalance degree coefficient is $0.1 \leqslant |k_i| < 0.5$ is post services and tourism, and there are higher integrity function of economic technology structure and better aggregation quality for these industries. The only industry whose unbalance degree coefficient is $|k_i| \geqslant 1$ is aircraft, and there is lowest integrity function of economic technology structure and poorest aggregation quality for this industry. Unbalance degree coefficient of consumption of the other industries is $0.5 \leqslant |k_i| < 1$, and there are lower integrity function of economic technology structure and poorer aggregation quality for these industries.

As far as unbalance degree coefficient of investment is concerned, the only industry whose unbalance degree coefficient is $0 \leqslant |k_i| < 0.1$ is aircraft, and there is highest integrity function of economic technology structure and best aggregation quality for this industry. Unbalance degree coefficient of consumption of the other industries is $|k_i| \geqslant 1$, and there are lowest integrity function of economic technology structure and poorest aggregation quality for these industries.

As far as unbalance degree coefficient of net export is concerned, industries whose unbalance degree coefficient is $0 \leqslant |k_i| < 0.1$ are public administration and other sectors and post services, and there are highest integrity function of economic technology structure and best aggregation quality for these industries. The only industry whose unbalance degree coefficient is $0.5 \leqslant |k_i| < 1$ is railway passenger transport, and there is lower integrity function of economic technology structure and poorer aggregation quality for this industry. The only industry whose unbalance degree coefficient is $|k_i| \geqslant 1$ is social welfare, and there are lowest integrity function of economic technology structure and poorest aggregation quality for this industry. Unbalance degree coefficient of consumption of the other industries is $0.1 \leqslant |k_i| < 0.5$, and there are higher integrity function of economic technology structure and better aggregation quality for these industries.

5.2.4.4   Analysis on production induced coefficient of whole industrial clusters in China

We can see from Table 5.45 that as far as production induced coefficient of final demand is concerned, the construction industrial cluster has a very high production induced degree. Industrial clusters with higher production induced degree are, respectively, the iron and steel industrial cluster, planting industrial cluster, diet industrial cluster, textile industrial cluster, commerce industrial cluster and the electronics industrial cluster.

For production induced coefficient of consumption, the degree is very high for the planting industrial cluster. Industrial clusters with higher production induced degree are respectively diet industrial cluster, textile industrial cluster, commerce industrial cluster and construction industrial cluster. There is an obvious difference in resident consumption and government consumption on industrial clusters. Industrial clusters with larger induced coefficient of resident consumption are respectively planting industrial cluster, diet industrial cluster, textile industrial cluster, commerce industrial cluster, construction industrial cluster and electronics industrial cluster, while industrial clusters with larger induced coefficient of government consumption are respectively administration industrial cluster, construction industrial cluster, petroleum industrial cluster, iron and steel industrial cluster and commerce industrial cluster. There is basically same rural consumption and town consumption on industrial clusters. Industrial clusters with larger induced coefficient of consumption are respectively planting industrial cluster, diet industrial cluster, textile industrial cluster and commerce industrial cluster.

For production induced coefficient of investment, the degree is extraordinarily high for construction industrial cluster and iron and steel industrial cluster, and investment has quite an impulse effect on the development of these clusters. Industrial clusters with higher production induced degree are commerce industrial cluster and electronics industrial cluster.

The induced coefficient of export has highest effect on construction industrial cluster and textile industrial cluster. Industrial clusters with higher export induced effect are electronics industrial cluster, commerce industrial cluster, iron and steel industrial cluster, planting industrial cluster and diet industrial cluster.

Table 5.45   Production induced coefficient of industrial clusters in China

Types of industrial clusters	Rural resident consumption	Town resident consumption	Resident consumption	Government consumption	Total consumption	Fixed cost	Stock increase	Total investment	Export	Final demand
Planting industrial cluster	0.7521	0.6324	0.6966	0.1262	0.6201	0.1235	0.6055	0.1727	0.2934	0.4113
Electronics industrial cluster	0.1817	0.2218	0.2003	0.1832	0.1980	0.2287	0.1861	0.2243	0.3498	0.2312
Commerce industrial cluster	0.2694	0.3021	0.2846	0.2155	0.2753	0.2266	0.2889	0.2329	0.3218	0.2677
Diet industrial cluster	0.5715	0.4965	0.5367	0.1147	0.4802	0.0793	0.4378	0.1159	0.2167	0.3106
Iron and steel industrial cluster	0.1631	0.1965	0.1786	0.2399	0.1868	0.8718	0.1899	0.8022	0.2977	0.4206
Petroleum industrial cluster	0.1275	0.1459	0.1360	0.2582	0.1524	0.1986	0.1371	0.1923	0.1821	0.1711
Nonferrous metal cluster	0.0674	0.0911	0.0784	0.1058	0.0821	0.1980	0.0834	0.1863	0.1625	0.1314
Construction industrial cluster	0.1938	0.2515	0.2205	0.3343	0.2358	1.0296	0.2717	0.9522	0.4101	0.5151
Textile industrial cluster	0.4627	0.3685	0.4190	0.1006	0.3763	0.1049	0.5043	0.1456	0.4006	0.2990
Chemistry industrial cluster	0.0900	0.0891	0.0896	0.0592	0.0855	0.0953	0.1755	0.1035	0.1476	0.1016
Administration industrial cluster	0.0157	0.0260	0.0205	0.5819	0.0957	0.0153	0.0107	0.0148	0.0388	0.0583

5.2.4.5   Analysis on production induced coefficient of each industrial cluster

(1) Planting industrial cluster.

We can see from Table 5.46 that as far as production induced coefficient of final demand is concerned, crop cultivation has very high production induced degree. Industries with higher production induced degree are livestock and livestock products and grain mill products, vegetable oil and forage.

For production induced coefficient of consumption, the degree is very high for crop cultivation. Industries with higher production induced degree are livestock and livestock products and grain mill products, vegetable oil and forage.

For production induced coefficient of investment, industries with higher total investment induced coefficient are agriculture, forestry, animal husbandry and fishing machinery, crop cultivation and livestock and livestock products; investment has a very high impulse effect on development of these industries.

Production induced coefficient of export has the highest effect on crop cultivation, and the industries with the higher production induced effect are livestock and livestock products.

(2) Electronics industrial cluster.

We can see from Table 5.47 that those industries whose production induced coefficient of final demand is higher are respectively wholesale and retail trade, electronic element and device, electronic appliances, other social services and other electronic and communication equipment; final demand has quite an impulsive effect on the development of these industries.

Industries whose production induced coefficient of consumption is higher are respectively wholesale and retail trade, electronic element and device, electronic appliances and other social services; consumption has quite an impulsive effect on development of these industries.

Table 5.46   Production induced coefficient of planting industrial cluster

Items	Total consumption	Total investment	Net export	Total demand
Crop cultivation	0.6661	0.4431	0.6367	0.6412
Livestock and livestock products	0.3222	0.3430	0.2187	0.3185
Other agriculture products	0.0444	0.0031	0.0257	0.0390
Natural gas products	0.0083	0.0058	0.0296	0.0092
Grain mill products, vegetable oil and forage	0.2076	0.1951	0.1816	0.2048
Sugar refining	0.0185	0.0079	0.0232	0.0177
Slaughtering, meat processing, eggs and dairy products	0.0510	0.0174	0.1429	0.0528
Other food products	0.1493	0.0667	0.1427	0.1404
Wines, spirits and liquors	0.0735	0.0300	0.0179	0.0658
Nonalcoholic beverage	0.0355	0.0150	0.0596	0.0348
Tobacco products	0.0699	0.0456	0.0494	0.0662
Hemp textiles	0.0022	0.0049	0.0327	0.0042
Other textiles	0.0008	0.0217	0.1460	0.0113
Chemical fertilizers	0.1131	0.0848	0.1295	0.1111
Chemical pesticides	0.0227	0.0176	0.0558	0.0240
Agriculture, forestry, animal husbandry and fishing machinery	0.0147	0.5191	0.0292	0.0680
Storage and warehousing	0.0007	0.0006	0.0006	0.0007
Technical services for agriculture, forestry, livestock and fishing	0.0181	0.0089	0.0109	0.0167

Industries whose production induced coefficient of investment is higher are respectively wholesale and retail trade, other electronic and communication equipment, electronic element and device and other social services; investment has quite an impulsive effect on development of these industries.

Industries whose production induced coefficient of export is higher are respectively wholesale and retail trade, electronic element and device, electronic computer, electronic appliances and other social services; export has quite an impulsive effect on the development of these industries.

Table 5.47   Production induced coefficient of electronics industrial cluster

Items	Total consumption	Total investment	Net export	Total demand
Glass and glass products	0.1111	0.0875	0.1755	0.1353
Electronic computer	0.1090	0.1320	0.2654	0.1811
Electronic appliances	0.3576	0.1910	0.2413	0.2792
Electronic element and device	0.3618	0.2632	0.6145	0.4558
Other electronic and communication equipment	0.1028	0.6123	0.1686	0.2160
Cultural and office equipment	0.0325	0.0203	0.0734	0.0483
Wholesale and retail trade	1.6658	1.2135	1.0371	1.3161
Other social services	0.3196	0.2285	0.2160	0.2592
Scientific research	0.0802	0.0061	0.0061	0.0355

(3) Commerce industrial cluster.

We can see from Table 5.48 that industries whose production induced coefficient of final demand is higher are respectively wholesale and retail trade, paper and products, tobacco products, printing and record medium reproduction, real estate and electronic appliances; final demand has quite an impulsive effect on the development of these industries.

Industries whose production induced coefficient of consumption is higher are respectively wholesale and retail trade, paper and products, tobacco products, real estate and printing and record medium reproduction; consumption has quite an impulsive effect on development of these industries.

Table 5.48   Production induced coefficient of commerce industrial cluster

Items	Total consumption	Total investment	Net export	Total demand
Forestry	0.0264	0.1383	0.0611	0.0491
Logging and transport of timber and bamboo	0.0075	0.0023	0.0075	0.0068
Nonalcoholic beverage	0.1313	0.0688	0.0756	0.1096
Tobacco products	0.4026	0.2346	0.1542	0.3199
Paper and products	0.4730	0.4796	0.4146	0.4594
Printing and record medium reproduction	0.2259	0.2285	0.2288	0.2269
Cultural goods	0.0134	0.0093	0.0237	0.0154
Chemical pesticides	0.0449	0.0589	0.0687	0.0525
Rubber products	0.0358	0.0344	0.1706	0.0690
Electronic appliances	0.1936	0.1889	0.3193	0.2241
Storage and warehousing	0.0115	0.0139	0.0144	0.0126
Wholesale and retail trade	0.9519	1.1870	1.2280	1.0499
Real estate	0.2675	0.2587	0.1020	0.2254
Recreational services	0.0164	0.0148	0.0705	0.0296
Culture and arts, radio, film and television services	0.1072	0.0008	0.0126	0.0704

Industries whose production induced coefficient of investment is higher are respectively wholesale and retail trade, paper and products, real estate, tobacco products and printing and record medium reproduction; investment has quite an impulsive effect on development of these industries.

Industries whose production induced coefficient of export is higher are respectively wholesale and retail trade, paper and products, electronic appliances and printing and record medium reproduction; and export has quite an impulsive effect on the development of these industries.

(4) Diet industrial cluster.

We can see from Table 5.49 that industries whose production induced coefficient of final demand is higher are respectively crop cultivation, grain mill products, vegetable oil and forage and other food products; final demand has quite an impulsive effect on the development of these industries.

Industries whose production induced coefficient of consumption is higher are respectively crop cultivation, grain mill products, vegetable oil and forage and other food products; consumption has quite an impulsive effect on development of these industries.

Industries whose production induced coefficient of investment is higher are respectively crop cultivation, grain mill products, vegetable oil and forage and other food products; investment has quite an impulsive effect on development of these industries.

Industries whose export induced coefficient is higher are respectively crop cultivation, slaughtering, meat processing, eggs and dairy products, grain mill products, vegetable oil and forage, prepared fish and seafood and other food products; export has quite an impulsive effect on the development of these industries.

Table 5.49 Production induced coefficient of diet industrial cluster

Items	Total consumption	Total investment	Net export	Total demand
Crop cultivation	0.7353	0.8027	0.5696	0.7274
Fishery	0.1216	0.0794	0.1262	0.1197
Grain mill products, vegetable oil and forage	0.2251	0.3472	0.1997	0.2297
Slaughtering, meat processing, eggs and dairy products	0.1127	0.0713	0.2489	0.1199
Prepared fish and seafood	0.0361	0.0317	0.1880	0.0463
Other food products	0.1936	0.1725	0.1748	0.1912
Wines, spirits and liquors	0.0973	0.0744	0.0320	0.0916
Nonalcoholic beverage	0.0470	0.0375	0.0699	0.0481
Tobacco products	0.0900	0.1153	0.0597	0.0892
Eating and drinking places	0.0736	0.0044	0.1058	0.0722
Technical services for agriculture, forestry, livestock and fishing	0.0196	0.0139	0.0103	0.0186

(5) Iron and steel industrial cluster.

We can see from Table 5.50 that industries whose production induced coefficient of final demand is higher are respectively steel processing, construction, metal products, other special industrial equipment and electricity production and supply; final demand has quite an impulsive effect on the development of these industries.

Table 5.50 Production induced coefficient of iron and steel industrial cluster

Type	Total consumption	Total investment	Net export	Total demand
Coal mining and processing	0.2782	0.1688	0.1846	0.1757
Ferrous ore mining	0.0499	0.1431	0.1295	0.1372
Coking	0.0211	0.0628	0.1168	0.0641
Iron-smelting	0.0406	0.1089	0.1833	0.1101
Steel-smelting	0.0584	0.1743	0.1206	0.1645
Steel processing	0.2824	0.9575	0.5637	0.8956
Alloy iron smelting	0.0216	0.0671	0.0894	0.0661
Metal products	0.3619	0.5735	0.6567	0.5675
Other special industrial equipment	0.0725	0.2401	0.1903	0.2279
Bicycle	0.1725	0.0018	0.0640	0.0151
Electricity production and supply	0.4661	0.2032	0.1649	0.2149
Steam and hot water production and supply	0.0125	0.0050	0.0019	0.0052
Gas production and supply	0.0709	0.0042	0.0026	0.0077
Water production and supply	0.0857	0.0144	0.0062	0.0177
Construction	0.0051	0.8988	0.0222	0.7937

Industries whose production induced coefficient of consumption is higher are respectively electricity production and supply, metal products, steel processing and coal mining and processing; consumption has quite an impulsive effect on development of these industries.

Industries whose production induced coefficient of investment is higher are respectively steel processing, construction, metal products, and other special industrial equipment and electricity production and supply; investment has quite an impulsive effect on development of these industries.

Industries whose production induced coefficient of export is higher are respectively metal products, steel processing, other special industrial equipment, coal mining and processing, iron-smelting and electricity production and supply; export has quite an impulsive effect on the development of these industries.

(6) Petroleum industrial cluster.

We can see from Table 5.51 that industries whose production induced coefficient of final demand is higher are respectively petroleum refining, public services, electricity production and supply and crude petroleum products; final demand has quite an impulsive effect on the development of these industries.

Table 5.51   Production induced coefficient of petroleum industrial cluster

Items	Total consumption	Total investment	Net export	Total demand
Crude petroleum products	0.1883	−0.0206	0.4307	0.2295
Petroleum refining	0.4975	0.4254	0.3999	0.4670
Railroad transport equipment	0.0116	0.6186	0.0259	0.0692
Electricity production and supply	0.3540	0.0276	0.1081	0.2642
Railway freight transport	0.0646	0.1124	0.0720	0.0707
Highway freight transport	0.0790	0.1964	0.1047	0.0958
Pipeline transport	0.0039	0.0014	0.0067	0.0044
Water freight transport	0.0104	0.0161	0.1055	0.0344
Railway passenger transport	0.0587	0.0021	0.0040	0.0402
Highway passenger transport	0.1528	0.0026	0.0040	0.1027
Water passenger transport	0.0312	0.0007	0.0395	0.0305
Public services	0.4912	0.0076	0.1967	0.3754
Social welfare	0.0482	0.0002	0.0001	0.0320

Industries whose production induced coefficient of consumption is higher are respectively petroleum refining, public services, electricity production and supply and crude petroleum products; consumption has quite an impulsive effect on development of these industries.

Industries whose production induced coefficient of investment is higher are respectively railroad transport equipment, petroleum refining and highway freight transport; investment has quite an impulsive effect on development of these industries.

Industries whose production induced coefficient of export is higher are respectively crude petroleum products, petroleum refining and public services; export has quite an impulsive effect on the development of these industries.

(7) Nonferrous metal cluster.

We can see from Table 5.52 that is industries whose production induced coefficient of final demand is higher are respectively other electric machinery and equipment, non-ferrous metal smelting and nonferrous metal processing; final demand has quite an impulsive effect on the development of these industries.

Table 5.52   Production induced coefficient of nonferrous metal cluster

Items	Total consumption	Total investment	Net export	Total demand
Nonferrous metal mining	0.0852	0.1702	0.1134	0.1189
Nonferrous metal smelting	0.4985	0.5908	0.5762	0.5628
Nonferrous metal processing	0.4613	0.4797	0.6013	0.5472
Other electric machinery and equipment	1.0732	1.3773	0.9584	1.0673
Scrap and waste	0.0137	0.0161	0.0158	0.0154
Telecommunication	0.8157	0.0259	0.1473	0.2636

Industries whose production induced coefficient of consumption is higher are respectively other electric machinery and equipment, telecommunication, nonferrous metal smelting, and nonferreus metal processing; consumption has quite an impulsive effect on development of these industries.

Industries whose production induced coefficient of investment is higher are respectively other electric machinery and equipment, nonferrous metal smelting, and nonferrous metal processing; investment has quite an impulsive effect on development of these industries.

Industries whose production induced coefficient of export is higher are respectively other electric machinery and equipment, nonferrous metal processing, and nonferrous metal smelting; export has quite an impulsive effect on the development of these industries.

(8) Construction industrial cluster.

We can see from Table 5.53 that industries whose production induced coefficient of final demand is higher are respectively construction, steel processing, metal products, other electric machinery and equipment, bricks, tiles, lime and lightweight building materials and cement; final demand has quite an impulsive effect on the development of these industries.

Industries whose production induced coefficient of consumption is higher are respectively steel processing, metal products, bricks, tiles, lime and lightweight building materials, furniture and products of wood, bamboo, cane, palm, straw, etc., other electric machinery and equipment, geological prospecting and water conservancy; consumption has quite an impulsive effect on development of these industries.

Table 5.53   Production induced coefficient of construction industrial cluster

Items	Total consumption	Total investment	Net export	Total demand
Nonmetal minerals and other ore mining	0.1264	0.1684	0.0944	0.1573
Logging and transport of timber and bamboo	0.0195	0.0196	0.0165	0.0193
Sawmills and fiberboard	0.1880	0.0688	0.1368	0.0859
Furniture and products of wood, bamboo, cane, palm, straw, etc.	0.2558	0.0414	0.1641	0.0722
Cement	0.1071	0.2493	0.0396	0.2159
Cement and asbestos products	0.0572	0.1155	0.0103	0.0998
Bricks, tiles, lime and lightweight building materials	0.2866	0.2730	0.0684	0.2534
Pottery, china and earthenware	0.0532	0.0583	0.0844	0.0605
Fireproof products	0.0594	0.0914	0.0835	0.0879
Steel processing	0.6206	0.7283	0.9301	0.7396
Metal products	0.4410	0.3658	0.6354	0.3997
Other electric machinery and equipment	0.2392	0.2650	0.5454	0.2913
Instruments, meters and other measuring equipment	0.0564	0.0436	0.1699	0.0575
Construction	0.0546	0.9651	0.0302	0.7922
Storage and warehousing	0.0006	0.0025	0.0006	0.0022
Telecommunication	0.1797	0.0763	0.1130	0.0889
Geological prospecting and water conservancy	0.2191	0.0193	0.0007	0.0344

Industries whose production induced coefficient of investment is higher are respectively construction, steel processing, metal products, bricks, tiles, lime and lightweight building materials, other electric machinery and equipment and cement; investment has quite an impulsive effect on development of these industries.

Industries whose production induced coefficient of export is higher are respectively steel processing, metal products and other electric machinery and equipment; export has quite an impulsive effect on the development of these industries.

(9) Textile industrial cluster.

We can see from Table 5.54 that is industries whose production induced coefficient of final demand is higher are respectively crop cultivation, wearing apparel and cotton textiles; final demand has quite an impulsive effect on the development of these industries.

Table 5.54   **Production induced coefficient of textile industrial cluster**

Items	Total consumption	Total investment	Net export	Total demand
Crop cultivation	0.7809	0.4200	0.1991	0.5660
Cotton textiles	0.1484	0.3538	0.4625	0.2652
Wool textiles	0.0822	0.1562	0.1901	0.1226
Hemp textiles	0.0079	0.0148	0.0198	0.0122
Silk textiles	0.0713	0.1595	0.1365	0.0992
Knitted mills	0.0333	0.0635	0.2684	0.1107
Other textiles	0.0485	0.1296	0.1577	0.0899
Wearing apparel	0.2441	0.2420	0.3858	0.2891
Chemical fibers	0.0583	0.1719	0.1943	0.1109
Rubber products	0.0196	0.0246	0.1016	0.0462

Industries whose production induced coefficient of consumption is higher are respectively crop cultivation and wearing apparel; consumption has quite an impulsive effect on development of these industries.

Industries whose production-induced coefficient of investment is higher are respectively crop cultivation, cotton textiles and wearing apparel; investment has quite an impulsive effect on development of these industries.

Industries whose production-induced coefficient of export is higher are respectively cotton textiles, wearing apparel and knitted mills; export has quite an impulsive effect on the development of these industries.

(10) Chemistry industrial cluster.

We can see from Table 5.55 that is industries whose production induced coefficient of final demand is higher are respectively plastic products and other chemical products; final demand has quite an impulsive effect on the development of these industries.

Table 5.55   **Production induced coefficient of chemistry industrial cluster**

Items	Total consumption	Total investment	Net export	Total demand
Other chemical products	0.4696	0.6188	0.5600	0.5608
Chemical fibers	0.1177	0.2251	0.1834	0.1838
Plastic products	0.9682	0.5611	0.7041	0.7088
Other manufacturing products	0.5522	0.0762	0.2866	0.2749

Industries whose production induced coefficient of consumption is higher are respectively plastic products and other manufacturing products; consumption has quite an impulsive effect on development of these industries.

Industries whose production induced coefficient of investment is higher are respectively other chemical products and plastic products; investment has quite an impulsive effect on development of these industries.

Industries whose production-induced coefficient of export is higher are respectively plastic products and other chemical products; export has quite an impulsive effect on the development of these industries.

(11) Administration industrial cluster.

We can see from Table 5.56 that industries whose production induced coefficient of final demand is higher are respectively public administration and other sectors and hotels; final demand has quite an impulsive effect on the development of these industries.

Industries whose production induced coefficient of consumption is higher is respectively public administration and other sectors and hotels; consumption has quite an impulsive effect on development of these industries.

The only industry whose production induced coefficient of investment is bigger is aircraft; investment has quite an impulsive effect on development of this industry.

Industries whose production-induced coefficient of export is higher are respectively tourism, air passenger transport, hotels and aircraft; export has quite an impulsive effect on the development of these industries.

Table 5.56 **Production induced coefficient of administration industrial cluster**

Items	Total consumption	Total investment	Net export	Total demand
Aircraft	0.0362	1.0024	0.2266	0.0966
Post services	0.0393	0.0000	0.0151	0.0353
Rail way passenger transport	0.0731	0.0000	0.0935	0.0717
Air passenger transport	0.0862	0.0000	0.4394	0.1152
Hotels	0.1863	0.0000	0.3109	0.1896
Tourism	0.0586	0.0000	0.5495	0.1015
Social welfare	0.0172	0.0000	0.0001	0.0149
Public administration and other sectors	0.8643	0.0000	0.0097	0.7467

5.2.4.6 Analysis on dependence degree coefficient of final demand of whole industrial clusters in China

Dependence degree coefficients of final demand for each industrial cluster are shown in Table 5.57. In general, there is highest dependence degree on consumption demand for industrial clusters in China in 1997, whose average dependence degree is 0.495, and whose average dependence degrees on investment and export are respectively 0.342 and 0.163. Industrial clusters can be divided into consumption dependence type, investment dependence type and export dependence type.

Table 5.57 **Dependence degree coefficient of final demand of industrial cluster in China**

Types of industrial clusters	Rural resident consumption	Town resident consumption	Resident consumption	Government consumption	Total consumption	Fixed capital	Inventory increase	Total investment	Export
Planting industrial cluster	0.417	0.303	0.720	0.020	0.740	0.095	0.053	0.148	0.112
Electronics industrial cluster	0.179	0.189	0.368	0.052	0.420	0.312	0.029	0.341	0.238
Commerce industrial cluster	0.229	0.222	0.452	0.053	0.505	0.267	0.039	0.306	0.189
Diet industrial cluster	0.419	0.315	0.735	0.024	0.759	0.081	0.051	0.131	0.110
Iron and steel industrial cluster	0.088	0.092	0.180	0.038	0.218	0.654	0.016	0.670	0.112
Petroleum industrial cluster	0.170	0.168	0.338	0.099	0.437	0.366	0.029	0.395	0.168
Nonferrous metal cluster	0.117	0.137	0.254	0.053	0.307	0.476	0.023	0.498	0.195
Construction industrial cluster	0.086	0.096	0.182	0.043	0.225	0.631	0.019	0.650	0.125
Textile industrial cluster	0.353	0.243	0.596	0.022	0.618	0.111	0.061	0.171	0.211
Chemistry industrial cluster	0.202	0.173	0.375	0.038	0.413	0.296	0.062	0.358	0.229
Administration industrial cluster	0.061	0.088	0.149	0.656	0.806	0.083	0.007	0.089	0.105

Industrial clusters belonging to consumption dependence type are respectively administration industrial cluster, diet industrial cluster, planting industrial cluster, textile industrial cluster and commerce industrial cluster, with dependence degree coefficients of consumption

> 0.500. As far as total consumption is concerned, dependence degree of resident consumption of industrial clusters in China (average dependence degree is 0.395) is far higher than dependence degree of government consumption (average dependence degree is 0.100). As far as resident consumption is concerned, dependence degree of rural resident consumption (average dependence degree is 0.211) is higher than that of town resident consumption (average dependence degree is 0.184) for industrial clusters in China.

Clusters belonging to investment dependence type are respectively iron and steel industrial cluster, construction industrial cluster, nonferrous metal cluster, petroleum industrial cluster and chemistry industrial cluster, with dependence degree coefficients of consumption > 0.342.

Clusters belonging to export dependence type are respectively electronics industrial cluster, chemistry industrial cluster, textile industrial cluster, nonferrous metal cluster, commerce industrial cluster and petroleum industrial cluster, whose dependence degree coefficients of consumption > 0.163.

5.2.4.7   Analysis on dependence degree coefficient of final demand of each industrial cluster

(1) Planting industrial cluster.

We can see from Table 5.58 that dependence degree of the planting industrial cluster on consumption demand is highest, with average dependence degree of 0.757; average dependence degrees of the planting industrial cluster on investment and export are respectively 0.115 and 0.128.

Table 5.58   Dependent degree coefficient of final demand of planting industrial cluster

Items	Total consumption	Total investment	Export
Crop cultivation	0.8712	0.0719	0.0569
Livestock and livestock products	0.8486	0.1121	0.0393
Other agriculture products	0.9541	0.0082	0.0377
Natural gas products	0.7504	0.0660	0.1837
Grain mill products, vegetable oil and forage	0.8501	0.0991	0.0508
Sugar refining	0.8781	0.0467	0.0752
Slaughtering, meat processing, eggs and dairy products	0.8106	0.0343	0.1551
Other food products	0.8923	0.0495	0.0582
Wines, spirits and liquors	0.9370	0.0474	0.0156
Nonalcoholic beverage	0.8568	0.0450	0.0982
Tobacco products	0.8855	0.0717	0.0427
Hemp textiles	0.4340	0.1212	0.4449
Other textiles	0.0626	0.1993	0.7382
Chemical fertilizers	0.8538	0.0794	0.0668
Chemical pesticides	0.7911	0.0761	0.1329
Agriculture, forestry, animal husbandry and fishing machinery	0.1813	0.7942	0.0245
Storage and warehousing	0.8647	0.0832	0.0520
Technical services for agriculture, forestry, livestock and fishing	0.9071	0.0556	0.0373

Except for hemp textiles, agriculture, forestry, animal husbandry and fishing machinery and other textiles, most industries in the planting industrial cluster are highly dependent on consumption, with dependence degree on consumption demand above 0.757. Industries which are dependent on investment demand are agriculture, forestry, animal husbandry and fishing machinery, other textiles and hemp textiles, with dependence degree on investment demand above 0.115. Industries which are dependent on export are other textiles, hemp textiles, natural gas products, slaughtering, meat processing, eggs and dairy products and chemical pesticides, with dependence degree on export demand above 0.128.

(2) Electronics industrial cluster.

We can see from Table 5.59 that the average dependence degree of the electronics industrial cluster on consumption demand is 0.415; the average dependence degrees of the electronics industrial cluster on investment and export are respectively 0.145 and 0.440.

Table 5.59   Dependent degree coefficient of final demand of electronics industrial cluster

Items	Total consumption	Total investment	Export
Glass and glass products	0.3262	0.1072	0.5666
Electronic computer	0.2391	0.1208	0.6400
Electronic appliances	0.5090	0.1134	0.3776
Electronic element and device	0.3154	0.0957	0.5889
Other electronic and communication equipment	0.1892	0.4699	0.3409
Cultural and office equipment	0.2669	0.0696	0.6635
Wholesale and retail trade	0.5029	0.1528	0.3443
Other social services	0.4899	0.1461	0.3640
Scientific research	0.8969	0.0283	0.0748

Industries which depend on consumption demand are scientific research, electronic appliances, wholesale and retail trade and other social services, with dependence degree on consumption demand above 0.415. Industries which depend on investment demand are other electronic and communication equipment, wholesale and retail trade and other social services, with dependence degree on investment demand above 0.145. Industries which depend on export are cultural and office equipment, electronic computer, electronic element and device glass and glass products, with dependence degree on export demand above 0.440.

(3) Commerce industrial cluster.

We can see from Table 5.60 that dependence degree of commerce industrial cluster on consumption demand is highest, with average dependence degree of 0.598; average dependence degrees of the commerce industrial cluster on investment and export are respectively 0.114 and 0.289.

Table 5.60   Dependent degree coefficient of final demand of commerce industrial cluster

Items	Total consumption	Total investment	Export
Forestry	0.3368	0.3551	0.3081
Logging and transport of timber and bamboo	0.6868	0.0419	0.2713
Nonalcoholic beverage	0.7502	0.0791	0.1707
Tobacco products	0.7882	0.0925	0.1193
Paper and products	0.6449	0.1317	0.2234
Printing and record medium reproduction	0.6234	0.1270	0.2496
Cultural goods	0.5436	0.0758	0.3806
Chemical pesticides	0.5352	0.1414	0.3234
Rubber products	0.3251	0.0629	0.6120
Electronic appliances	0.5410	0.1063	0.3527
Storage and warehousing	0.5760	0.1401	0.2840
Wholesale and retail trade	0.5679	0.1426	0.2895
Real estate	0.7432	0.1448	0.1120
Recreational services	0.3470	0.0631	0.5899
Culture and arts, radio, film and television services	0.9542	0.0014	0.0443

Industries which depend on demand are culture and arts, radio, film and television services, tobacco products, nonalcoholic beverage, real estate, logging and transport of timber and bamboo, paper and products and printing and record medium reproduction, with dependence degree on consumption demand above 0.598. Industries which depend on investment demand are forestry, real estate, wholesale and retail trade, chemical pesticides, storage and warehousing, paper and products and printing and record medium reproduction, with dependence degree on investment demand above 0.114. Industries which are dependent on export are rubber products, recreational services, cultural goods, electronic appliances, chemical pesticides, forestry and wholesale and retail trade, with dependence degree on export demand above 0.289.

(4) Diet industrial cluster.

We can see from Table 5.61 that dependence degree of diet industrial cluster on consumption demand is highest, with average dependence degree of 0.867; average dependence degrees of the diet industrial cluster on investment and export are respectively 0.044 and 0.089.

Table 5.61   Dependent degree coefficient of final demand of diet industrial cluster

Items	Total consumption	Total investment	Export
Crop cultivation	0.8882	0.0577	0.0540
Fishery	0.8925	0.0347	0.0728
Grain mill products, vegetable oil and forage	0.8609	0.0791	0.0600
Slaughtering, meat processing, eggs and dairy products	0.8257	0.0311	0.1432
Prepared fish and seafood	0.6841	0.0358	0.2801
Other food products	0.8897	0.0472	0.0631
Wines, spirits and liquors	0.9334	0.0425	0.0241
Nonalcoholic beverages	0.8589	0.0408	0.1003
Tobacco products	0.8861	0.0676	0.0462
Eating and drinking places	0.8956	0.0032	0.1012
Technical services for agriculture, forestry, livestock and fishing	0.9230	0.0389	0.0381

Except for grain mill products, vegetable oil and forage, nonalcoholic beverage, slaughtering, meat processing, eggs and dairy products and prepared fish and seafood, the other industries in the diet industrial cluster are highly dependent on consumption demand, with dependence degree on consumption demand above 0.867. Industries which depend on investment demand are grain mill products, vegetable oil and forage, tobacco products, crop cultivation and other food products, with dependence degree on investment demand above 0.044. Industries which are dependent on export are prepared fish and seafood, slaughtering, meat processing, eggs and dairy products, eating and drinking places and nonalcoholic beverages, with dependence degree on export demand above 0.089.

(5) Iron and steel industrial cluster.

We can see from Table 5.62 that dependence degree of with iron and steel industrial cluster on investment demand is highest, with average dependence degree of 0.805; average dependence degrees of the iron and steel industrial cluster on consumption and export are respectively 0.124 and 0.071.

Except for water production and supply, gas production and supply and bicycle, the other industries in the iron and steel industrial cluster are highly dependent on investment, with dependence degree on investment demand above 0.805. Industries which depend on consumption demand are bicycle, gas production and supply, water production and supply and steam and hot water production and supply, with dependence degree on consumption demand above 0.124. Industries which are dependent on export are bicycle, coking, iron-smelting, alloy iron smelting and metal products, with dependence degree on export demand above 0.071.

(6) Petroleum industrial cluster.

We can see from Table 5.63 that dependence degree of the petroleum industrial cluster on consumption demand is highest, with average dependence degree of 0.670; average dependence degrees of the Petroleum industrial cluster on investment and export are respectively 0.099 and 0.231.

Table 5.62   Dependent degree coefficient of final demand of iron and steel industrial cluster

Items	Total consumption	Total investment	Export
Coal mining and processing	0.0852	0.8464	0.0684
Ferrous ore mining	0.0196	0.9190	0.0614
Coking	0.0177	0.8637	0.1186
Iron-smelting	0.0199	0.8718	0.1083
Steel-smelting	0.0191	0.9332	0.0477
Steel processing	0.0170	0.9421	0.0410
Alloy iron-smelting	0.0176	0.8944	0.0880
Metal products	0.0343	0.8904	0.0753
Other special industrial equipment	0.0171	0.9286	0.0543
Bicycle	0.6160	0.1076	0.2764
Electricity production and supply	0.1167	0.8334	0.0499
Steam and hot water production and supply	0.1298	0.8467	0.0235
Gas production and supply	0.4942	0.4842	0.0216
Water production and supply	0.2603	0.7169	0.0228
Construction	0.0003	0.9978	0.0018

Table 5.63   Dependent degree coefficient of final demand of petroleum industrial cluster

Items	Total consumption	Total investment	Export
Crude petroleum products	0.5448	−0.0080	0.4632
Petroleum refining	0.7074	0.0812	0.2114
Railroad transport equipment	0.1109	0.7968	0.0923
Electricity production and supply	0.8897	0.0093	0.1010
Railway freight transport	0.6069	0.1417	0.2514
Highway freight transport	0.5476	0.1827	0.2697
Pipeline transport	0.5960	0.0276	0.3764
Water freight transport	0.2005	0.0417	0.7577
Railway passenger transport	0.9705	0.0046	0.0249
Highway passenger transport	0.9880	0.0023	0.0097
Water passenger transport	0.6787	0.0021	0.3192
Public services	0.8688	0.0018	0.1294
Social welfare	0.9983	0.0006	0.0011

Industries which depend on consumption demand are social welfare, highway passenger transport, railway passenger transport, electricity production and supply, public services, petroleum refining and water passenger transport, with dependence degree on consumption demand above 0.670. Industries which depend on investment demand are railroad transport equipment, highway freight transport and railway freight transport, with dependence degree on investment demand above 0.099. Industries which are dependent on export are water freight transport, crude petroleum products, pipeline transport, water passenger transport, highway freight transport and railway freight transport, with dependence degree on export demand above 0.231.

(7) Nonferrous metal cluster.

We can see from Table 5.64 that dependence degree of the nonferrous metal industrial cluster on export is highest, with average dependence degree of 0.544. Average dependence degrees of the nonferrous metal industrial cluster on consumption and investment are respectively 0.261 and 0.195.

Only telecommunication depends on consumption demand with dependence degree on consumption demand above 0.261. Industries which depend on investment demand are nonferrous metal mining, other electric machinery and equipment, nonferrous metal smelting and scrap and waste, with dependence degree on investment demand above 0.195. Industries which are dependent on export are nonferrous metal processing , scrap and waste, nonferrous metal smelting and nonferrous metal mining, with dependence degree on export demand above 0.544.

Table 5.64   Dependent degree coefficient of final demand of nonferrous metal cluster

Items	Total consumption	Total investment	Export
Nonferrous metal mining	0.1510	0.2891	0.5598
Nonferrous metal smelting	0.1867	0.2122	0.6011
Nonferrous metal processing	0.1777	0.1772	0.6451
Other electric machinery and equipment	0.2120	0.2608	0.5272
Scrap and waste	0.1869	0.2112	0.6019
Telecommunication	0.6522	0.0199	0.3280

(8) Construction industrial cluster.

We can see from Table 5.65 that dependence degree of the construction industrial cluster on investment demand is highest, with average dependence degree of 0.776; average dependence degrees of the construction industrial cluster on consumption and export are, respectively, 0.119 and 0.105.

Industries which depend on consumption demand are geological prospecting and water conservancy, furniture and products of wood, bamboo, cane, palm, straw, etc., sawmills and fiberboard and telecommunication, with dependence degree on consumption demand above 0.119. Industries which depend on investment demand are construction, storage and warehousing, cement and asbestos products, cement, bricks, tiles, lime and lightweight building materials, nonmetal minerals and other ore mining, fireproof products, logging and transport of timber and bamboo, steel processing and pottery, china and earthenware, with dependence degree on investment demand above 0.776. Industries which are dependent on export are instruments, meters and other measuring equipment, furniture and products of wood, bamboo, cane, palm, straw, etc., other electric machinery and equipment, sawmills and fiberboard, metal products, pottery, china and earthenware, telecommunication and steel processing, with dependence degree on export demand above 0.105.

Table 5.65   Dependent degree coefficient of final demand of construction industrial cluster

Items	Total consumption	Total investment	Export
Nonmetal minerals and other ore mining	0.0686	0.8703	0.0610
Logging and transport of timber and bamboo	0.0866	0.8266	0.0868
Sawmills and fiberboard	0.1869	0.6512	0.1619
Furniture and products of wood, bamboo, cane, palm, straw, etc.	0.3027	0.4661	0.2312
Cement	0.0424	0.9390	0.0187
Cement and asbestos products	0.0489	0.9405	0.0105
Bricks, tiles, lime and lightweight building materials	0.0966	0.8759	0.0274
Pottery, china and earthenware	0.0751	0.7831	0.1418
Fireproof products	0.0577	0.8457	0.0966
Steel processing	0.0717	0.8005	0.1279
Metal products	0.0943	0.7441	0.1616
Other electric machinery and equipment	0.0702	0.7395	0.1903
Instruments, meters and other measuring equipment	0.0837	0.6160	0.3002
Construction	0.0059	0.9902	0.0039
Storage and warehousing	0.0220	0.9483	0.0297
Telecommunication	0.1727	0.6981	0.1292
Geological prospecting and water conservancy	0.5435	0.4546	0.0019

(9) Textile industrial cluster.

We can see from Table 5.66 that dependence degree of the textile industrial cluster on export is highest, with average dependence degree of 0.514; average dependence degrees of the textile industrial cluster on consumption and investment are respectively 0.396 and 0.090.

Table 5.66 **Dependent degree coefficient of final demand of textile industrial cluster**

Items	Total consumption	Total investment	Export
Crop cultivation	0.8276	0.0602	0.1122
Cotton textiles	0.3355	0.1082	0.5562
Wool textiles	0.4019	0.1034	0.4947
Hemp textiles	0.3858	0.0978	0.5164
Silk textiles	0.4309	0.1304	0.4387
Knitted mills	0.1803	0.0465	0.7731
Other textiles	0.3236	0.1170	0.5595
Wearing apparel	0.5064	0.0679	0.4257
Chemical fibers	0.3153	0.1258	0.5589
Rubber products	0.2547	0.0432	0.7021

Industries which depend on consumption demand are crop cultivation, wearing apparel, silk textiles and wool textiles, with dependence degree on consumption demand above 0.396. Industries which depend on investment demand are silk textiles, chemical fibers, other textiles, cotton textiles, wool textiles and hemp textiles, with dependence degree on investment demand above 0.090. Industries which are dependent on export are knitted mills, rubber products, other textiles, chemical fibers, cotton textiles and hemp textiles, with dependence degree on export demand above 0.514.

(10) Chemistry industrial cluster.

We can see from Table 5.67 that dependence degree of the chemistry industrial cluster on export is highest, with average dependence degree of 0.607; average dependence degrees of the chemistry industrial cluster on consumption and investment are, respectively, 0.184 and 0.209.

Table 5.67 **Dependent degree coefficient of final demand of chemistry industrial cluster**

Items	Total consumption	Total investment	Export
Other chemical products	0.1267	0.2719	0.6013
Chemical fibers	0.0969	0.3019	0.6012
Plastic products	0.2067	0.1951	0.5982
Other manufacturing products	0.3040	0.0683	0.6277

Industries which depend on consumption demand are other manufacturing products and plastic products, with dependence degree on consumption demand above 0.184. Industries which depend on investment demand are chemical fibers and other chemical products, with dependence degree on investment demand above 0.209. The only industries that are dependent on export have other manufacturing products with dependence on the export demand above 0.607.

(11) Administration industrial cluster.

We can see from Table 5.68 that dependence degree of the administration industrial cluster on consumption demand is highest, with average dependence degree of 0.769; average dependence degrees of the administration industrial cluster on investment and export are, respectively, 0.057 and 0.174.

Industries which depend on consumption demand are social welfare, public administration and other sectors, post services, railway passenger transport and hotels, with dependence degree on consumption demand above 0.769. Only aircraft depends on investment demand, and its dependence degree on investment demand is 0.459. Industries which are dependent on export are tourism, air passenger transport and aircraft, whose dependence degree on export demand above 0.174.

Table 5.68   Dependent degree coefficient of final demand of administration industrial cluster

Items	Total consumption	Total investment	Export
Aircraft	0.3233	0.4590	0.2177
Post services	0.9603	0	0.0397
Railway passenger transport	0.8791	0	0.1209
Air passenger transport	0.6459	0	0.3541
Hotels	0.8479	0	0.1521
Tourism	0.4977	0	0.5023
Social welfare	0.9992	0	0.0008
Public administration and other sectors	0.9988	0	0.0012

5.3   ANALYSIS OF BENEFITS OF INDUSTRIAL CLUSTERS IN CHINA

5.3.1   Analysis of benefits of whole industrial clusters

We can see from Table 5.69 that as far as value added rate is concerned, the administration industrial cluster has the highest total benefit. Industrial clusters which have higher total benefits are respectively commerce industrial cluster, petroleum industrial cluster, textile industrial cluster, diet industrial cluster, planting industrial cluster and electronics industrial cluster. The chemistry industrial cluster has the lowest total benefit.

As far as cost output rate is concerned, commerce industrial cluster has highest cost output benefit. Industrial clusters which have higher cost output benefit are respectively petroleum industrial cluster, nonferrous metal cluster and electronics industrial cluster. The administration industrial cluster has the lowest cost output benefit.

As far as intermediate input-output rate is concerned, the commerce industrial cluster has lowest consumption of intermediate use in formation of unit product; it has highest intermediate input-output benefit. Industrial clusters which have higher intermediate input-output benefit are respectively petroleum industrial cluster, diet industrial cluster, administration industrial cluster, textile industrial cluster and planting industrial cluster. The chemistry industrial cluster has the lowest intermediate input-output benefit.

Table 5.69   Benefit table of industrial clusters in China

Types of industrial clusters	Value added rate	Cost output rate	Intermediate input-output rate	Capital consumption output rate	Labor consumption output rate
Planting industrial cluster	0.3194	1.0990	1.8179	30.2058	3.0607
Electronics industrial cluster	0.3130	1.1879	1.6834	20.8704	5.0037
Commerce industrial cluster	0.3639	1.2162	1.8822	12.0816	4.8033
Diet industrial cluster	0.3466	1.1265	1.8642	29.9421	3.1453
Iron and steel industrial cluster	0.2430	1.0984	1.4320	23.3365	5.9090
Petroleum industrial cluster	0.3631	1.2078	1.8743	8.2756	5.7617
Nonferrous metal cluster	0.2347	1.1905	1.5041	12.4966	10.5134
Construction industrial cluster	0.2229	1.1044	1.4186	23.8349	6.3062
Textile industrial cluster	0.3474	1.1007	1.8325	28.9190	3.0470
Chemistry industrial cluster	0.1985	1.1197	1.3520	22.0791	9.2471
Administrate cluster	0.4490	1.0738	1.8578	12.7726	3.1777

As far as consumption of capital factor use is concerned, the planting industrial cluster has lowest consumption in formation of unit product; it has highest capital consumption output benefit. Industrial clusters which have higher capital consumption output benefit are respectively diet industrial cluster, textile industrial cluster, construction industrial cluster, iron and steel industrial cluster, chemistry industrial cluster and electronics industrial cluster. The petroleum industrial cluster has the lowest capital consumption output benefit.

As far as consumption of labor factor use is concerned, the nonferrous metal cluster has lowest consumption in formation of unit product; and it has highest labor consumption output benefit. Industrial clusters which have higher labor consumption output benefit are respectively chemistry industrial cluster, construction industrial cluster, iron and steel industrial cluster and petroleum industrial cluster. The textile industrial cluster has the lowest labor consumption output benefit.

5.3.2 Analysis of benefit of each industrial cluster

5.3.2.1 Planting industrial cluster

We can see from Table 5.70 that as far as value added rate is concerned, crop cultivation has highest total benefit. Industries which have higher total benefits are respectively other agriculture products, tobacco products, natural gas products, technical services for agriculture, forestry, livestock and fishing and livestock and livestock products. Slaughtering, meat processing, eggs and dairy products have the lowest total benefit.

Table 5.70   Benefit table of planting industrial cluster

Items	Value added rate	Cost output rate	Intermediate input-output rate	Capital consumption output rate	Labor consumption output rate
Crop cultivation	0.6496	1.0617	2.8539	41.8321	1.7619
Livestock and livestock products	0.4961	1.0209	1.9843	49.2109	2.1964
Other agriculture products	0.5715	1.0798	2.3339	48.2813	2.0969
Natural gas products	0.5415	1.1710	2.1808	4.4414	5.8723
Grain mill products, vegetable oil and forage	0.1868	1.1163	1.2298	28.1022	21.2336
Sugar refining	0.1867	1.0576	1.2296	19.6136	12.3038
Slaughtering, meat processing, eggs and dairy products	0.1528	1.0549	1.1804	30.7455	14.6565
Other food products	0.3131	1.1783	1.4559	27.1235	8.0007
Wines, spirits and liquors	0.3294	1.2253	1.4913	19.6032	10.5717
Nonalcoholic beverage	0.2966	1.1925	1.4218	27.4631	10.1214
Tobacco products	0.5584	1.9454	2.2646	35.4403	22.6047
Hemp textiles	0.2791	1.0766	1.3872	21.7930	6.1696
Other textiles	0.2589	1.1633	1.3493	24.3926	12.9013
Chemical fertilizers	0.2241	1.0663	1.2888	19.2340	9.0979
Chemical pesticides	0.2299	1.1007	1.2985	22.2232	10.7032
Agriculture, forestry, animal husbandry and fishing machinery	0.2180	1.0785	1.2787	31.6225	8.8026
Storage and warehousing	0.3351	0.9783	1.5041	11.2293	3.7281
Technical services for agriculture, forestry, livestock and fishing	0.5058	1.0636	2.0234	2.8665	10.2925

As far as cost output rate is concerned, tobacco products has highest cost output benefit. Industries which have higher cost output benefit are respectively wines, spirits and liquors, nonalcoholic beverages, other food products, natural gas products and other textiles. Storage and warehousing have the lowest cost output benefit.

As far as intermediate input-output rate is concerned, crop cultivation has lowest consumption of intermediate use in formation of unit product, it has highest intermediate input-output benefit. Industries which have higher intermediate input-output benefit are respectively other agriculture products, tobacco products, natural gas products, technical services for agriculture, forestry, livestock and fishing and livestock and livestock products. Slaughtering, meat processing, eggs and dairy products have the lowest intermediate input-output benefit.

As far as consumption of capital factor use is concerned, livestock and livestock products has lowest consumption in formation of unit product; it has highest capital consumption output benefit. Industries which have higher capital consumption output benefit are respectively other agriculture products, crop cultivation, tobacco products, agriculture, forestry, animal husbandry and fishing machinery, slaughtering, meat processing, eggs and dairy products, grain mill products, vegetable oil and forage, nonalcoholic beverage and other food products. Technical services for agriculture, forestry, livestock and fishing have the lowest capital consumption output benefit.

As far as consumption of labor factor use is concerned, tobacco products have lowest consumption in formation of unit product, and it has highest labor consumption output benefit. Industries which have higher labor consumption output benefit are, respectively, grain mill products, vegetable oil and forage, slaughtering, meat processing, eggs and dairy products, other textiles, sugar refining, chemical pesticides, wines, spirits and liquors, technical services for agriculture, forestry, livestock and fishing and nonalcoholic beverage. Crop cultivation has the lowest labor consumption output benefit.

5.3.2.2   Electronics industrial cluster

We can see from Table 5.71 that as far as value added rate is concerned, wholesale and retail trade has highest total benefit. Industries which have higher total benefits are respectively scientific research, glass and glass products and electronic element and device. Electronic appliances have the lowest total benefit.

As far as cost output rate is concerned, wholesale and retail trade has highest cost output benefit. Industries which have higher cost output benefit are respectively electronic element and device, electronic computer, other electronic and communication equipment and glass and glass products. Scientific research have the lowest cost output benefit.

As far as intermediate input-output rate is concerned, wholesale and retail trade has lowest consumption of intermediate use in formation of unit product; it has highest intermediate input-output benefit. There is only scientific research which has higher intermediate input-output benefit. Electronic appliances have the lowest intermediate input-output benefit.

Table 5.71   Benefit table of electronics industrial cluster

Items	Value added rate	Cost output rate	Intermediate input-output rate	Capital consumption output rate	Labor consumption output rate
Glass and glass products	0.3073	1.1294	1.4436	17.9232	7.3017
Electronic computer	0.2341	1.1425	1.3056	47.1023	11.3531
Electronic appliances	0.1978	1.0908	1.2465	48.4949	10.6500
Electronic element and device	0.2965	1.1693	1.4214	22.5217	9.3217
Other electronic and communication equipment	0.2734	1.1356	1.3763	21.7037	9.2665
Cultural and office equipment	0.1982	1.1025	1.2472	49.9900	11.7348
Wholesale and retail trade	0.5100	1.2500	2.0409	17.8796	3.9354
Other social services	0.2515	1.0562	1.3360	27.6884	6.1652
Scientific research	0.3878	1.0346	1.6334	18.6263	3.3262

As far as consumption of capital factor use is concerned, cultural and office equipment has lowest consumption in formation of unit product; it has highest capital consumption output benefit. Industries which have higher capital consumption output benefit are respectively electronic appliances and electronic computer. Wholesale and retail trade have the lowest capital consumption output benefit.

As far as consumption of labor factor use is concerned, cultural and office equipment has lowest consumption in formation of unit product; it has highest labor consumption output benefit. Industries which have higher labor consumption output benefit are respectively electronic computer, electronic appliances, electronic element and device and other electronic and communication equipment. Scientific research has the lowest labor consumption output benefit.

5.3.2.3　Commerce industrial cluster

We can see from Table 5.72 that as far as value added rate is concerned, real estate has highest total benefit. Industries which have higher total benefits are respectively forestry, logging and transport of timber and bamboo, tobacco products, wholesale and retail trade, culture and arts, radio, film and television services and recreational services. Electronic appliances have the lowest total benefit.

Table 5.72　Benefit table of commerce industrial cluster

Items	Value added rate	Cost output rate	Intermediate input-output rate	Capital consumption output rate	Labor consumption output rate
Forestry	0.7330	1.0608	3.7457	35.2275	1.5449
Logging and transport of timber and bamboo	0.6210	1.3454	2.6386	13.3325	3.4570
Nonalcoholic beverage	0.2966	1.1925	1.4218	27.4631	10.1214
Tobacco products	0.5584	1.9454	2.2646	35.4403	22.6047
Paper and products	0.2937	1.0948	1.4159	24.9750	5.9843
Printing and record medium reproduction	0.3738	1.1799	1.5970	21.4240	5.7242
Cultural goods	0.4109	1.3088	1.6976	45.3515	6.5383
Chemical pesticides	0.2299	1.1007	1.2985	22.2232	10.7032
Rubber products	0.2497	1.1200	1.3328	38.9296	8.5606
Electronic appliances	0.1978	1.0908	1.2465	48.4949	10.6500
Storage and warehousing	0.3351	0.9783	1.5041	11.2293	3.7281
Wholesale and retail trade	0.5100	1.2500	2.0409	17.8796	3.9354
Real estate	0.7591	1.1885	4.1505	2.0830	8.3056
Recreational services	0.4495	1.0965	1.8165	8.0922	4.2033
Culture and arts, radio, film and television services	0.4886	1.0943	1.9554	7.6051	3.6913

As far as cost output rate is concerned, tobacco products has highest cost output benefit. Industries which have higher cost output benefit are respectively logging and transport of timber and bamboo, cultural goods and wholesale and retail trade. Storage and warehousing have the lowest cost output benefit.

As far as intermediate input-output rate is concerned, real estate has lowest consumption of intermediate use in formation of unit product; it has highest intermediate input-output benefit. Industries which have higher intermediate input-output benefit are respectively forestry, logging and transport of timber and bamboo, tobacco products and wholesale and retail trade. Electronic appliances have the lowest intermediate input-output benefit.

As far as consumption of capital factor use is concerned, electronic appliances has lowest consumption in formation of unit product; it has highest capital consumption output benefit. Industries which have higher capital consumption output benefit are respectively cultural goods, rubber products, tobacco products, forestry, nonalcoholic beverage and paper and products. Real estate has the lowest capital consumption output benefit.

As far as consumption of labor factor use is concerned, tobacco products has lowest consumption in formation of unit product; it has highest labor consumption output benefit. Industries which have higher labor consumption output benefit are respectively chemical pesticides, electronic appliances, nonalcoholic beverage, rubber products and real estate. Forestry has the lowest labor consumption output benefit.

5.3.2.4   Diet industrial cluster

We can see from Table 5.73 that as far as value added rate is concerned, crop cultivation has highest total benefit. Industries which have higher total benefits are respectively fishery, tobacco products and technical services for agriculture, forestry, livestock and fishing. Slaughtering, meat processing, eggs and dairy products has lowest total benefit.

Table 5.73   Benefit table of diet industrial cluster

Items	Value added rate	Cost output rate	Intermediate input-output rate	Capital consumption output rate	Labor consumption output rate
Crop cultivation	0.6496	1.0617	2.8539	41.8321	1.7619
Fishery	0.6062	1.0686	2.5394	29.7935	1.9668
Grain mill products, vegetable oil and forage	0.1868	1.1163	1.2298	28.1022	21.2336
Slaughtering, meat processing, eggs and dairy products	0.1528	1.0549	1.1804	30.7455	14.6565
Prepared fish and seafood	0.2510	1.1429	1.3352	21.3739	12.6217
Other food products	0.3131	1.1783	1.4559	27.1235	8.0007
Wines, spirits and liquors	0.3294	1.2253	1.4913	19.6032	10.5717
Non-alcoholic beverage	0.2966	1.1925	1.4218	27.4631	10.1214
Tobacco products	0.5584	1.9454	2.2646	35.4403	22.6047
Eating and drinking places	0.3561	1.1896	1.5529	73.4452	5.4621
Technical services for agriculture, forestry, livestock and fishing	0.5058	1.0636	2.0234	2.8665	10.2925

As far as cost output rate is concerned, tobacco products have the highest cost output benefit. Only wines, spirits and liquors have higher cost output benefit. Slaughtering, meat processing, eggs and dairy products have the lowest cost output benefit.

As far as intermediate input-output rate is concerned, crop cultivation has lowest consumption of intermediate use in formation of unit product; it has highest intermediate input-output benefit. Industries which have higher intermediate input-output benefit are respectively fishery, tobacco products and technical services for agriculture, forestry, livestock and fishing. Slaughtering, meat processing, eggs and dairy products have the lowest intermediate input-output benefit.

As far as consumption of capital factor use is concerned, eating and drinking places has lowest consumption in formation of unit product; it has highest capital consumption output benefit. Industries which have higher capital consumption output benefit are respectively crop cultivation, tobacco products and slaughtering, meat processing, eggs and dairy products. Technical services for agriculture, forestry, livestock and fishing have the lowest capital consumption output benefit.

As far as consumption of labor factor use is concerned, tobacco products has lowest consumption in formation of unit product; it has highest labor consumption output benefit. Industries which have higher labor consumption output benefit are respectively grain mill products, vegetable oil and forage, slaughtering, meat processing, eggs and dairy products and prepared fish and seafood. Crop cultivation has the lowest labor consumption output benefit.

5.3.2.5   Iron and steel industrial cluster

We can see from Table 5.74 that as far as value added rate is concerned, coal mining and processing has highest total benefit. Industries which have higher total benefits are respectively water production and supply, electricity production and supply, ferrous ore mining, other special industrial equipment and steam and hot water production and supply. The bicycle industry has the lowest total benefit.

Table 5.74   **Benefit table of iron and steel industrial cluster**

Items	Value added rate	Cost output rate	Intermediate input-output rate	Capital consumption output rate	Labor consumption output rate
Coal mining and processing	0.5136	1.0748	2.0558	10.6882	2.8538
Ferrous ore mining	0.3367	1.1033	1.5077	12.4721	6.1378
Coking	0.2298	1.0793	1.2983	22.3429	8.9629
Iron-smelting	0.2227	1.0572	1.2865	21.8321	8.1455
Steel smelting	0.2370	1.0412	1.3107	18.0936	7.0320
Steel-processing	0.2095	1.0760	1.2649	23.1549	10.4529
Alloy iron-smelting	0.2138	1.0753	1.2720	34.0788	8.7382
Metal products	0.2334	1.0915	1.3044	29.9068	8.6093
Other special industrial equipment	0.3248	1.1654	1.4810	28.5434	6.7639
Bicycle	0.1882	1.0686	1.2319	44.5235	9.8396
Electricity production and supply	0.4361	1.2394	1.7733	7.4223	9.2393
Steam and hot water production and supply	0.3144	1.0390	1.4586	9.4806	5.8353
Gas production and supply	0.2635	0.9778	1.3578	12.1866	4.8978
Water production and supply	0.5005	1.2080	2.0019	7.1475	5.3079
Construction	0.2875	1.0777	1.4034	60.5960	5.0278

As far as cost output rate is concerned, electricity production and supply has highest cost output benefit. Industries which have higher cost output benefit are, respectively, water production and supply, other special industrial equipment and ferrous ore mining. Gas production and supply have the lowest cost output benefit.

As far as intermediate input-output rate is concerned, coal mining and processing has lowest consumption of intermediate use in formation of unit product; it has highest intermediate input-output benefit. Industries which have higher intermediate input-output benefit, are, respectively water production and supply, electricity production and supply, ferrous ore mining and other special industrial equipment. The bicycle industry has the lowest intermediate input-output benefit.

As far as consumption of capital factor use is concerned, construction has lowest consumption in formation of unit product; it has highest capital consumption output benefit. Industries which have higher capital consumption output benefit are respectively bicycle, alloy iron smelting, metal products, other special industrial equipment and steel processing. Water production and supply have the lowest capital consumption output benefit.

As far as consumption of labor factor use is concerned, steel processing has lowest consumption in formation of unit product; it has highest labor consumption output benefit. Industries which have higher labor consumption output benefit are respectively bicycle, electricity production and supply, coking, alloy iron smelting, metal products and iron-smelting. Coal mining and processing have the lowest labor consumption output benefit.

5.3.2.6   Petroleum industrial cluster

We can see from Table 5.75 that as far as value added rate is concerned, crude petroleum products has highest total benefit. Industries which have higher total benefits are respectively railway freight transport, railway passenger transport, highway passenger transport, highway freight transport and public services. Petroleum refining has the lowest total benefit.

As far as cost output rate is concerned, crude petroleum products has highest cost output benefit. Industries which have higher cost output benefit are respectively railway freight transport, railway passenger transport, electricity production and supply and highway passenger transport. Public services has the lowest cost output benefit.

As far as intermediate input-output rate is concerned, crude petroleum products has lowest consumption of intermediate use in formation of unit product; it has highest intermediate input-output benefit. Industries which have higher intermediate input-output benefit are respectively railway freight transport, railway passenger transport, highway passenger transport and highway freight transport. Petroleum refining has the lowest intermediate input-output benefit.

Table 5.75   Benefit table of petroleum industrial cluster

Items	Value added rate	Cost output rate	Intermediate input-output rate	Capital consumption output rate	Labor consumption output rate
Crude petroleum products	0.7522	1.7111	4.0358	5.0366	7.2411
Petroleum refining	0.2196	1.1739	1.2814	31.6362	25.0780
Railroad transport equipment	0.2627	1.0542	1.3563	20.1302	6.1876
Electricity production and supply	0.4361	1.2394	1.7733	7.4223	9.2393
Railway freight transport	0.6711	1.4565	3.0404	6.5902	4.8565
Highway freight transport	0.5447	1.0677	2.1962	8.0969	2.7956
Pipeline transport	0.4592	1.1799	1.8493	5.0576	9.1720
Water freight transport	0.3572	1.0886	1.5558	8.2779	6.4504
Railway passenger transport	0.6364	1.2414	2.7501	7.4314	3.2534
Highway passenger transport	0.6251	1.2263	2.6672	6.2746	3.5566
Water passenger transport	0.3194	0.9698	1.4694	5.3918	6.0574
Public services	0.4963	0.9504	1.9852	6.4898	2.5360
Social welfare	0.4757	1.0951	1.9071	80.7954	2.6565

As far as consumption of capital factor use is concerned, social welfare has lowest consumption in formation of unit product; it has highest capital consumption output benefit. Industries which have higher capital consumption output benefit are respectively petroleum refining and railroad transport equipment. Crude petroleum products have the lowest capital consumption output benefit.

As far as consumption of labor factor use is concerned, petroleum refining has lowest consumption in formation of unit product; it has highest labor consumption output benefit. Industries which have higher labor consumption output benefit are respectively electricity production and supply, pipeline transport and crude petroleum products. Public services have the lowest labor consumption output benefit.

5.3.2.7   Nonferrous metal cluster

We can see from Table 5.76 that as far as value added rate is concerned, scrap and waste has highest total benefit. Only telecommunication has higher total benefits. Nonferrous metal processing has lowest total benefit.

Table 5.76   Benefit table of nonferrous metal cluster

Items	Value added rate	Cost output rate	Intermediate input-output rate	Capital consumption output rate	Labor consumption output rate
Nonferrous metal mining	0.3611	1.0950	1.5653	11.7313	5.2881
Nonferrous metal smelting	0.1953	1.0714	1.2427	29.6849	10.5294
Nonferrous metal processing	0.1538	1.0671	1.1817	28.6403	17.8756
Other electric machinery and equipment	0.1971	1.0723	1.2454	30.4347	10.3289
Scrap and waste	1.0000				
Telecommunication	0.5929	1.3228	2.4564	3.9556	10.4111

As far as cost output rate is concerned, telecommunication has highest cost output benefit. Only nonferrous metal mining has higher cost output benefit. Nonferrous metal processing has the lowest cost output benefit.

As far as intermediate input-output rate is concerned, telecommunication has lowest consumption of intermediate use in formation of unit product; it has highest intermediate input-output benefit. Only nonferrous metal mining has higher intermediate input-output benefit. Nonferrous metal processing has the lowest intermediate input-output benefit.

As far as consumption of capital factor use is concerned, other electric machinery and equipment has lowest consumption in formation of unit product; it has highest capital consumption output benefit. Industries which have higher capital consumption output benefit are nonferrous metal smelting and nonferrous metal processing. Telecommunication has the lowest capital consumption output benefit.

As far as consumption of labor factor use is concerned, nonferrous metal processing has lowest consumption in formation of unit product; it has highest labor consumption output benefit. Only nonferrous metal smelting has higher labor consumption output benefit. Nonferrous metal mining has the lowest labor consumption output benefit.

5.3.2.8 Construction industrial cluster

We can see from Table 5.77 that as far as value added rate is concerned, logging and transport of timber and bamboo has highest total benefit. Industries which have higher total benefits are respectively telecommunication, geological prospecting and water conservancy, nonmetal minerals and other ore mining, bricks, tiles, lime and lightweight building materials, pottery, china and earthenware and instruments, meters and other measuring equipment. Other electric machinery and equipment have the lowest total benefit.

As far as cost output rate is concerned, logging and transport of timber and bamboo has highest cost output benefit. Industries which have higher cost output benefit are respectively telecommunication, geological prospecting and water conservancy, bricks, tiles, lime and lightweight building materials, instruments, meters and other measuring equipment, pottery, china and earthenware and nonmetal minerals and other ore mining. Storage and warehousing have the lowest cost output benefit.

As far as intermediate input-output rate is concerned, logging and transport of timber and bamboo has lowest consumption of intermediate use in formation of unit product; it has highest intermediate input-output benefit. Industries which have higher intermediate input-output benefit are respectively telecommunication, geological prospecting and water conservancy, nonmetal minerals and other ore mining and bricks, tiles, lime and lightweight building materials. Other electric machinery and equipment has lowest intermediate input-output benefit.

As far as consumption of capital factor use is concerned, construction has lowest consumption in formation of unit product; it has highest capital consumption output benefit. Industries which have higher capital consumption output benefit are respectively furniture and products of wood, bamboo, cane, palm, straw, other electric machinery and equipment, metal products, instruments, meters and other measuring equipment, cement and asbestos products and bricks, tiles, lime and lightweight building materials. Telecommunication has the lowest capital consumption output benefit.

As far as consumption of labor factor use is concerned, steel processing has lowest consumption in formation of unit product; it has highest labor consumption output benefit. Industries which have higher labor consumption output benefit are respectively telecommunication, other electric machinery and equipment, metal products, cement, cement and asbestos products, fireproof products, furniture and products of wood, bamboo, cane, palm, straw, and sawmills and fiberboard. Logging and transport of timber and bamboo have the lowest labor consumption output benefit.

Table 5.77   Benefit table of construction industrial cluster

Items	Value added rate	Cost output rate	Intermediate input-output rate	Capital consumption output rate	Labor consumption output rate
Nonmetal minerals and other ore mining	0.4009	1.1452	1.6692	21.2665	4.4035
Logging and transport of timber and bamboo	0.6210	1.3454	2.6386	13.3325	3.4570
Sawmills and fiberboard	0.3051	1.1363	1.4390	21.6961	7.1923
Furniture and products of wood, bamboo, cane, palm, straw	0.2630	1.1202	1.3569	41.3470	7.6006
Cement	0.2404	1.0515	1.3165	15.3760	7.9125
Cement and asbestos products	0.2675	1.1150	1.3653	29.1522	7.6881
Bricks, tiles, lime and lightweight building materials	0.3956	1.2011	1.6547	24.6343	5.3294
Pottery, china and earthenware	0.3778	1.1711	1.6071	18.6387	5.6181
Fireproof products	0.2846	1.1187	1.3978	20.7573	7.6758
Steel processing	0.2095	1.0760	1.2649	23.1549	10.4529
Metal products	0.2334	1.0915	1.3044	29.9068	8.6093
Other electric machinery and equipment	0.1971	1.0723	1.2454	30.4347	10.3289
Instruments, meters and other measuring equipment	0.3631	1.1798	1.5700	29.6027	5.6532
Construction	0.2875	1.0777	1.4034	60.5960	5.0278
Storage and warehousing	0.3351	0.9783	1.5041	11.2293	3.7281
Telecommunication	0.5929	1.3228	2.4564	3.9556	10.4111
Geological prospecting and water conservancy	0.5657	1.2560	2.3024	4.7629	6.5850

5.3.2.9   Textile industrial cluster

We can see from Table 5.78 that as far as value added rate is concerned, crop cultivation has highest total benefit. Industries which have higher total benefits are wearing apparel and wool textiles. Chemical fibers have the lowest total benefit.

Table 5.78   Benefit table of textile industrial cluster

Items	Value added rate	Cost output rate	Intermediate input-output rate	Capital consumption output rate	Labor consumption output rate
Crop cultivation	0.6496	1.0617	2.8539	41.8321	1.7619
Cotton textiles	0.2828	1.1016	1.3943	15.7493	7.8698
Wool textiles	0.3347	1.2452	1.5031	24.9684	10.2310
Hemp textiles	0.2791	1.0766	1.3872	21.7930	6.1696
Silk textiles	0.2486	1.1237	1.3309	33.2034	9.2208
Knitted mills	0.2813	1.1381	1.3913	24.1517	8.4392
Other textiles	0.2589	1.1633	1.3493	24.3926	12.9013
Wearing apparel	0.3613	1.1474	1.5656	32.9056	4.9406
Chemical fibers	0.2328	1.1142	1.3035	17.1643	13.8790
Rubber products	0.2497	1.1200	1.3328	38.9296	8.5606

As far as cost output rate is concerned, wool textiles have highest cost output benefit. Industries which have higher cost output benefit are other textiles, wearing apparel and knitted mills. Crop cultivation has the lowest cost output benefit.

As far as intermediate input-output rate is concerned, crop cultivation has lowest consumption of intermediate use in formation of unit product; it has highest intermediate input-output benefit. Only wearing apparel has a higher intermediate input-output benefit. Chemical fibers have the lowest intermediate input-output benefit.

As far as consumption of capital factor use is concerned, crop cultivation has lowest consumption in formation of unit product; it has highest capital consumption output benefit. Industries which have higher capital consumption output benefit are rubber products, silk textiles and wearing apparel. Cotton textiles has the lowest capital consumption output benefit.

As far as consumption of labor factor use is concerned, chemical fibers has lowest consumption in formation of unit product; it has highest labor consumption output benefit. Industries which have higher labor consumption output benefit are respectively other textiles, wool textiles, silk textiles, rubber products and knitted mills. Crop cultivation has the lowest labor consumption output benefit.

5.3.2.10 Chemistry industrial cluster

We can see from Table 5.79 that as far as value added rate is concerned, other manufacturing products has highest total benefit. Only other chemical products has higher total benefits. Chemical fibers has the lowest total benefit.

Table 5.79 Benefit table of chemistry industrial cluster

Items	Value added rate	Cost output rate	Intermediate input-output rate	Capital consumption output rate	Labor consumption output rate
Other chemical products	0.2568	1.1099	1.3456	26.6490	8.3139
Chemical fibers	0.2328	1.1142	1.3035	17.1643	13.8790
Plastic products	0.2433	1.1113	1.3215	21.1817	10.4254
Other manufacturing products	0.3428	1.1678	1.5216	25.4943	6.2540

As far as cost output rate is concerned, other manufacturing products has highest cost output benefit. Only chemical fibers has higher cost output benefit. Other chemical products have the lowest cost output benefit.

As far as intermediate input-output rate is concerned, other manufacturing products has lowest consumption of intermediate use in formation of unit product; it has highest intermediate input-output benefit. Only other chemical products has higher intermediate input-output benefit. Chemical fibers have the lowest intermediate input-output benefit.

As far as consumption of capital factor use is concerned, other chemical products has lowest consumption in formation of unit product; it has highest capital consumption output benefit. Only other manufacturing products has higher capital consumption output benefit. Chemical fibers have the lowest capital consumption output benefit.

As far as consumption of labor factor use is concerned, chemical fibers has lowest consumption in formation of unit product; it has highest labor consumption output benefit. Only plastic products has higher labor consumption output benefit. Other manufacturing products have the lowest labor consumption output benefit.

5.3.2.11 Administration industrial cluster

We can see from Table 5.80 that as far as value added rate is concerned, railway passenger transport has highest total benefit. Industries which have higher total benefits are hotels and social welfare. Tourism has the lowest total benefit.

As far as cost output rate is concerned, hotels has highest cost output benefit. Industries which have higher cost output benefit are railway passenger transport and air passenger transport. Public administration and other sectors have the lowest cost output benefit.

As far as intermediate input-output rate is concerned, railway passenger transport has lowest consumption of intermediate use in formation of unit product; it has highest intermediate input-output benefit. Only hotels have higher intermediate input-output benefit. Tourism has the lowest intermediate input-output benefit.

Table 5.80   Benefit table of administration industrial cluster

Items	Value added rate	Cost output rate	Intermediate input-output rate	Capital consumption output rate	Labor consumption output rate
Aircraft	0.4239	1.1208	1.7357	15.6439	3.9659
Post services	0.4144	1.0404	1.7077	16.6033	3.1712
Railway passenger transport	0.6364	1.2414	2.7501	7.4314	3.2534
Air passenger transport	0.4000	1.2077	1.6666	7.3719	10.8284
Hotels	0.6142	1.4620	2.5918	9.0672	5.3232
Tourism	0.2204	1.0766	1.2827	42.2792	7.9598
Social welfare	0.4757	1.0951	1.9071	80.7954	2.6565
Public administration and other sectors	0.4510	1.0060	1.8216	14.1423	2.6710

As far as consumption of capital factor use is concerned, social welfare has lowest consumption in formation of unit product; it has highest capital consumption output benefit. Only tourism has higher capital consumption output benefit. Air passenger transport has the lowest capital consumption output benefit.

As far as consumption of labor factor use is concerned, air passenger transport has lowest consumption in formation of unit product; it has highest labor consumption output benefit. Industries which have higher labor consumption output benefit are tourism and hotels. Social welfare has the lowest labor consumption output benefit.

CHAPTER 6

Developing Policies of Industrial Clusters in China

There are 96 industries in 11 industrial clusters in China in 1997, accounting for 77.4% of all industries. Total output amount is up to 16603.591 billion yuan, accounting for 83.1% of total output of all industries. Industrial clusters have been important sub-systems in the Chinese national economic system, the working status of which directly affects healthy development of the Chinese national economy. Structural and functional analyses of about industrial clusters provide abundant raw data and new thinkings. During the period of the 11th five-year planning, in order to carry out scientific development, propel the transition of economic growth, and facilitate healthy and sustainable development of the national economy, the policies of industrial clusters should be stressed in the development of industrial policies in China. This chapter discusses focal points of industrial cluster policies and cut-in points and path of industrial cluster policies.

6.1 FOCAL POINTS OF INDUSTRIAL CLUSTER POLICIES

6.1.1 Industrial development should specially attach importance to developing policies of industrial clusters

We have entered into the stage when industrial competitiveness is closely associated with the industrial cluster. As far as the whole country is concerned, industry development is not an abstract concept, and it must be established in concrete regions. The location choice of industrial development with competitiveness to a greater degree depends on the condition, size, level and degree of industrial clustering in the region. Our country is in a fast-growing period as a whole, with many industries still having much growth space, but whether this opportunity of growth belongs to one given region is a great uncertainty, especially in those regions with a definite development foundation. In recent years, some industries are experiencing rapid development, while industrial centers are relocating dependence degree of industrial competitiveness on industrial clusters are being strengthened. Therefore, in drawing up the 11th five-year plan of national economy and social development, when we work on regional industrial development planning, industrial development should specifically attach importance to developing policies of the industrial cluster, resolutely get rid of 'big and complete' planning tendency, avoid balanced development of three industries, and avoid being reckless about developing a series of hot industries such as the automobile industry, or high technology industries (Liu, 2003). The emphasis of industry development should focus on different industrial clusters and improve international competitiveness of key industries through the development of industrial clusters.

6.1.2 Developing policies of industrial clusters should give prominence to characteristics of dynamic changes of industrial clusters

Our country has entered a period of rapid change in economic structures and economic growth; economic link patterns among industries have become more complicated; development and evolution of industrial clusters show remarkable and continuous changes. Because of this developing policies of industrial clusters should give prominence to the characteristics of dynamic changes of industrial clusters, and put forward developing policies which promote the development of different industrial clusters.

With the dynamic development of industrial clusters, if the ratio of interindustry input of an industrial cluster to total output becomes large with time, the industrial cluster is termed an interindustry input-oriented industrial cluster; otherwise, it is called a primary input-oriented industrial cluster. If the ratio of interindustry output of industrial clusters to total output becomes large with time, the industrial cluster is termed an interindustry output-oriented industrial cluster; otherwise, it is called a final demand-oriented industrial cluster. Developing policies of industrial clusters should be adjusted in a timely manner according to the status of dynamic changes of industrial clusters in order to facilitate the development of different types of industrial clusters. For interindustry input and interindustry output-oriented industrial clusters, the emphasis of developing policies is to ensure stability and expedite economic links among industries within an industrial cluster, thereby improving the development of industrial cluster specialization. For primary input and final demand-oriented industrial clusters, the core of development policies is to facilitate the course (memberships of industrial clusters) of 'localization' of industries outside an industrial cluster, and enforce economic links among industrial clusters or between industrial clusters and nonindustrial clusters, thereby strengthening the ability of the industrial system to resist exterior shocks.

6.1.3   Adjustment of regional industrial structures should follow developing laws of industrial clusters

Industrial development follows the 'Industrial Cluster Law'; that is, service industries have the tendency to concentrate in given regions (metropolitan centers). Adapting to the industrial distribution, general manufacturing industries sensitive to costs tend to amass in the periphery regions of metropolises. This arrangement can not only reduce production costs of manufacturing industries but also bring into play roles of metropolises in reducing transaction costs of manufacturing industries. Industrial Cluster Law causes structures of counties (cities) around metropolises to take on normal patterns of an 'excessively high' proportion of second industries.

In the last five-year plans, industrial development planning of each region has assumed the industrial structure of 'third-second-first' as an object of high-grade industrial structures. For the sake of realizing the object, practices which violate regional industrial division of labor and spatial clustering laws are taken in the adjustment of industrial structures, unilaterally emphasizing developing third industries. The economic structure of a county (city) is a nonindependent open system and an important part of a national economic system; its form of industrial structure is mainly dependent on economic links among industries and the resulting requests of competitive advantages of an industrial cluster. If we mechanically apply the evolving law of three industries for the adjustment of economic structure of the whole country to nonindependent economic structures of counties (cities), and require to realize ring upon ring the form of industrial structure of 'Third-Second-First', then this creates separate artificially intrinsic economic links among industries, and cancels economic cooperation and complementarity among regions (Liu, 2004). The results are necessarily similarly developing patterns of all regions and identical industrial structures of all regions, thereby inducing a new round of construction of redundant projects and overlapping investments.

6.1.4   The core of industrial cluster policies is industrial competition policy

Industrial cluster policies are advantaged tools by which governments adjust industrial structures. They should include two aspects: industrial structure policies and industrial organization policies. The main contents of industrial structure policies include choosing key industrial clusters, central industries and nodal industries, mostly embodying relationships among industries within industrial clusters or among industrial clusters; the main contents of industrial organization policies are to create fair competition environments for enterprises, mostly embodying relationships among enterprises within industries. The former mostly embodies strategic orientations of governments, whose purposes are to speed up resource transfer among industries and change of industry structures by strengthening economic links among industries or among industrial clusters. The latter mostly embodies essential the characteristics of a market economy and speeds up resource transfer among enterprises within industries and change of industry structures by strengthening economic links among enterprises. By always striving for perfection and the development of a socialist market economy system, the strength of industrial structure policies will decline gradually, while that of industrial organization policies will be gradually reinforced. The core of industrial cluster policies is necessarily to facilitate competitive policies taking competitions as objects.

6.2 CUT-IN POINTS AND THE PATH OF INDUSTRIAL CLUSTER POLICIES

6.2.1 Vigorously increase technological matrix levels of industrial clusters

Increasing technological matrix levels is in fact reducing levels of material consumptions and driving technological progresses of industries. Increasing technological matrix levels will not only advance transitions of economic growth modes, but also enhance transforming capacities of industrial structures, thereby facilitating constant, stable and rapid growth of a national economy. When we develop energetically industrial clusters of low material consumptions such as commerce industrial cluster, petroleum industrial cluster and textile industrial cluster, we should pay attention simultaneously to driving industrial clusters of high material consumptions such as chemistry industrial cluster, construction industrial cluster and iron and steel industrial cluster, reducing their levels of material consumptions. Industrial clusters such as administration industrial cluster, chemistry industrial cluster and nonferrous metal industrial cluster with high technological matrix levels become important industrial clusters which propel adjustments of industrial structures and transitions of economic growth patterns for China, while industrial clusters such as construction industrial cluster, iron and steel industrial cluster and planting industrial cluster with low technological matrix levels should be speeded up in technological transformation, containing increasing nuclear technologies and key technologies, advancing technological equipment and product quality, improving the levels of technology introduction, and absorption and assimilation, boosting technological development and technological innovation of industries.

6.2.2 Take great pains to upgrade the whole function of industrial structure relationships

Increasing the technological economy structure relationship degrees of industrial clusters means enlarging transaction sizes, increasing transaction links and augmenting transaction quantities among industries within industrial clusters. This is a universal tendency in the period of accelerated development of industrialization, it is an objective need of specialization development of industrial clusters, and it is an inevitable tendency to interfuse the development among industrial clusters and speed up their integration as well as. Upgrading the whole function of industrial structure relationships can improve efficiency of resourses on the one hand and can create external economy effects on the other hand. It can not only facilitate coordination of relationship among industries within industrial clusters, but also facilitate coordination of relationship among industrial clusters, thereby intensifying the whole function of industrial structure relationships of industrial clusters and ensuring stable economic growth. Upgrading the whole function of industrial structure relationships of industrial clusters should pay special attention to the industrial clusters which are influenced greatly by demands; for example, the administration industrial cluster is influenced greatly by consumption demands, the iron and steel industrial cluster is influenced greatly by investment demands, industrial clusters such as textile industrial cluster, commerce industrial cluster and electronics industrial cluster are influenced greatly by net exports. When industrial structures are adjusted in China, these industrial clusters should be energetically supported, the objective being to further enhance the whole quality of industrial clusters.

6.2.3   Energetically develop demand pull type and demand dependence type industrial clusters

Industrial clusters pulled greatly by demands, for example, industrial clusters affected greatly by consumption pulls such as planting industrial cluster and diet industrial cluster, industrial clusters affected greatly by investment pulls such as construction industrial cluster and iron and steel industrial cluster, and industrial clusters affected greatly by export pulls such as construction industrial cluster and textile industrial cluster should pay special attention to right pull actions of industrial clusters for a national economy. Industrial clusters depending greatly on demand degree, for example, industrial clusters depending on consumptions such as administration industrial cluster, diet industrial cluster and planting industrial cluster, industrial clusters depending on investments such as iron and steel industrial cluster and construction industrial cluster, industrial clusters depending on exports as electronics industrial cluster, chemistry industrial cluster and textile industrial cluster, should pay special attention to adapting to the demands, having the courage to create new demands and exert driving actions of these industrial clusters for a national economy.

6.2.4   Central and nodal industries should be key industries in the adjustment and upgrade of industrial structures

An industrial cluster is in fact a functional network connecting structure modes which integrate many different industries through close economic links among industries surrounding central industries. Thus, an industrial cluster is a concept covering wide industry scope, contains many united industries and is a physical distribution chain, service chain, information chain and capital chain connecting the upper reaches and the lower reaches industries. According to Porter's value chain theory, value actions of industries in industrial clusters are links that may produce added value; the interrelationship and value chain of any industry exists in the value system of industrial clusters and is composed of interconnecting value chains of many industries. According to the 'bullwhip effect' theory of Drezner et al. (1996), misinterpreting information on industries has an enlarged phenomenon from lower reaches to upper reaches in a value chain. This effect exhibits uncertainty of demands, which in turn results from distortions during transferring information on demand changes and then proceeds to be the result of distortion and enlargement. In view of the special positions of central industries in a value chain of an industrial cluster, adjustment and upgrade of industrial structure of a region should be implemented through enforcing demand management of central industries and reducing or eliminating the 'bullwhip effect.' Industrial chains are accelerated to climb toward high ends and new competitive advantages are created by increasing value added to central industries.

An industrial system of a region is actually composed of different industrial clusters, and the tie of links among industrial clusters is nodal industries in industrial clusters. Types and quantities of nodal industries reflect the open degrees of industrial clusters on the one hand, and reflect capacities of regional industry systems against external shocks on the other hand.

Therefore, that reinforcing cultivation and support of nodal industries of a region in the industrial structure adjustment of the region not only make for increasing competitiveness of industrial clusters of a region, but also favor stable and continuous development of the industrial system of the region. In the aspect of the 11th-year industrial development planning, what deserves attention is those industries which are both central industries and nodal industries. These industries play important roles in industrial development and adjustment and optimizing of industrial structures. In industrial clusters in China, such industries are crop cultivation, grain mill products, vegetable oil and forage, wholesale and retail trade, steel processing, social welfare and construction.

6.2.5   Develop the driving industry and leading industry greatly

At the present time, the industrial development of China is at the stage of relying mainly on quantitive explosion; that is enlarging output size of industries makes for elevating economic general quantity. On the basis of this judgment, as far as industrial development of China is concerned, pushing industrial sectors of industrial clusters should be developed vigorously because these sectors can increase supplies, stimulate demands and impel the development of other industries. Pulling industrial sectors of industrial clusters should also be developed vigorously because these sectors will produce more ultimate products which will increase demands for products of other industrial sectors and increase more size of development of other industrial sections by effecting production processes. Furthermore, industrial development of China should attach specific importance to and vigorously develop these industrial sectors which are greater pushing actions and pulling actions, for example, grain mill products, vegetable oil and forage, fishery and sugar refining in planting industrial cluster; electronic element and device, electronic computer and electronic appliances in electronics industrial cluster; other textiles, cotton textiles and hemp textiles in textile industrial cluster, coking, iron-smelting, steel-smelting, electricity production and supply and steel processing in iron and steel industrial cluster; pipeline transport and petroleum refining in petroleum industrial cluster; nonferrous metal smelting, nonferrous metal processing and other electric machinery and equipment in nonferrous metal industrial cluster; telecommunication and cement and asbestos products in construction industrial cluster; logging and transport of timber and bamboo, paper and products and printing and record medium reproduction in commerce industrial cluster; crop cultivation, grain mill products, vegetable oil and forage and fishery in diet industrial cluster; and air passenger transport and railway passenger transport in administration industrial cluster.

References

Campbell J. 1970. The relevance of input-output analysis and digraph concepts to growth pole theory. Unpublished PhD dissertation, University of Washington.

Cheng Jianquan. 2001. Urban systematic engineering. Wuhan: Wuhan University Press.

Czamanski S. 1971. Some empirical evidence of the strengths of linkages between groups of related industries in urban-regional complexes. Papers, Regional Science Association, 27: 136-150.

Czamanski S. 1974. Study of clustering of industries. Halifax, Nova Scotia: Dalhousie University.

De Bresson C, Hu Xiaoping. 1999. Identifying clusters of innovation activity: a new approach and a toolbox. *In*: Boosting innovation: the cluster approach. Paris: OECD.

Drezner Z, Ryan J K, Simchi-Levi D. 1996. Quantifying the bullwhip effect: the impact of forecasting, lead time and information. Working Paper. Northwestern University.

Kier L B, Hall L H. 1986. Molecular connectivity in structure-activity analysis. Letchworth: Research Studies Press.

Liu Shijin. 2003 Industrial agglomerations and their meanings for economic development. Industrial economic dynamic of China.

Liu Zhibiao. 2004. Some strategic considerations about adjustments of industrial structures of China. The Second Session of Symposia on Cooperation and Development Forum of Middle and Small Businesses in 'Chang Triangle.'

Malmberg A, Solvell O, Zander I. 1996. Spatial clustering, local accumulation of knowledge and firm competitiveness. Geografiska Annaler, 78B: 85-97.

McCarty H H, Hock J C, Knos D S. 1956. The measurement of association in industrial geography. Iowa City: State University of Iowa Press.

Roepke H, Adams D, Wisemen R. 1974. A new approach to the identification of industrial complex using input-output data. Journal of Regional Science, 14: 15-29.

Shirasu T. 1980. Changes in the structure of industrial clusters in a growth economy: a case study of Japan, 1960-1970. New York: Cornell University Press.

Slater P B. 1977. The determination of groups of functionally integrated industries in the United States using a 1967 inter-industry flow table. Empirical Economics, 2: 1-9.

Streit M E. 1969. Spatial association and economic linkages between industries. Journal of Regional Science, 9: 177-188.

Su Dongshui. 2000. Industrial economics. Beijing: Higher Education Press.

Wang Chaorui. 1997. Graph theory. Beijing: Beijing University of Science and Technology Press.

Xu Lu, Hu Changyu. 2000. Graph theory of applying chemistry. Beijing: Science Press.

Zhang Wenchang, Jin Fengjun, Rong Chaohe et al. 1992. Spatial transport links. Beijing: Chinese Railway Press.

Zhu Yingming. 2003. On industrial agglomeration. Beijing: Economic Science Press.

Zhu Yingming. 2004. Analysis on economic space of urban agglomerations. Beijing: Science Press.

APPENDIX 1

Industrial Composing and Codes of Chinese National Economy

Codes	Industries	Codes	Industries	Codes	Industries
001	Crop cultivation	032	Paper and products	063	Metalworking machinery
002	Forestry	033	Printing and record medium reproduction	064	Other general industrial machinery
003	Livestock and livestock products	034	Cultural goods	065	Agriculture, forestry, animal husbandry and fishing machinery
004	Fishery	035	Toys, sporting and athletic and recreation products	066	Other special industrial equipment
005	Other agriculture products	036	Petroleum refining	067	Railroad transport equipment
006	Coal mining and processing	037	Coking	068	Motor vehicles
007	Crude petroleum products	038	Raw chemical materials	069	Ship building
008	Natural gas products	039	Chemical fertilizers	070	Aircraft
009	Ferrous ore mining	040	Chemical pesticides	071	Bicycle
010	Nonferrous metal mining	041	Organic chemical products	072	Other transport machinery
011	Salt mining	042	Chemical products for daily use	073	Generators
012	Nonmetal minerals and other ore mining	043	Other chemical products	074	Household electric appliances
013	Logging and transport of timber and bamboo	044	Medical and pharmaceutical products	075	Other electric machinery and equipment
014	Grain mill products, vegetable oil and forage	045	Chemical fibers	076	Electronic computer
015	Sugar refining	046	Rubber products	077	Electronic appliances
016	Slaughtering, meat processing, eggs and dairy products	047	Plastic products	078	Electronic element and device
017	Prepared fish and seafood	048	Cement	079	Other electronic and communication equipment
018	Other food products	049	Cement and asbestos products	080	Instruments, meters and other measuring equipment
019	Wines, spirits and liquors	050	Bricks, tiles, lime and lightweight building materials	081	Cultural and office equipment
020	Nonalcoholic beverage	051	Glass and glass products	082	Maintenance and repair of machinery and equipment
021	Tobacco products	052	Pottery, china and earthenware	083	Arts and crafts products
022	Cotton textiles	053	Fireproof products	084	Other manufacturing products
023	Wool textiles	054	Other nonmetallic mineral products	085	Scrap and waste
024	Hemp textiles	055	Iron-smelting	086	Electricity production and supply
025	Silk textiles	056	Steel-smelting	087	Steam and hot water production and supply
026	Knitted mills	057	Steel processing	088	Gas production and supply
027	Other textiles	058	Alloy iron smelting	089	Water production and supply
028	Wearing apparel	059	Nonferrous metal smelting	090	Construction
029	Leather, furs, down and related products	060	Nonferrous metal processing	091	Railway freight transport
030	Sawmills and fiberboard	061	Metal products	092	Highway freight transport
031	Furniture and products of wood, bamboo, cane, palm, straw, etc.	062	Boiler, engines and turbine	093	Pipeline transport

Continued

Codes	Industries	Codes	Industries	Codes	Industries
094	Water freight transport	105	Air passenger transport	116	Sports services
095	Air freight transport	106	Finance	117	Social welfare
096	Transportation not elsewhere classified and auxiliary body	107	Insurance	118	Education services
097	Storage and warehousing	108	Real estate	119	Culture and arts, radio, film and television services
098	Post services	109	Public services	120	Scientific research
099	Telecommunication	110	Resident services	121	General technical service
100	Wholesale and retail trade	111	Hotels	122	Technical services for agriculture, forestry, livestock and fishing
101	Eating and drinking places	112	Tourism	123	Geological prospecting and water conservancy
102	Railway passenger transport	113	Recreational services	124	Public administration and other sectors
103	Highway passenger transport	114	Other social services		
104	Water passenger transport	115	Health services		

APPENDIX 2

Input-output Tables of China in 1997 (Intermediate Input and Use Part)

(Units: ten thousand yuan)

Output / Input	1	2	3	4	5	6	7	8	9	10	11	12	13	14
1	11625925.2	86883.6	16009712	1219357	1435993.6	0	0	0	0	0	0	0	0	25568510
2	79414.5	575835.4	53168.2	6399.2	322608.5	90732.6	13.4	1.8	809.4	937.4	50.9	755.4	393489.7	5860.5
3	851461.3	7347.9	3808535.1	0	0	0	0	0	0	0	0	0	0	429983.8
4	0	0	0	1398660.1	32236.2	147095.5	0	0	0	0	0	0	0	427364.3
5	804703.2	31944.7	479120.3	50258.9	761891.7	622926.8	88086.2	1164	13184.8	32914	22588.9	4964.4	3580.4	6089.5
6	0	7352.6	141857.7	27115.8	44103.6	0	61196.7	5812.7	7268.7	43092.2	0	27815.4	16607.2	7245.9
7	0	0	95841.1	0	0	0	239.1	0	23.4	1270.9	0	69370.6	679.2	1429.9
8	0	0	0	0	0	343.4	0	30077	0	91.1	0	0	0	0
9	0	0	0	0	0	0	0	0	343707.6	3233	0	0	0	0
10	0	0	0	0	0	0	0	0	0	1708743.1	0	0	0	0
11	0	0	0	0	0	0	1079.9	44	24.9	188	8619	0	0	0
12	0	14892.8	9458.1	7291.6	26509.5	28637.8	4245.8	1360.2	1693.6	3754.4	2816.7	999177.8	49.5	27019.8
13	22524.4	15167.4	30841.6	29980.1	54607.2	966663.9	6173.5	527.4	11942	27096.9	408.3	11772	46013.4	462.3
14	1481247.7	0	12513839	2270132.8	39333.3	0	0	0	0	0	0	0	0	3764757.3
15	0	0	0	0	0	0	0	0	0	0	0	0	0	25035.8
16	0	0	0	0	0	0	0	0	0	0	0	0	0	82525
17	0	0	0	0	0	0	0	0	0	0	0	0	0	597714.2
18	0	0	0	0	0	0	0	0	0	0	0	0	0	120212.2
19	24661.8	1545.2	9438	5298.1	8314.3	3936.3	7593.3	0	0	7662.9	0	0	0	0
20	0	0	0	0	0	0	0	0	0	0	0	0	0	0
21	0	0	0	0	0	0	0	0	0	0	0	0	0	0
22	19167.8	1275.6	13672.1	2058.7	58963.2	54116.1	20371.7	2286.3	910.6	20806.7	592.5	7115.4	922.4	57716.8
23	1450.5	27.3	184.7	548.1	17952.6	8348.6	8688.8	506.6	268.3	2841.5	69.2	2813.8	123.9	1390.9
24	46457.1	2004.6	114706	3155.2	93295.5	2270.2	1857.1	123.1	172.1	782.8	404.3	49860.9	109.3	16393.8
25	0	0	622.2	1100	59574.9	4963.1	4089.9	313.6	99.4	1442.9	89.7	1842.5	18.9	2176.7
26	4060.2	864.3	0	0	4978.3	3887.9	2136.1	338.4	819.6	5431.6	80.1	1671.8	279.7	843.1
27	0	0	0	0	81518.3	0	0	0	0	0	0	0	0	0
28	16424.4	5273.6	8642.1	4579.3	54943.4	95673.4	60956.9	5493.6	15455.6	47196.8	3562	41047.2	8243.4	37279.5
29	13773.6	859.1	66779.8	809.9	6554.8	29908.9	25685	1769.1	3441.5	25706	1006	5925.5	1434.5	4009.8
30	21469.4	8812.8	46348.1	40169.7	20502.6	60236.9	21413.8	1979.5	5715	26248.8	875.3	10095.8	4824.5	1649.6
31	82609.7	2736.6	15294.9	29477.5	63911.1	19826.1	7473.4	1053.7	3647	6046.7	559.5	10622.4	1162	1685.5
32	33196.4	4933.4	4592.2	9628.6	72480.7	10915.6	4763	216.8	666	1451.4	70.2	4123.6	142.3	29129
33	15877.3	3671.3	4006.3	6554.7	18687.4	8930.9	8102.3	630.7	1191.9	3846.1	198.2	1870.8	527.1	4336.8
34	46198.4	5877.5	11754	18117.7	7075.2	13632.1	10147.6	834.1	1118.4	2942.9	290.1	2586.8	194.3	1471.4
35	4430	1150.5	1993	3779.9	1272.5	1185.8	0	0	0	2159.7	0	4647.2	0	0
36	1123765.8	53204.8	159668	664480.4	87382.4	306522.3	282424.2	45516.4	85714	132922.3	10055.9	582573.2	64394.2	90913

Continued

Input \ Output	1	2	3	4	5	6	7	8	9	10	11	12	13	14
37	0	0	0	0	0	0	0	0	11854.6	1488.6	0	0	0	0
38	1957.3	200.9	1908.8	1083.3	1501.7	28056.2	84402.3	10091.1	4336.9	128877.1	3663.4	17553.4	306.8	117487.8
39	13707775.9	276179	3385.7	0	0	0	0	0	0	0	0	0	0	14944.2
40	2383211	81885.7	0	0	0	0	0	0	0	0	0	0	0	0
41	15825.2	1567.4	2400.1	2699.2	7285.3	41801.1	95254	42115.9	2381.5	39582.7	923.5	33600.7	1667.2	44572.6
42	8834.6	1484.9	5416.9	2339	2784.4	29045.9	26164.6	1573.3	4241	21894.6	1290.1	20235.4	1302.8	32255.8
43	5577.6	446.3	24943.4	3605.3	3774.5	250148.1	155512.4	1258.2	79333.2	200645.9	4335.1	567622	3028	299442.4
44	16910.5	9815.9	308966.5	52794.3	3128.6	6087.1	35.7	2.7	146.2	1599	492.9	198.7	407.4	148283.4
45	0	0	0	59205.6	3461.2	0	0	0	0	0	0	0	0	0
46	66603.5	4852.9	19102.7	13625.5	13997.7	346416.1	48134.6	4859.9	25803.5	112836.7	33989	50655.2	65078	6447.9
47	949229.5	21347	25927.2	100042.9	40086.9	54026.7	22403.1	1822.4	8247	66390.3	64733	503646.3	3320.6	226864.4
48	74560.2	10276.2	44819.6	21859.6	32833.4	83600.2	35653.1	7570	8464.9	29455.8	2012.3	28285.9	2464.5	2291.4
49	75456.3	6956.4	21146.3	26245.7	10478.5	61060.1	13288.9	2823.8	5823.4	17793.2	463.6	21560.5	166.6	878.8
50	91665	8712.8	84096.8	29739.9	12691.1	73229.5	72632.4	11735.3	7000.6	39072.3	1461.4	22247.7	3564.2	3663.8
51	7373.1	1324.1	3518.6	2825.9	31370.2	33141.2	12087.8	921.7	5217.1	7845.4	898.7	19417.9	1996	4543.9
52	1625.1	348.1	1587.6	508.2	1936.3	47939.3	14935.1	1542.3	7758.7	11977.3	4353	32552.4	2935	14618.3
53	1136.1	318.8	398	391.7	1115.1	6215.9	3276.4	279.5	546.3	535.7	135	3609.8	67.3	142.1
54	1041.9	262.9	446.6	398.5	0	5944.3	7417.9	188.8	1546.5	3753.9	60.7	23724.9	166.3	230
55	0	0	0	0	0	98900	20268.3	3280.4	13473	2307.2	13	1622.5	4.3	0
56	13583.1	0	0	0	16800.1	972783.2	81048.4	15077.5	31270.2	14860.3	427.6	36659	148.6	0
57	0	1187.3	1427	2766	0	13976.5	13380.2	1218.8	10496.1	43647.5	2564	6767.3	1985.1	2913.7
58	0	0	0	0	0	10694.6	5873.8	255	480.8	9015.9	252.3	1583.6	71.8	0
59	0	0	1.5	0	295	93617.4	13574	797.9	1048.5	18945.2	13.4	332.6	14.4	0
60	828.8	24.8	34257.8	330.2	137631	460659.2	112404.5	9569.9	104579	5509.4	149.4	3035.8	87.6	304.5
61	477919.3	37905.3	0	43835.7	0	21201.2	13222.3	1674.4	119.4	232411.2	6443.6	118721.7	46138.8	45093.3
62	0	0	0	0	0	71716.5	10225.4	1063.1	8998.2	4379.3	270.8	5090.8	1927.3	4877.5
63	0	0	0	0	54809.1	483206	215266.8	35389.9	98509.2	34872.8	1424.5	29368.3	600.1	4037.9
64	289943.4	27359.9	44921.5	171586	78773.4	6668.4	904.8	74.9	590.7	205389.9	14518.9	205001.9	46383.4	37553.4
65	1268501.4	48991.8	162690.3	318369.6	7829.6	376710.6	301991.8	44734.9	291367.2	2855.8	1884.2	17187.2	52263.8	0
66	26551.1	0	4048.4	12144	0	7787.3	3534.6	501.3	489.5	96392.1	8662.6	638656.9	17834.2	29151.7
67	0	0	0	0	8229.4	125152	126434.4	16119.5	125493	569.6	9.7	38.7	193.5	0
68	26289	9392	24973.9	13549.4	1630.8	0	0	0	0	58559.2	6829.2	324874.8	53988.8	87196.1
69	5072	219.6	10	528550.1	0	0	0	0	0	0	0	0	0	0
70	0	0	0	0	0	0	0	0	0	0	0	0	0	0
71	3116.6	381.5	360.4	995.1	1110.3	9296	4714.2	748.7	1923.9	10399	459.6	9554.1	1129.6	0
72	19922.7	12312.4	19279.6	31207	12743.4	37878.5	36999.6	3644.8	13332.3	16740.6	1759.9	8842.1	1341.8	1754
73	28589.9	1201.8	15261.8	7213.1	1006.3	6573.5	2997.1	745.7	1410.6	6734	188.6	2103.2	370.5	7788.3
74	10200.7	249.3	777.6	12115.3	11967.9	511989.8	143486.9	13911.9	32004.4	40532.6	4775.2	61826	9498.4	964.5
75	36267.9	6710.7	14086	5850.2	2899.1	7662.3	7388.6	1463.8	1934.1	2532.7	369.1	4529.4	942.4	22864.1
76	725.7	246.5	20.2	477.3	803.4	27758.3	2199.9	639.3	2464.5	2152.5	366.9	17508.4	673.1	1332.7
77	2650.1	296.1	751.7	1977.2	206.7	10761.3	33269.5	2225.6	1174.3	26038.3	259.4	6537.8	34.9	1939.1
78	98	14.1	0.2	73.9	19	15768.1	28689	5321.4	2942.9	6936.8	620.2	10266.6	910	2400.1
79	3374.6	298.4	657.9	4649.9	1407.7	47606.4	82878.8	16214.2	17838.2	31554.6	1725.6	25584.6	1612.9	3401.3
80	2045.6	288	3238.1	1506.7	446.7	0	0	0	0	0	0	0	0	12341.1

Continued

Input \ Output	1	2	3	4	5	6	7	8	9	10	11	12	13	14
81	5138.7	621	538.7	730.7	97.9	16288.1	4683.8	795.3	3058.5	3255	558.1	6674.5	1485.6	1818.2
82	260632.8	18647.6	39674.3	164871.9	36493.3	113636.1	30366.2	2039.5	12095	36109	3284.6	65749.4	3364.4	13576.7
83	161144.8	11406.4	83926.8	29361.3	14822.8	46356.1	28372.5	1888.3	5501	13563.7	1600	21279.3	4474.7	47274.4
84	63055.5	863.8	22061.9	9698.3	3151.6	328267.9	72879.5	8724.8	23349	44436.4	6397.1	81175.6	2196.9	30285.7
85						2459.6	151.5	10.2	44	296.3	58.3	49.9	0.6	238881.5
86	1286665.7	41374.3	143823.3	130680.6	185489.4	1052623.3	437175	57979.1	332978.9	638077.1	61612.7	711373.2	28488.2	236599
87	88.4	38.4	15916.4	1116.2		9904.8	1889.5	345.2	165.1	929.4	734.5	454.4	736.8	2511.1
88						2774.3	160.6	11	128.4	36.4	34.7	56.1	8.7	245.9
89	25562.6	883.2	4736.8	3799.5	2175.6	29939.7	12548.4	1195.2	8554.7	6787.4	1309.4	26234.8	945.3	11920
90	293143.9	20479.2	132738.2	27511.5	15665	54615.9	26746.2	2110.9	11586.1	17209.8	2102.3	33932.9	3377.7	11926.7
91	270958.4	15925.9	98106.7	34560.8	28159.7	132451.4	43848.9	3623.7	37603.8	90023.4	3476.3	326828.8	12671.9	96818.4
92	771358.6	120254.6	330230.4	216376.5	85634.4	191484.3	25420.3	2721.3	72562.4	302681.8	22020.5	354806.6	24518.9	303380.5
93	4145	408.9	1550.8	2384.2	319.5	1136.2	1658.6	951.8	526.9	507.4	37.4	2925.5	238.5	358.7
94	32741.1	1393.7	20377.7	9131.5	2920.6	14427.4	4233.4	446.9	19383.6	13196.1	1038.1	29866	1522.8	18074.5
95	2107.2	221.4	1997.9	3229.9	740.7	1851.4	1055.8	136.9	478.5	655.8	90	1550.2	157.5	813.6
96	97913.3	4667.8	60559.5	20574.2	8774.6	75451.9	11681.3	1238.1	16936.5	95925.1	4455.6	48754.8	4696	87297.3
97	76629.2	2782.1	34646.1	28847	4646.4	114.4	3.7		1.3	493.9	27.8	29	0.5	617.5
98	22021	2480	17294	8600.4	2459.2	82.5	20.8	1.6		5.5	0.9	1.9		31.6
99	34013.2	5711.6	19422.2	14000.1	3211.3	138619	72665.4	4557.7	18613.9	13583.2	1309.3	621190.8	5084	34432
100	1782734.7	112412.6	1845792.2	344955.9	259829.4	630403.1	155208.4	17678.4	83590.9	239388.9	22813.9	429865.6	65678.7	1207997.3
101	70893.7	9820.2	22446.3	14097.8	11928.8	146509	16331.2	1779.2	37987.9	48738.8	6935.7	167118.1	16360.3	69867.6
102	42563	3142.8	17182.3	12340.4	3858.5	145.7	76.4	4.8	19.6	14.3	1.4	20518.1	5.3	36.2
103	79389.7	13803.7	21815.1	17543.9	19758.7							34546.5		
104	98904.9	4852.1	16229.3	43230.7	9259.3	137.9	72.4	4.5	18.5	13.6	1.3	6752.6	5.1	34.3
105	4837.4	558.7	1278.5	2598.3	514	2281.8	1196.4	75	306.4	223.8	21.5	13448.7	83.5	567.5
106	521007.3	54795.9	248192.2	94432	66089.3	177200.5	136620.2	9196	21597.3	105380.8	9989.8	189714.4	23142.5	227839.9
107	88076	9397.9	22588.6	45725.4	6818.8	48928.9	51616.6	3756.2	6669.4	14297.7	1825	28627	8244.3	18392.2
108	32284.5	7479.5	9445.9	6721.2	3086.4	18591.6	574.8	126	521.5	7107.6	183.5	13453.3	594.3	10034.6
109	117362.1	21073.9	32499.7	17060.6	16617.5	107.9	56.5	3.5	14.5	10.6	1	5343.2	4	26.8
110	237478.3	28122.1	30281.3	29889.2	5065	137902.4	55579.9	4141.7	20470.6	32059.3	10112.2	72927.1	37473	52156.8
111	95902.5	8552.3	35118.5	17227.5	28082.6	5130.5	1536.7	95.4	445	4472.2	30.4	77169.6	150.2	1175.4
112												1079.6		
113	511.1	105.3	305.5	193.9	137.5	15445.7	1908.9	209.4	4140.3	5742.5	811.5	19750.9	1902.6	8277.2
114	218578.2	23856.4	14977.6	30066	15949.3	142952.1	7697.3	711.9	3993.2	33896.4	3511.7	20500.7	3703.9	75343.9
115	9873.1	2514.7	3918.3	3419.1		69103.5	2958.9	204.4	2006.1	3375.9	169.6	4341.8	1678.2	1391.7
116														
117														
118	96094.4	10621.5	24083.8	22808.6	6455.5	69489.6	3511	432.9	2804.9	11620.6	1506.9	11142.7	4107.3	6968.1
119	29251.3	3244.7	9002	4171.8	3013.5	567.6	143.3	11.3	293.6	37.8	6.4	7828.5		217.3
120	2297.5	733.9	515.9	1111.9	177.3	4300.3	1773.1	237.7		2121.9	26.7	1994.1	30.2	2015.8
121	94502.6	8850.8	29456.3	38285.3	6874.7	33031.5	5658.3	910.4	5493.8	19378.7	283.5	25873.5	12977.5	9264.1
122	1617084.6	211887	524408.1	267972.5	64048.2									
123	310558.4	13020.1	7886.2	13966	5369.8	144984.6	72889.6	5212	13366.2	35709.6		38391.4		
124														
Total	44900786.1	2204867.2	38419927.5	9006707.2	4825701.2	10834977.2	3772658.2	498945.8	2206917.5	5507278.1	393516.1	8244577.7	1137531.8	35490256.4

Continued

Input \ Output	15	16	17	18	19	20	21	22	23	24	25	26	27	28
1	1722741.6	17319	1905.2	4524737.9	4311710.4	1999860.3	2212267	2441975.3	2505.7	594254.6	130265.3	9094.2	3766038.2	14923.4
2	527.5	933.2	0	62526.3	1668.5	3037.2	137	30.7	0.8	46	2131.8		44.9	2214.4
3	0	9974326.8	0	2009587.1	0	66853.2	0	0	2503391.2	1517.7	1252248.6	65166.1	741966.1	5784.8
4		0	4535951.9	442211.1	0	0	0	0	0	0	0	0	0	0
5	1155.4	888.7	10714	730028.5	7406.6	58446.8	91.2	498	6.5	34.5	75882.6	138.4	55	166.4
6	44305.7	16675.7	3403.3	211314.9	133216.4	17549.7	20267.9	151513.2	42875.4	5008.5	53741.8	68530.6	20222.7	38914.8
7	1031.8	0	0	0	0	0	0	0	0	0	0	0	0	0
8	72.5	0	0	483	1222.3	74.5	270	1335.9	174.5	0	0	0	0	0
9	0	0	0	0	0	0	0	0	0	0	0	0	0	0
10	0	0	0	0	0	0	0	0	0	0	0	0	0	0
11	0	0	0	66387.6	0	0	0	0	0	0	0	0	0	0
12		0	0	0	0	0	0	0	0	0	0	0	0	0
13	480.8	240	7717.3	3768.1	2560	627.5	766.3	1507.2	472	94.3	316.8	350.6	356.4	1273.4
14	8722	25712.6	106294.1	3362943.3	990862	273860	0	0	0	0	0	0	0	0
15	245413.9	5982.3	1038.6	1018951.5	83486.4	549461.6	0	0	0	0	0	0	0	0
16		471042.3	2887.9	443402	680.4	48.3	0	0	0	0	0	0	0	0
17	0	0	147989.4	101538.8	501.1	643.9	0	0	0	0	0	0	0	0
18	7076	99354.3	11516.3	1015543	147136.4	180219.2	12467.6	13058.3	31.2	95	140.1	45.4	18.7	53.9
19	0	3652.8	380.3	36820.6	1312847.9	217037.7	18594.6	0	0	0	0	0	0	0
20	0	28433	0	63081.4	27591.8	300109.7	0	0	0	0	0	0	0	0
21	0	0	0	0	0	0	1748027.5	0	0	0	0	0	0	0
22	1572.1	604.4	701.6	3916.4	2403.8	513.4	67438.5	12346460	170114.6	21280.9	176774.6	2729762.8	374440.7	6773986.5
23	387.9	185.6	264.9	1023.6	914.4	297.7	353.3	107404.2	3025074.4	26589.8	16473.2	1627163	8654.4	4146116.4
24	2077.1	73.2	61.3	5313.6	3430.5	834.6	4472.2	33838.9	8752.4	69923.9	3517.5	19987.9	11040.6	649251.9
25	395	106.8	180.1	1938.5	2313.6	606.6	166.4	55135	83012.3	2780.4	7072914.6	566289.1	25573.6	3120982.7
26	729.2	97.7	57.2	537.3	788.7		193.4	30241.4	15137.5	675	3588.2	404019.1	6942.6	166721.7
27	0	0	0	0	0	0	0	3997533.3	1048945.9	195185.3	59402.8	55516.6	1192145.4	2032201.4
28	2989.4	5675.4	2597.6	22992.3	15494.4	7639.6	17011.9	185256.9	6194.2	6806.7	30755.6	22510.6	6017.5	509283.3
29	855.7	3803.6	626.4	5194.8	54223.8	2725.6	1101.6	52955.5	48153.7	157.4	1157.8	8117.2	1291.2	165440.4
30	587.3	545.6	747	10134	6128.2	7010.3	1056.9	7420.8	2680.5	265.6	698.5	1368.3	3027	3096.2
31	300.3	1506	2197.6	8673.5	8606.3	3831.7	1406.9	5430.3	1059.9	160.9	714	514	1069.7	5150.1
32	1785.6	77049.5	37524.1	632180.2	546778.8	270504.7	596417.8	93589.7	16938.6	3130.5	5832.9	70032.8	35601.6	230700.5
33	282.9	5022.7	352.6	202282.2	161093.6	97631	349407	4309.5	3888.2	249.9	2100.4	3694.7	1643.8	9510.5
34	267.3	517	2953.2	1812.7	3332.4	291.6	583.8	2774	2379.7	295.8	750.5	1750.2	642.8	4425.1
35	0	35.8	0	0	8.2	0	7.9	19.5	877.4	0	17.1	16.4	2.4	64.1
36	13119.5	15989.4	20382.3	90618.3	33427.4	29620.8	7706.1	60111.8	26996.4	7961.5	8078.3	29476.7	19648.3	51066.5
37	1959.2	0	0	0	0	0	0	0	0	0	0	0	0	0
38	9602.2	1368	1437.2	137180.3	31651.1	11995.8	30744.3	54665.2	9001.9	3121.7	10003.4	26096.5	20813.3	40310.8
39	0	0	0	22761.1	1658.8	176.9	0	0	0	0	0	0	0	0
40	0	0	0	0	0	0	0	0	0	0	0	0	0	0
41	8834.6	1484.9	5416.9	2339	2784.4	29045.9	26164.6	1573.3	4241	21894.6	1290.1	20235.4	1302.8	32255.8
42	5577.6	446.3	24943.4	3605.3	3774.5	250148.1	155512.4	1258.2	79333.2	200645.9	4335.4	567622	3028	299442.4

Continued

Input \ Output	15	16	17	18	19	20	21	22	23	24	25	26	27	28
43	1997.4	5598.6	14038	123110.3	27619.3	75701.2	27606.5	50870.8	1470.4	1151.3	45372.7	4956.2	5285	30613
44	154.1	2426.8	5.5	2438.2	4217.6	9106.6	130.9	560.2	59.4	12.5	2.5	5.2	717.3	16.9
45	0	0	0	0	2769	5458.4	363919.7	3019729.8	392080	68563.7	1021773.2	1100149.4	346509.6	2508382.5
46	2317.2	2456.3	2613.3	87103.2	11818	9046.7	5950.8	48047.6	4337.1	1099.4	3012.5	7944.3	4293.2	163287.3
47	24998.9	95184.4	15428.5	732678.1	73460.7	296809.8	43278.8	98941.7	26029.7	3659.4	4767.7	57484.9	55195.8	271375.2
48	1175.1	1140.6	539.7	5525.3	62132	1173.6	1822.3	5467.8	2243.9	295.3	727	1666.1	1196.8	4042.1
49	775.9	509.8	339.6	3416.1	4050.9	479.6	252.2	1721.1	1166	127.4	370.7	564.6	245.9	911.2
50	3939.2	1403.9	553.6	6744.5	8660.2	1441.3	2501.3	7014	2381.4	347.5	847.3	2090.7	1509.6	5785.8
51	851.3	413.8	724.5	55848.7	606934.5	30365.5	2070.6	12437.6	1373.7	199	2249.1	2915.2	1399.8	7706.3
52	1841.6	2918.7	1451.2	17967.3	37300.8	29459.6	4480	45402.3	5014	499.5	3607	11007.8	6653.2	26132.7
53	528.8	9.9	0.6	739.6	431.3	95.2	9.7	241.7	23.4	4.1	33.2	1038.7	3.1	93
54	285.1	9.3	0.5	1218.8	6274.9	165.9	42.6	241.8	514.3	2.6	288.9	24.9	9891.3	1090.8
55	0	0	0	0	0	0	0	0	0	0	0	0	0	0
56	0	6	0	0	0	0	0	0	0	0	0	0	0	0
57	3357.1	300.5	461	11783.3	5808.9	2539.9	711.9	5952.6	1626.2	458.6	904.4	750.5	2931.9	8756.2
58	0	0	0	0	0	0	0	0	0	0	0	0	0	0
59	0	0	0	0	0	0	0	0	0	0	0	0	0	0
60	534.3	87.9	373.2	2741.1	4452.1	14495.2	3073.3	1490.8	1652.4	50.8	13.9	85.5	531.2	1110.2
61	6098.7	9979.3	25536.2	278306.2	256172.2	354400.8	28958	80340.3	28680.3	3703	17824	19538.8	24972.4	117002.2
62	570.9	225.8	6.7	2637.9	1586.7	150.4	382.4	541.5	320.9	75.3	402.8	55.6	28.2	2890
63	1510	873	696.6	6853.1	5421.3	4945.2	2668.6	13418.6	3589.8	1090	2733.6	3188.5	17120.6	7286.2
64	19157.5	10850.5	7812.9	68537.2	54764.3	20242.6	62365	113127.7	21560.8	4631.5	15418.4	20924.1	33191.7	30686.8
65	0	0	0	0	0	0	0	0	0	0	0	0	0	0
66	14399.4	6429.6	4296.4	46800.7	48313.8	9964.7	42878.8	787203.7	25438.5	4891.4	7935.1	13790.9	15573.5	28425.5
67	0	0.9	0	0	2.8	0	0.3	0	0	0	0	0	0	0
68	12531.3	10329.2	2815.1	76489.9	29887	12988.3	11845	36507.7	6119.4	1840.1	9987.1	11328.6	8821.8	28417.5
69	0	0	0	0	0	0	0	0	0	0	0	0	0	0
70	0	0	0	0	0	0	0	0	0	0	0	0	0	0
71	0	0	0	0	0	0	0	0	0	0	0	0	0	0
72	267.8	468.7	204.2	1481.8	2860.1	473.7	941.7	1520.3	276.8	67.5	220.5	647	256.3	1349.3
73	954.7	1362.1	1671.8	7148.9	4777.6	1091.2	546	63793.4	12867.8	186.9	938.3	2318.9	2224.5	10037.4
74	206.1	932.9	207.4	1333.3	1141.9	611.9	817.5	1681	932.9	102.8	1004	677.2	4044.7	1732.5
75	9344.6	4329.3	3360.9	27978.3	25232.3	8235.9	18672.7	175635.3	15009.1	1475.2	6285.4	8968.4	9391.4	22205.1
76	428.2	558.2	77.8	1968.5	1696.9	686.5	1191	2732.6	380.3	99.9	306.3	947.9	358.8	1630.2
77	600.4	1411	428.5	3635.4	2317.2	1871.1	886.6	3869.5	692.9	124.4	1653.8	1921.8	1367.9	6417.3
78	629.4	256.9	454.4	1597.2	1521	481.1	2108.5	9270.2	2920.4	184.8	476	1313.7	1530.6	1320.1
79	980.7	707.5	513.5	3034.7	2801.9	1026.5	5399	10240	3147.1	254.7	686.9	2119.4	1347.6	2623.2
80	4235.6	3036.6	2988	15767.8	18499.8	7407.3	7940.9	25288.9	4113.7	396.1	3021.9	6636	3344.8	13818.4
81	577.2	859.1	134.1	2837.3	2366.2	1052.4	1730.5	3489.2	570.3	168	416.9	1465.2	556.5	2530.6
82	4838.9	4007	4457.9	24734.5	17073.8	6636.5	30023	30125.4	3593.3	1094.2	4442.3	5591.8	11498.4	10494.1
83	3515	12899	7345.4	33819.4	17693.5	10176.1	23959.7	55220.7	12661.8	1807.3	12007	14042.9	10767.3	41887.3
84	5641.6	23992.5	37067.8	114950.9	75706.2	25130.9	11960.4	305352.3	39555.7	10647.4	29831.1	53334.4	20038.3	375700.4

Continued

Input \ Output	15	16	17	18	19	20	21	22	23	24	25	26	27	28
85	802.3	213.4	0.2	3138.3	5686	131.6	282.6	2175.2	73.2	9.7	1195.9	110.8	779.5	384
86	55262.8	56061.2	48165.2	323079.6	237352.2	54061.9	35737.2	582058.1	29241	27382.5	7387.4	54921.6	23852.4	97526.7
87	2674.3	689.6	119.2	23410.7	14246.2	2783.3	1817.8	36436	5619.8	331.2	4097.8	5743.6	1866.6	5051.2
88	52.4	45.8	839.8	2400.2	2005.2	703.6	1314.4	4182.5	421.2	21.1	504.9	241.8	1177.9	1337.9
89	2275.7	6783.6	5946.8	32847.1	16039.6	10561.8	3059.9	49213.6	8814.5	2187.9	10424.3	11456.6	3682.3	24080.9
90	2735.8	9851	2251.7	20247.8	12057.9	7306.7	7505.6	46168	3180.6	678	3677.9	1899.2	4437.4	16576.6
91	11143.9	8743.4	12018.4	108908.1	48352.2	28943.7	13023.1	60501.5	16179	2704	17540.6	17300.8	23255.8	51680.4
92	36697	43586.8	44355.4	207325.6	98366.4	120631.8	39914.9	185419.7	39079.3	12272.9	35629.4	353578.8	58441.7	174860
93	61.7	58.4	75.5	356.1	266.6	129	38.7	324	113.1	29	38.1	132.8	76.6	193
94	1839.4	2492.1	2464.3	15237.7	6436.8	5012.3	2489.8	8237.9	2429.9	553.8	1736.3	12009.2	2991.7	7406.7
95	80	1223	166.8	1693.7	5269	718.6	293	1698.4	881	40.1	917.6	589.2	303.4	2046.7
96	8595.7	7293.2	2647.9	39194.6	27351.3	11881.7	16413.8	29230.3	6077.1	2781.5	8190.2	19390.9	15865.6	19988.1
97	139.9	93.1	302.1	4347.6	758.2	272.1	843.1	1008.5	285.1	5	94.1	538.9	97.2	3430.9
98	2.7	0	0.9	41.3	9.6	147.5	16.4	13.9	1.8	11	11	6.1	40.7	14.5
99	20225.5	15406.6	11489.9	73048.4	35739.8	8097.1	24194.4	70796.7	35135.8	1596.7	36711.7	157312.1	6111.9	155447.5
100	91757.8	1111294.7	399400.9	1114801.1	592339	339996.9	407437.3	1717095.4	663613.6	56263.4	692868.8	560564.2	382655.2	1602112.5
101	6890.4	16879.2	10037.7	79603.9	55365.7	37695.3	15625	416908.7	18495.6	6940.6	26298.1	48296.6	40924.5	103753.2
102	21.3	16.2	12.1	76.8	37.6	8.5	25.4	74.4	36.9	1.7	38.6	5180.2	6.4	163.5
103	0	0	0	0	0	0	0	0	0	0	0	8708.5	0	0
104	20.1	15.3	11.5	72.9	35.6	8	24.1	70.6	35	0.2	36.7	1712.3	5.6	154.7
105	333	253.7	188.9	1203.2	588.7	133.3	398.3	1165.3	578.4	2.6	604.4	3506.8	100.6	2559.9
106	31517	42499.2	31081.4	171161.3	88922.7	33444.9	38498	278863.8	81478.9	12540.8	94265.3	125221.4	70476.7	162320.7
107	4603.1	3609.2	2459.9	25218.7	17657	9570.6	9256.9	49612.6	8343	1987.1	5858.1	13633.9	12589.2	45654.1
108	865.6	17398.3	1280.7	21173.1	7138	17503.8	3062.4	10726.2	2815	952.6	4250.7	6864.6	31526.8	47617.6
109	15.7	15	9	56.9	27.8	6.3	18.8	55.1	27.3	0.1	28.5	1352.8	4.8	121
110	8588.9	10138.2	3630.2	46989.8	27776.6	9705	17570.4	91527.5	8568.4	1213.3	6162	8987.5	8300.9	48533.7
111	661.2	2071.7	258	11607	2671.9	460.2	967.2	1898.4	1105.4	22.3	837.4	19579.6	448	6345.4
112	0	0	0	0	0	0	0	0	0	0	0	271.8	0	0
113	810.5	1995.5	1178.3	9406.5	6529.3	4004.6	1848.1	49538.1	2176.2	799.8	3108.7	5733.8	4849.8	12181.8
114	5768.9	24606.5	7550.6	557794.8	342317.8	266272.1	23814.5	143718.8	39060.5	1161.1	12916.2	58982.5	9308.2	300643.8
115	777.9	4451.8	396	846.3	4983.4	1091.8	1019	3387.3	476.5	102.9	250.2	254.4	343	2137.8
116	0	0	0	0	0	0	0	0	0	0	0	0	0	0
117	0	0	0	0	0	0	0	0	0	0	0	0	0	0
118	1857.3	2111.6	861.6	9643.4	8495.5	2487.8	3533.4	12362.9	4603.3	715.2	6199.1	4029.2	3320.6	16071.6
119	18.9	0.4	6.2	284.3	65.9	1015.3	113.2	96	12.6	0.3	76	2012.4	280.5	99.6
120	202.1	91.7	126	2192.4	2846.8	1689.5	1283	497.8	225.6	20.8	419.3	178.9	311.3	1533.5
121	746.4	2541	2190.8	22977.4	11816.8	37977.1	2220.7	8751.7	1931	65.5	1631.6	11656.4	1918.9	10589
122	0	0	0	0	0	0	0	0	0	0	0	0	0	0
123	0	0	0	0	0	0	0	0	0	0	0	0	0	0
124	0	0	0	0	0	0	0	0	0	0	0	0	0	0
Total	2497948.9	12359480	5649108.4	20016199.1	10982673.1	6094380.6	6575037.8	29252609.3	8722303.1	11185369.2	11162167	8819613.4	7591991.1	24703134.8

Continued

Input \ Output	29	30	31	32	33	34	35	36	37	38	39	40	41	42
1	583.7	4214.9	96822.4	1008628.7	0	0	0	0	0	0	0	0	58405.7	30185.3
2	612.5	645399.5	61044.8	190353.3	0	406	97478.3	22.7	249.1	54.2	187.7	415	5340.2	4156.3
3	2146763.4	0	0	0	0	3939.7	1206.6	0	0	0	0	0	519735.9	14706.8
4	0.6	49.6	0	0	0	0	0	0	0	0	0	0	0	107.6
5	8350.2	218000.4	184959.7	624837.5	16106.9	16172.9	41325.6	181379.6	1073072.7	0	339	2699.1	12610.2	32295.3
6	17484.3	102927.4	25406.7	246204.1		3498	16136.2	15462999	114812.4	403953	897977.8	69931.9	396192.7	61388.5
7	0	0	0	0		44.9	133.1	17954.4	2256.6	41219.4	370776.2	3323.1	518377.8	7840.4
8	0	0	0	476.6	219.8	0	0	0	0	12711.1	310193	1059.6	131767.7	277.8
9	0	0	0	0		0	0	0	0	0	0	0	0	0
10	0	0	0	0		0	0	0	0	300828.8	36294.4	3321.8	30634.2	0
11	0	0	0	0		0	0	0	0	657642	117012.1	9170.1	21555.3	2406.7
12	0	0	0	0		0	5973.2	0	0	344993.2	860766.8	14448.6	20487.9	66890.8
13	346.3	676484.4	72426.9	1390513		194.8	353.1	3695	2889.3	1451.9	2459.1	578.8	2574.7	2734.4
14	0	0	0	0		0	0	0	0	0	0	18842.3	179661.4	247820.9
15	0	0	0	6071		0	0	0	0	452.9	125.7	216.3	2700.9	7179
16	2410947.5	0	0	0		0	0	0	0	0	0	0	5156.5	221149.8
17	0	0	0	0		0	0	0	0	0	0	0	0	3.3
18	459	343.2	0	0		0	0	0	0	0	9619.2	57405	2004.4	3591.6
19	0	0	0	0		0	0	0	0	0	0	14798.8	237663	56156.6
20	0	0	0	0		0	0	0	0	0	0	0	0	0
21	1504102.7	203974.4	892262	286909.8	19643.9	131914.5	296995.8	4996.7	703.6	56656.4	82993.9	1597.2	84530.9	19402
22	92486	15888.1	123442.3	509261.7	1800.7	99.3	424062.6	1067	162	904	65384.7	147.7	3226.8	365.9
23	15210.3	69.3	20591.5	2807.8	241.7	13.9	615.9	360	43.3	398.8	1033.5	48.6	751.4	132.5
24	524113.1	184.8	22444.6	282313.8	7390.9	1303.6	23313.1	790.5	106.1	1199.5	780.4	107.9	2129.4	226.5
25	154597.7	84	142856.4	5911.5	2904.8	492.2	37872.7		57.4	1055.3	955.8	158.3	885.2	190.9
26	173685	0	155806.4	65505.5		108	168296.4						0	0
27	201789.1	45818.2	24069.3	7545.7	15657.5	2324	248096.8	41848.3	6059.6	31182.9	34348.1	6070.5	34755.8	7975.3
28	5842676.5	95860.7	402165.9	70948.3	3440	1522.3	171791.5	20682.3	1894.3	4707.7	6913.6	1354.1	7071	5236.3
29	1609503	33784.8	281815.4	89780.4	1838.8	57971	101516	8475.6	1129.2	4408	4735.4	2064.9	7157.7	8217.4
30	4143	1609503	826822.1	345883.4	5449	6257.3	18148.5	7371.9	1450.2	3398.4	3996.8	1215.2	7651.9	1456.8
31	2474.4	140135	174790.1	5570642.1		328650.1	174634.5	5911.7	74.5	11405.8	2518	55987.8	46571	467529.6
32	136821.1	812.6	4810.7	3114228.6		38277.3	27131.5	4875.7	391.9	1646.3	16931.2	9272.7	7338.3	48247.8
33	9647.3	2229.6	869.5	594463.2	8694.6	39242.3	2098.6	2663.4	431	1154.7	1294.4	449.4	1603.3	1463.8
34	1522.6			5127.9	28.8	53.9	223397.2	8.3		51.3	28.9		166.2	0
35	38685.1	40141.8	46270	149760.6	42613.4	4550.2	38442.2	1353548.4	22024.5	120047.5	220386.9	29893.8	964268.1	190918.7
36	0	0	0	0								1639.2	27776.5	
37									23469.6	50297.7	134360.1			
38	116836.9	36632.4	7139.4	725790.1	11880.6	11876.1	24118.2	95554.7	3770.7	1495851.3	435980.3	543460.6	1446203.8	537073
39		44423.6	0	24295.3	0	0			0	106001.3	1487906.7	10790.9	60877.1	1165.1
40	0								0			692846.8	12001.6	58572.7
41	188488.1	375802.7	210878.6	263477.7	302768.2	36844.3	101006.3	193543.4	1326.3	442407.1	45734.1	603521.8	6855895.7	498643.9
42	14440.3	13243.4	10105.5	19564.1	32155.2	25818.3	30431.4	16917.5	2089.8	16949.4	92752.3	24898.6	292131.1	1056911.3

Continued

Input \ Output	29	30	31	32	33	34	35	36	37	38	39	40	41	42
43	113722.7	51159	44999.5	54819.2	134305.3	88200.6	140393.5	237616.5	7656.9	56949	21641	50038.3	780853.9	37729.2
44	68.4	52.3	180.8	5527.2	760.4	2.8	0.1	274.2	11.8	311.7	12413	4865	40944.8	8598.9
45	51390.2	1738.7	43066.8	173115.6	5792.3	2254.7	394153.9	0	0	12588.7	16856	1257.3	7526.5	8710
46	345530.5	10225	20587.8	67692.9	43612.1	13493.7	296609.6	53001.3	15097.9	18827.3	44185	15772.7	18364	100194.3
47	284592.4	40639.4	174314.5	192254.1	179684.4	85976.2	536195.9	53134.6	3160.5	167668.6	1551420.4	104857.9	76742.4	418462.5
48	1348.1	35410.2	3952.7	15402.6	2150.9	240.4	3196.6	36451.5	2365.6	18228.2	10530.3	4242.9	43611	3032.2
49	199.7	20825.5	5795.3	7016.2	417.3	44	2675.9	26443.9	1224	9793.4	3551.7	2407.6	32652.5	8338.6
50	1891.2	11227	5325.4	41243.8	2920.1	407.4	3022.9	45366.5	2002.3	78021.8	40883.3	13694.2	105408.5	17896.2
51	5740.8	2910.2	21322.8	24598.3	8907	13783.5	37116.6	10410.2	1899.1	10589.1	7967.3	46919.5	36751.7	32078.6
52	10380.4	7556.5	21577	26113.4	34590.9	2245.3	1432.5	23490.9	4076.7	15708.1	12059.7	4048.4	27351.8	4669.3
53	414.7	11208.4	23926.8	2715.5	634.4	143.3	11782.7	12190.5	56949.4	37295.4	4059.8	5627	19629.4	2246.3
54	2865.3	19888.8	2857.6	8090.8	286.4	3321.6	7035.4	54653.1	63.7	129814.7	8350.1	6298.6	100255.9	3514.1
55	0	0	0	0	0	0	0	0	0	0	0	223.9	5157.2	210.8
56	0	0	0	0	0	0	12565	0	0	1330.9	180.4	89.8	1068.4	38.6
57	1993.5	4640.4	447485.1	18249	4576.6	4202.4	56910.7	96754.5	4336.7	13846.7	22920.8	3962	24340.6	1629.9
58	0	0	0	0	0	0	22441.4	0	0	2490.3	4876.1	1666.1	3150.6	445.8
59	0	0	0	0	0	0	104354.1	0	0	32430.4	13774.3	10063.5	28299.1	7876.1
60	1525.5	729.8	52519.6	34699.7	9374.2	25193.1	12806.8	5991.8	510.6	8503.1	11435.3	713.3	8000.6	11030
61	160075	200920.9	652215.2	176026.1	58576.9	105376.2	355062.3	80808.5	6232.7	63970	55598.9	48524.9	504884.6	49164.2
62	7580.6	1145.7	68.6	2635.4	190.8	477.1	0.4	5003.3	173.2	2105.1	2357.8	1085.1	1519.4	458.8
63	13157.7	3204.8	14945.3	15929.9	10363	3247.5	3032.4	4550.1	2248.3	12813.3	10575.6	2697.5	17474.5	1984.4
64	35737.1	32519.9	71468	228588.8	56820.3		54857.7	250764.8	73347.1	107564.5	291508.6	23317.8	156402.4	18592.6
65	0	0	0	0	0	0	0	0	0	0	0	0	0	0
66	26832.5	39859.6	16608.3	174285.2	46533.2	1856.7	11094.6	199369.1	68141.4	112041.3	153703.2	20921.1	119692.6	11822.5
67	0	0	0	0	0	0	0	0	0	0	0	0	0	0
68	20664.7	28573	40549.4	164654.6	42332.6	5052.8	15875.8	102707.7	7449.9	52122.4	55281.3	23094.2	73532.4	37150.3
69	0	0	0	0	0	0	0	0	0	0	0	0	0	0
70	0	0	0	0	0	0	0	0	0	0	0	0	0	0
71	0	0	0	0	0	0	0	0	0	0	0	0	0	0
72	342.2	552.2	1624.9	2616.5	2589.3	87	175.5	5126.6	2594	1647.3	2387.5	701	6697.9	661.4
73	3685.2	1672.7	2577.6	24762.7	6577.8	239.8	21922.5	12127	691.1	10537.8	7746.2	1693.7	9175.3	1327.3
74	295.7	368.7	1084	5887.9	1018.9	80.7	2043.5	3638.3	285.3	1509.7	2175.3	821.6	1764.1	316.2
75	15273.6	20389.6	13213.4	93225.5	29372	1382.6	26560.9	181694.3	6957.1	66620.4	84194.9	13873.7	61905	8634.6
76	613.9	713.2	2399.7	3737.2	2596.6	126.7	264	9018.6	672.6	1979.8	3600.3	804.1	3081.6	767.3
77	4113.4	1960.2	5322.8	11187.9	9744	406.4	8776.9	3020	1499.7	6350.4	3320.4	1288.7	5667.3	1316.9
78	1707.4	1033.2	2664.5	4423.1	981.7	189.1	337142.5	9124.5	429.3	2797.9	5018.7	958.1	4469.3	718.8
79	1319.1	1560.6	3132.1	26585	7295.2	486.7	2457.4	32781.6	3258.4	4188.7	6788.9	2096.1	5648.5	1335.2
80	8199.3	4589.7	7342.8	33305.9	15282.9	1158.6	3428.7	53109.1	3705.5	42229.9	52459.6	9609.2	48444.5	6397.8
81	669.5	1106	3765.6	5423.6	15035.5	202	524.7	7722.6	1432.8	2872.7	5049.5	1205.6	3752.3	1146.5
82	10238.2	6488	11586.5	26466	27105.9	1472.7	3263.9	120174.3	9031.1	26420.4	42239.9	5320.9	65247.4	9497.5
83	19718.5	13063.6	20984.6	36052.2	14132.6	3269.6	10440.6	55657.3	5053.6	19885.3	20988	6415.7	34960.9	13197.5
84	96748.4	22815.8	112869.7	487576.6	88727.9	6744.3	75334.9	58703.2	10241.6	50725.2	57717.9	21449	74662.7	21424.6

Continued

Input\Output	29	30	31	32	33	34	35	36	37	38	39	40	41	42
85	1022.2	4859.6	326.3	876653.2	253.1	21.4	738.4	16080.7	22.7	6605.1	8751.7	2710.3	1336.3	26190.6
86	49370	152328.9	155666	826235	94594.4	14416.4	75021.9	383037.6	204745.2	1620263.3	924502	104850.3	658183.2	54778.8
87	1751.8	4352.8	2165.2	56122	2122.7	610.8	544	29788.1	12330.3	40953.7	24520.9	7086.6	51864.6	3787.2
88	231.6	657.3	47.3	1082.6	319	94.6	83.4	366.9	9836	8357.5	9129.3	1131.6	8275	19665.6
89	7515.6	9440.6	14653	60863.9	8544.2	1297.7	4420.1	42534	3711.2	34908.3	30005	26666.8	49758.3	6502.4
90	17503.6	3575.1	8183	15720.8	20614.8	2308.7	4997.1	30925.6	1418.5	12649.9	20288	3979.7	22499.3	10176
91	38177.2	49860.6	44243.8	177790.4	30887.1	3744.9	20906.9	98952.9	87338.4	145851.3	170688.8	53092.2	184115.2	36208.8
92	108119.8	99039.3	125797	192017.2	59441.6	18064.3	78626.7	73698	46448.5	153421.2	212835.3	42420.8	197309.7	83447.6
93	167	190.2	190.1	627.9	192.5	21.9	153.3	1330.9	1330.9	1348.8	12190.1	269.8	12760.7	872.1
94	6426.5	6706.1	8330	14390.3	3284.8	817.2	3837.1	162289.9	8851.2	20242.5	24408.9	3545.5	21893.1	5789.3
95	5469.5	443.4	1268.9	1641.5	784.6	572.2	1833.7	101022.8	124.2	690.7	1647.1	629.6	1580.8	1148.8
96	16611.2	15441	21691	37939.7	8716.3	2086.6	6176.7	113876.3	43767.3	47007.3	69513.1	9209.8	49599.8	13630.4
97	41.6	61.7	1088.1	898.4	459.9	263.8	320.2	838	41	528.4	393.8	12.3	3354.1	799.5
98	2.6	1.5	24.6	146.8	18.7	0.1	0.2	19.9	0.2	15.4	41.4	0	29.4	8.9
99	152062.6	38709.7	152741.4	40141.7	18317.2	5892.9	48030	111608.1	2794	40715.2	30371.6	23210.4	67293.7	28818.1
100	1600283.3	578972.7	924874	1272681.8	602104.1	111400.7	373890.2	813746.6	268377.4	404400.5	625962.7	163770.6	791301.8	290783.4
101	40022.3	32737.8	69553.4	217922.1	57621	18515.8	25501.6	33061.1	21965.5	39359.6	56590.2	35871.2	106103	39163.3
102	159.9	40.7	473.6	42.2	19.4	6.2	50.5	117.3	2.9	42.8	31.9	24.5	70.8	30.3
103	0	0	770.2	0	0	0	0	0	0	0	0	0	0	0
104	151.4	38.3	165.6	39.9	18.2	5.8	47.9	111	2.8	40.5	30.2	23.1	67	28.7
105	2503.5	638	544.4	660	302.3	97.1	790.5	1837.1	46	670.4	500	382.3	1108	474.7
106	90992.1	71727.1	93420.1	215505.9	71146.4	12386.3	41307.6	228975	22529.9	122659.9	199244.3	32427.2	239023.6	40715.1
107	11225.2	12371.7	18512.5	54368.2	17023.8	2933.2	6811.6	52539.1	5794.6	16027.2	25524.3	8552.1	40595.2	10635.4
108	18854.9	7471.9	50889.5	11935.9	15297	2002.2	26256.7	4599.7	78	16415	1985.1	1634	8704.5	16577
109	104.4	30.2	129.9	31.1	14.3	4.6	37.4	86.8	2.2	31.7	23.6	18.1	52.4	22.4
110	15042.4	16660	22487.2	67478.4	26300.4	3915.3	11350.1	140825.7	17527.4	58452.5	70930.4	17423.5	88217.8	17261.2
111	4418.2	917.1	3030.5	3394.8	1448.8	180.3	1282.8	3513.5	68.9	1001.8	726.1	565.9	3205.9	5876.6
112			23.4											0
113	4682.1	3857.5	8236.4	25812.3	6779.7	2174	3004.6	3873.6	2602.1	4643.3	6668.3	4236.7	12519.1	4632.5
114	102776.6	21804.1	176051.6	83810.5	28719	10196.8	25048.6	48564.3	2785.4	18410.9	29547.2	42978.2	55938.1	354425
115	2435.3	1996.3	765.6	23938.3	672.1	39.1	182.4	4139.8	586.4	2319.9	2478	876	2837	291.1
116	0	0	0	0	0	0	0	0	0	0	0	0	0	0
117	0	0	0	0	0	0	0	0	0	0	0	0	0	0
118	5555.8	8051.2	4475.1	26467.6	8633.5	1461.9	4916.8	10653.1	1553	8891.9	10748.6	5653.9	11885.9	4961.1
119	15.7	10.2	343.8	1010.5	128.8	0.8	1.2	136.8	1.4	105.9	284.8	0.2	202.4	60.9
120	1897.5	783.4	366.1	6143.7	810	117.5	296.5	15767.4	279.6	2340.3	1309.6	1508.1	7295.7	5461.6
121	31582.6	3769.4	13476.7	66125.3	6207.7	1171.6	4998.9	33199.5	1147.3	13386.2	9283.4	9553.1	38656.7	9554.6
122	0	0	0	0	0	0	0	0	0	0	0	0	0	0
123	0	0	0	0	0	0	0	0	0	0	0	0	0	0
124	0	0	0	0	0	0	0	0	0	0	0	0	0	0
Total	17193078.1	16054301.8	10096140.2	18467421.1	6138286.4	1299764.1	5629900.7	21817764.1	2329375	8158975.2	10484131.8	3248676.8	17304591.1	5608015.4

Continued

Input\Output	43	44	45	46	47	48	49	50	51	52	53	54	55	56
1	51356.3	671107.6	63029.7	4565.7	0	0	0	169553.1	0	0	0	0	0	0
2	1173351.8	3772.9		2038622.7	34.8	0	10883.4	20072.1	3557.2	1242.3	3281.3	8523.7	0	0
3	0	685945.7		0	0	0	0		0	0	0	0	0	0
4	12047	22272		0	0	0	0		0	0	0	0	0	0
5	1305828.7	307521.7	71852.8	43731.4	56289.4	1598453.7	14080.2	3981.9	15920.6	4032.8	2237.3	813.8	181581.3	138453.9
6	258580.7	34409.7	43488.6	90734.8	9085.1	3178	116761.9	1222114.5	222445.9	132399.6	240491.7	91224.3	26990.4	23347.6
7	647079.1	5578.8	4515.1	8055.6	207.8	652	107.8	4539.5	151018	12228.7	2005.2	10712.7	411.6	1449.7
8	42459.9	1036.9	0	295.8	0	0	23.9	1325.8	24169.7	25018.4	435.2	17504	806574.5	365778.6
9	0	0	0	0	0	0	0	609	0	0	0	0	0	103198.6
10	80431.7	0	0	0	0	7028	0	18501.8	0	0	0	4731.2	6540.2	0
11	32870.6	2525.7	0	0	0	0	452.4	66298.2	2202.5	598	26310	34828.1	0	0
12	61189.8	2201.6	0	0	0	1266622.5	570662.4	1677890.8	414243.9	1259364.5	727.1	784.2	220847.8	62434.8
13	2530.2	1098.7	0	0	0	3666.8	2405	2657.9	4143.2	0	473347.4	383152.8	2102.1	1806
14	26167.5	168067.1	0	0	0	0	0		0	0	0	0	0	0
15	3053.2	86474.9	0	0	0	0	0		0	0	0	0	0	0
16	571.7	28831.4	0	0	0	0	0		0	0	0	0	0	0
17	199.1	34428.5	0	0	0	0	0		0	0	0	0	0	0
18	6835.5	141232.6	5667	286.4	1295.1	0	0		0	0	0	0	0	0
19	93126	256168.8	3660.1	5925.8	2405.9	19132.1	0		3812.2	27240.8	12722.1	7803.9	0	0
20	0	0	0	0	0	0	0		0	0	0	0	0	0
21	0	0	0	0	0	0	0		0	0	0	0	0	0
22	266219.6	67710	189100.2	2184958.6	1119639.5	135571.7	136627.3	29205.8	42901.6	112578.6	13022.3	116014.5	2840.4	1773.5
23	1214.1	10495.4	24602.8	8696.9	223209.7	44333.7	112707.2	17582.8	1435	10519.6	8025.5	2200.8	1087.8	590.2
24	419.7	319.3	2169.2	841.7	1110.2	3560.9	584.4	741.9	2053.1	1514.9	74.4	281.7	272.2	163.5
25	773	842.6	21762	7554.4	1305.4	1423.2	1277	856.6	1277.6	11816.5	612.2	365	694.8	284.1
26	712.9	3064.1	1685.7	131495.5	114748.3	1376	452	574.3	488.8	400.3	7296.7	185.7	556	432.5
27	0	54561.3	687676	134869.2	126859.8	82199.4	31652.3	24440.5	0	0	8283.5	11082.6	0	0
28	28865.6	22817.3	9618.1	26678.5	35687	35140.2	15310.8	74751.8	17900.6	10233.6	25841.2	8413.3	11963.9	17362.9
29	8392.7	8726.7	2213.4	123734.9	40691.2	8418.6	4040.2	29921.5	5623.9	3687	39649.1	3473.4	2426.9	3164.7
30	11346.7	3375	2179.5	2738.4	4988.9	5331.9	9360.9	21546	37285.5	11679.3	7434	3192.4	4955.7	3956.1
31	6207.6	5819.9	1725.7	2842.3	13526.7		2930.9	49732.5	27986.2	21606.6	9826.8	39640.4	1521.5	1107.5
32	132045	462008.4	193340.5	99269.3	378435.5	1779351.3	7132.4	1403581.1	179927.8	245193.3	237054.9	24452.7	854.8	4019.1
33	7243.9	212307.2	1755.6	3133.3	11489.1	7106.5	593.5	3077.4	1385.5	11242.7	1070.7	8903.3	1852.6	653.8
34	1278.1	7711	718.2	2988	3616.1	2888.3	1772.7	3280.3	2386	9897.6	237.5	806.3	1145.4	878.9
35	12.7	0	0	0	1147.3	97.5	5.3	25.4	0	297.9		725.7	0	0
36	620809.4	19953.7	134098.3	103407.2	177013.4	193407.8	60897.1	1062298.8	355993.8	330656.1	61797	137482.8	59281	103057.9
37	20531.7	0	0	6421.9	0	0	0	29503.2	0	0	9379.7	187053.5	891558.3	169301.9
38	1138956.5	407626.4	149184.8	255769.7	128917.6	29479.1	7920	554515	463014.3	102578.6	16127	75034.8	4034.8	24243
39	191467.1	11412.1	0	0	0	0	0		0	0	0	0	0	0
40	3716.6	2870.6	0	0	0	0	0		0	0	0	0	0	0
41	1699565.8	722055.5	1284041.2	235757.7	2155584.5	45565.8	20559.4	92873.9	64858.3	166381	222181.7	31462.9	2906	5854.2
42	213176	59902.9	36790.6	55725.8	81518.8	38921.6	4792.6	355991.6	10511.4	5131	11975.7	93718.1	30112	3354.6

Continued

Input \ Output	43	44	45	46	47	48	49	50	51	52	53	54	55	56
43	1967422.7	119754	2839240.4	827441.1	7751438.6	68526.5	8085.6	977357.8	109035.4	32208.3	24902.2	104752.4	91914	12081.1
44	3811.8	3019639.4	293.5	160	460	270.5	5.8	363.4	283.4	291.9	6.3	8.3	69.2	144.3
45	19377	15044.2	2139971.1	332950.8	114639.7	5069.1	4346.1	60620.8	1454.3	1423.8	983.5	29155.6	17077.7	0
46	89886.7	109477.7	16598.2	887427.7	81293.8	115950.2	21111.5	65339.2	29657.8	15150.1	30974.3	14958.8	5619.5	19649.8
47	274570.1	182260	95618.8	59648.6	5402953.7	906094.6	25778.5	240617.2	80488.7	58162.4	51875.7	49157.5	6450.3	8195
48	20440.9	7744.3	5164.6	5880.8	7610.6	592007.1	2526490.6	552195.3	7887.9	52429.7	58242.8	36583.3	4821.1	7020.4
49	11366	3627.3	5081.5	3731.9	2783.8	123506.2	198406.8	72059.9	3947.3	29202	17719.7	14151.6	15938.9	5021
50	32689.8	21600.7	16493.7	5292.7	9414.4	182241.7	99636.7	2113572.4	41465	39180.5	66609.2	48063.5	2998.5	36931
51	49192.2	171948.9	3418.2	5890.7	15788.6	17183.3	6893.3	48608.4	492370.9	7673.7	11764.9	8509.1	45618	4491.6
52	20537.9	21348.1	7378	14194.1	25857.8	64102.2	10028.4	46061	26853.8	160321.9	85218.4	13127	140402.7	8088.1
53	7296.4	1238.8	3387.1	127.9	450.5	330018.8	79615	187230.9	142285.2	77432.7	261607.7	34549.3	11315	296085
54	42651.9	6305.3	8108.4	3255	36940.2	184738.7	860904.3	273461.1	103337.4	65760.2	516633.6	149354.4	72011.8	80640.5
55	241.9	120.2	17.3	9814	3280.2	105045.3	726.8	1944.4	606.4	517.3	183.3	619.8	112263.2	6890810
56	727.9	670.9	497.5	618.6	297	31778.4	29381.3	11851.9	715.5	901.3	72.7	211413.5	19955.1	310863.5
57	14884.6	40217	7775.3	90164.9	14374.2	395481.5	867660.2	600313	7689.8	11750.8	13123.3	8123.4	4628.5	72432
58	1578.5	1177.4	3458.5	2366.7	2113.9	21279.2	5107.2	4561.3	1029	3458.7	1285.9	5159.2	436.7	241870.7
59	125894.5	1369.8	1193.3	3276.7	1379.7	3002.8	745.6	10563.5	21119	3392.9	815.4	1038510		464711
60	31827.7	5656	2532.1	12834.4	8477.2	18576.1	13478	17590.8	8078	4136.8	7381	13484.3	7857.3	7562.7
61	150694.5	93878.4	25267.4	277116.8	163713.1	1224586.8	783699.5	225919.9	237459.3	85055.2	46726.2	61396.6	18658.1	615611
62	5872.7	6710	369.3	355.8	78.5	1200.2	124.4	467.4	2315	398.1	113.9	451.1	203.5	667.1
63	12512.8	5755.5	6063	64894.6	28457.8	88018.1	8317.2	43200.8	10418.7	26225.9	4822.7	11696.1	17768.1	67137.5
64	164181.5	59111.9	50321.7	43529.6	106899.2	1561264.7	78332.3	93021.7	49756	85602.8	39109.3	1146116	57444.3	117857.4
65	0	0	0	0	0	244179.4	30176.4	42504	48466.2	47114.5	34806.7	50748.5	0	0
66	140092.3	35367.7	40043.7	35563.7	58410.7	0	0	0	0	0	0	0	50379.1	97839.2
67	0	0	0	0	0	0	0	0	0	0	0	0	648.3	2305.6
68	67525.4	33903.2	29537.6	31560.3	83384.6	109431.4	23712.5	50205.8	32257.1	35979.2	31233.6	18458.2	35038.5	52518.1
69	0	0	0	0	0	0	0	0	0	0	0	0	0	0
70	0	0	0	0	0	0	0	0	0	0	0	0	0	0
71	0	0	0	0	0	0	0	0	0	0	0	0	0	0
72	3093.8	1397.1	972.8	1238.2	2406.8	5069.6	818.8	6162.7	1213.9	1043.3	472.2	833.8	3747.1	1341.5
73	19161.9	8011	5107.8	4667.5	50792.1	26655.9	4531.7	6006.4	5405.3	1684.4	2883.1	42906.2	4277.2	8224.1
74	1781.4	1192.4	759.7	638.1	1941.3	6058.9	751.7	1888.5	964	2381.9	340.5	732.6	519.2	1166.1
75	80349.9	25190.2	21666.1	24054.1	54500.6	123325	24344.9	47348.4	28716.3	35011	26617.7	53150.5	32383.7	46133.9
76	3313.6	4442.3	1484.1	1284.4	3480	7142.1	1159	1914.1	1846.9	998.2	697.7	785.9	846.3	1973.5
77	5012.2	3783.8	2113.2	5217	13425.7	11139.9	2006.4	18061.6	8110.7	5748	4109.3	5252.7	2173.4	3229.1
78	6522.3	3782.5	2946.5	2541.6	5279	22459.3	3632	2272.4	5339.5	4719	990.7	2865.9	6929	2717
79	5415.9	6181.3	4179.9	2909.9	7373.5	14717.2	4515.9	3346	2854	4831.7	1274	2390	7642.8	4661.7
80	50637.3	27522.9	17024.4	18552.1	41251	59230.2	7035.9	28219.5	16033	13580	10007.5	11764.9	11142.9	12114.8
81	4346.3	3191.7	2125.8	2947.9	5602.5	10154.5	1804.4	2708.1	2777.7	2104.7	1097.1	1225.4	928.3	1812.9
82	61446.6	28495.3	19351.7	14481.2	35731.4	36542.7	29574	28283.9	9789.2	14632.4	6587.1	12183.9	9767.2	20980.3
83	30030.4	22552.9	17208.2	18681.3	42537.6	36537	19568.9	40784.4	10713.3	13994.6	8567.1	12419.6	10122.2	12809.9
84	77110.9	79736.8	33856.2	68383.5	78472.8	137965.2	90179.6	218924.1	74361.6	76018.5	43747.5	35552.1	30367.9	114303.6

Output / Input	43	44	45	46	47	48	49	50	51	52	53	54	55	56
85	3193.4	85	1116.9	6279.7	37255.4	164732.2	5062.5	55772.3	34039.7	166.4	3425.5	19102.4	504018	597999.6
86	755988.6	319270.9	314480.5	238229.8	599874.5	1881146.4	255996.6	749697.8	212865.9	185358.1	247898.1	305284.6	230095.7	624737.2
87	41334.8	18988.4	11804.9	24655.2	4968.7	1943.5	2341.3	1699.6	2021	6663.5	1706.1	2683.6	2411.1	1835.2
88	9550.4	1370.3	353.4	5651.4	4383.8	734.1	244	1333	17464.4	13754.7	646.4	654.2	1546.3	6302.9
89	33096.2	26262.4	13343.9	13554.2	27352.2	17130	21384.1	17806.7	9377.1	12787.5	3856.8	8686.7	7548.4	24519.7
90	24125.5	14048	6648.8	6342.8	13177.7	19087.6	3608.9	21384.6	4710.1	6481.5	3790.9	7053	3025.6	4106.1
91	109456.8	40861.7	37829.4	49015	137345.7	285619.7	185877.7	407718.1	65404.6	128375.9	68594.6	68957.4	144487.6	61502.5
92	164960	107766.3	66601.6	110463.6	230849.8	487191.9	228754.2	368792.1	114964.9	152935.1	91531.2	135691.3	184177	83235.9
93	10098.8	279.1	1189.9	506.1	997.7	956.6	302.5	4151.3	2100.8	2100.8	350.1	1094.6	874.5	811.2
94	17605.9	5143.9	4739.8	6106.1	14829	37176.4	15782.9	40920.9	10216.7	16800.4	8606.2	9375.1	47043.3	22453.7
95	1525.2	6351.4	685.5	776.8	5205.2	2679.8	1104.3	1509.8	4382.3	523.8	633.1	1160.3	230	419.7
96	41145.4	16665.2	16034.9	15475.1	38764.6	135915.2	36195.4	111462.3	26710.2	31077.7	25835.3	25301.6	33895.3	34038.1
97	843	949.7	388.1	868.5	739.8	847.5	25.7	1307.1	891.7	84.8	9871.5	169	79.7	245.4
98	14.6	175.6	10.9	9.2	104.2	24.6	12.1	60	178.4	6.3	51.3	15.9	53	2.9
99	191193.3	42145.5	26056.2	44303.1	169962.2	20225.4	27463.5	56300.9	25298.5	17469.7	110605.2	80837.5	18836.9	12348.1
100	694599.9	804356.9	437374.1	561189.4	1051360.6	1240518.4	624506.6	1248562.7	355801	351525.6	297055.2	335549.9	273130.9	209789.5
101	75902.9	88476.3	30997.1	49734.6	125687.6	121393.9	115890.1	117839.8	39404.7	40814.3	47277.3	96059	14355.8	30608.4
102	451.3	44.3	27.4	46.6	178.7	21.3	28.9	5.9	26.6	18.3	116.4	85.1	0	13
103	617.8	0	0	0	0	0	0	0	0	0	0	0	0	0
104	199.8	42.1	26	44	169.2	20.2	27.3	5.6	25.3	17.5	109.6	80.5	1.9	12.1
105	1565.4	694.3	429.1	729.7	2798.1	332.5	451.7	92.7	415.9	286.8	1820.9	1331.3	31	204.2
106	188730.2	120268.6	116402.7	90169.5	243426.3	287103.5	73659.3	309945.2	86852.2	111055.2	58647.3	120825.1	49101.9	62288.1
107	32652.4	26262	13039.5	20135.2	41228.1	27704.7	13131.9	23472	12544.9	8725.5	7241.1	10538.3	5451.9	7937
108	12799.5	25222.1	4614.7	14702.4	44337	3852.1	2891.6	20934.1	6399.3	16371.7	5696.7	2216.2	682.2	1499.2
109	158.2	32.8	20.3	34.5	132.3	15.7	21.4	4.4	19.7	13.6	86.1	62.9	1.5	9.7
110	82196.1	31092.3	23593.2	24886	45764	72498.6	26960.8	50710	29214.6	29899.5	9358	24554	19778.2	38106.7
111	4455.8	7867.7	684.5	2679.8	3862.7	3723.9	4063.9	76380	724.8	4996.5	3601.4	3441.6	114.8	343.9
112	19.4	0	0	0	0	0	0	0	0	0	0	0	0	0
113	8952.6	10457.9	3663.4	5777.6	14789.3	14368.3	13700.9	13796.5	4634.8	4831	5532.7	11258.8	1692.4	3632.1
114	50445.2	424762.4	7758.1	29224.5	78978.9	52978.5	13041.8	90842.2	28464.8	32062.2	41478.7	17707.2	2916.9	7811.6
115	2114.2	1147.7	300	2042.8	3701.2	3640.7	1298.9	5777.4	1055.6	787.1	608.3	1502.3	2088.8	734.1
116	0	0	0	0	0	0	0	0	0	0	0	0	0	0
117	0	0	0	0	0	0	0	0	0	0	0	0	0	0
118	11442.6	10135.9	3500.8	10131	11025.6	19822.1	8387.8	10578.6	4522.3	4663.6	6109.1	5598.1	3319.1	4547.5
119	240.3	1208.7	75.2	63.5	717.2	169.1	83.3	411.5	1227.8	43.6	353.1	109.3	365.5	20.2
120	4993.9	8236.8	778.8	1499.8	11041.5	730.8	159.8	513.3	1296.8	1332.4	404.4	1451.8	105.2	503.1
121	27999.3	18593.6	7796.6	11508.4	21599.6	33383	5302.5	10111.9	2781.7	7597.1	5042.5	3547	2970.7	6350.1
122	0	0	0	0	0	0	0	0	0	0	0	0	0	0
123	0	0	0	0	0	0	0	0	0	0	0	0	0	0
124	0	0	0	0	0	0	0	0	0	0	0	0	0	0
Total	14358802.4	10894926.7	9523867.9	9809630.2	21938707.6	17141445.1	8694444.4	16908037.1	4789689.5	4727685.8	3927959	7201165.2	4677909.6	12537738.3

Continued

Input\Output	57	58	59	60	61	62	63	64	65	66	67	68	69	70
1	0	0	0	0	0	0	0	0	0	0	0	0	0	0
2	31319	0	0	0	21605.3	0	0	3095.5	1508.5	9538.3	0	18590.6	2669.5	0
3	0	0	0	0	0	0	0	0	0	0	0	0	0	0
4	0	0	0	0	0	0	0	0	0	0	0	0	0	0
5	0	0	0	0	19624.6	0	0	0	0	21223.4	0	0	0	0
6	1184534.1	39212.1	234545	27423.3	124706.8	23998.8	22485.9	194454.5	33260.5	103336.3	24120.5	87239.2	2702	4351.4
7	253834.9	1413.9	19745.8	1779.4	1915.5	832.4	3204.6	3603.4	1057.4	3758.4	735.9	2666.7	416.9	306.2
8	40441.4	32.8	3337.5	765.5	3309.7	4526.7	108.5	6296.7	817.7	5750.8	1202.5	9321.6	28.6	208.3
9	1779579.8	270944.7	45874.5	1765.4	751275.4	31721.9	5749.9	64867.6	11834.7	44420.7	7453.2	28837.5	3173	778.7
10	118444.4	626042.9	3976469.9	870301.3	89800.6	41371.2	4296.5	271953.9	2845	30348.8	2739.3	6359.5	835	1177.1
11	0	0	0	0	0	0	0	0	0	0	0	0	0	0
12	250012.8	39702.5	35060.6	2365.9	158350.4	11093.1	8759	39937	6025.1	9898.2	8275	8579.2	6547.2	243.9
13	9826.1	1435.2	7579	1810.1	15089.7	4524.1	7813.9	9100.2	868.8	5294.8	6140.6	2939	8238	499.2
14	0	0	0	0	0	0	0	0	0	0	0	0	0	0
15	0	0	0	0	0	0	0	0	0	0	0	0	0	0
16	0	0	0	0	0	0	0	0	0	0	0	0	0	0
17	0	0	0	0	0	0	0	0	0	0	0	0	0	0
18	0	0	0	0	0	0	0	0	0	0	0	0	0	0
19	0	0	0	0	0	0	0	0	0	0	0	0	0	0
20	0	0	0	0	0	0	0	0	0	0	0	0	0	0
21	0	0	0	0	0	0	0	0	0	0	0	0	0	0
22	30299.2	870.2	12183.9	17011.2	94640.5	16118.8	15959.8	112245.6	15527.6	261723.1	3760.7	129001.8	30045.9	1162.1
23	5015.6	91	1673.6	6220.1	25001.2	6235.3	4167	2080.8	569.1	21087.7	905.5	115848.7	9826.2	356.9
24	1387.3	388.9	449.6	426.2	23639.4	91.6	79.6	780.1	207.9	715.4	60.4	1198.7	943.7	84.3
25	2856.4	52.1	1002.7	7704.1	16523.1	258.1	243.2	16813.1	286.5	1745.7	165.7	14472.9	816.8	4072.6
26	2470.7	283.1	1011.2	681.8	3903.5	452	235.1	4615.1	333.5	1704.6	300.3	21230	315	2279.5
27	0	0	0	0	0	0	0	0	0	105479.2	0	37152.9	2031.1	220.7
28	83992.5	5498.1	29521.9	13739	87667.1	16206.9	7988.4	80987.9	15848.3	53790.9	7476.1	49769.7	8517	3195.8
29	16589.5	2465.6	4225.6	3879.6	23402.4	5205.1	2436.3	35695.2	32766	22894.1	826.2	101446.6	1353.5	326.8
30	29413.8	6418.7	9497.3	5337.4	122186.2	9168.5	9481.2	41578	5399.5	44101.4	14337.3	67647.8	13182.4	732
31	6347.3	2675.3	9381.8	12280	405262.1	15300.9	13305.1	63883.6	9094.5	61088.1	1423.8	42510	3600.7	1051.8
32	18276.8	683.6	8597.6	32400.4	329954.4	11104.2	3417.6	238183.6	3689.7	73743.3	811.6	95836.3	558.4	848.5
33	6674.3	344.4	1840	3872.5	20671.9	3519	1597.6	21791.9	3129.7	16989.6	700.5	16970.8	817.3	245.3
34	6426	256.6	1508.6	5218	7073.7	1097.9	848.6	6310.9	5418	5525.3	482.3	5172.7	668.5	417.2
35	27.9	117	186	17.7	515	27.8	24	29.5	9.1	50.1	100.8	200.9	24.6	0
36	796950.9	12719.2	176490.8	100310.4	279374.2	70334	55374.6	265287.4	60219.6	198831.7	28075.8	232941.3	30461.8	10911.6
37	453847	557767.2	42878	4280.3	70334	18084.7	18566.9	58924.4	11492.6	20814.6	4199.9	15551.4	2113	81.5
38	105874.3	155612	356904.5	63408.3	147279.4	13277.4	7671.9	75275.1	9472.7	48994.3	15875.9	100683	15727.5	41343
39	0	0	0	0	0	0	0	0	0	0	0	0	0	0
40	0	0	0	0	0	0	0	0	0	0	0	0	0	0
41	74072.5	14816	27598.2	36530.9	490768	23330.6	30151.9	1466454	80800	96342.8	28843.1	291428.6	61041.5	9086.7
42	32163.8	904.7	17241.7	4780.5	36545.3	6233	3078.3	38710.1	8008.2	34784.3	1350.5	17325.1	1054.3	573.4

Continued

Input \ Output	57	58	59	60	61	62	63	64	65	66	67	68	69	70
43	81983.1	6440.2	89425.8	3615.8	144078.3	11013.2	8016.1	73728.3	12808.5	110246.5	19150.2	61325.9	7177.4	9798.4
44	885	18	260.5	38.4	185	156.3	304.7	582.3	67.2	1353.7	41.6	792.8	1543.6	22.9
45	0	0	0	0	276966.6	45017	1405.5	4789.6	559.4	57029.4	1451.1	3128.1	1428	257.9
46	115699	4049.5	27774.9	9060.1	241882.8	38973.5	38058.5	473176.2	423241.6	233238.9	30664.9	959062.1	17829.4	5158.7
47	62464.8	9318.9	29903.8	18490.7	258082.9	29361.3	103818	759946.4	25768.7	186222.2	16852.5	529143	5615	2645.4
48	44218.2	47413	26362.5	6885.5	50942.6	77611	2487.3	15533.1	3175.3	17981.5	2478.7	18469.1	2848.2	603.7
49	24497	3525.5	7740.4	4806.2	14999.1	2380.6	447.4	6682.4	647.5	6580.2	8599.6	4767.3	978.7	123.5
50	88377.3	13250.5	24681.1	5831	177525.5	12125.8	4601.6	23936.7	4427.3	28778.5	5576.6	24953.8	9227.3	892.9
51	29754.5	1926.6	13781.4	10357.9	231937.4	8948.3	45514	42764.1	15529.9	29796.7	3111.5	189380.5	14842.8	1562.3
52	46662.7	3474	13294.6	7877	128384.8	18099.9	7358.5	65202.8	13318.2	45980.1	2891.6	33255.9	6161.3	1451.3
53	692941.6	13055	37714	8261.1	65062.4	60765.3	342812	57932.9	5289.5	35859.6	11019.2	69308.9	21321.6	394.5
54	151685.1	45533.6	135512.4	8519.3	295184.9	11229.9	10903.4	71205.4	18114.1	78267.2	8676.3	105692.5	8394.3	13176.1
55	1461347.8	1970.6	2460.2	699.7	1195416.6	83997.9	76468.1	473753.3	284580.6	152601.6	18408.4	77532.4	67717	149.3
56	2897893.4	11715.7	8423.8	1699.5	897849.9	44820.2	191709.3	509531.3	71342.7	416838.9	73975.2	815933.9	15050.4	796.8
57	7294700	123297.7	22168.5	21866.8	9960479	726608.8	408099.9	2824407.4	797365.7	3282937.6	293444.7	1360723.3	398744.4	22274.8
58	1006064.2	311668.9	36644.5	8776.3	256387.6	104538.4	54288.1	324619	15751.8	117433.4	8999.6	31550.4	6016.2	352.2
59	278888.8	215405.1	1644207.7	5047047.5	2122397.5	75224.8	173979.9	389965.4	106721.6	230093.1	9785.3	249181.7	26985.9	4098.5
60	63398.8	2959.6	64458.8	354644.4	1293556.3	65903.1	26051.1	247224.4	144889.4	294359.8	24042.6	169795	385918	17598.7
61	526478.6	27020.1	147059	63981.4	6497610.2	165688.8	580085.5	1576333.6	2297415	1001705.5	70694.5	371871.8	76643	11706.5
62	4148.2	91.5	2814.3	88.7	1969.9	1302864.5	6707.9	326162.8	813106.7	134607.9	32900.4	607974.2	281840.5	16143.7
63	260207.8	3149.5	16107.4	10001.9	326375.1	133611.8	361519.6	205273.1	298440.2	173148.7	15172.4	132203.8	10740.7	2206.1
64	784026	26445.6	133454.5	41825.3	377222.4	798577.8	929376	4379185.3	977434.7	1692587.7	160265.8	997379.2	320885.3	58713.2
65	0	0	0	0	0	0	3273.1	30012.7	1569981.3	22405.3	0	0	10140	0
66	738957.3	24284.2	105469.9	39465	165981.9	21664.4	13369.3	1333647	48005	1527826.5	3807.5	67364	139429.6	8088.1
67	18801.3	38.9	2249.8	50.4	583.6	187.1	6918.4	6298.5	3087.1	136.5	331215.8	2463.1	15.4	2.8
68	327368.6	15920.4	43043	22837.3	206146.2	95712.5	18859	261320.1	152801.8	130384.8	20954.9	10455796.7	44458.2	7652.3
69	0	0	0	0	0	0	0	0	0	0	0	0	212192.1	0
70	0	0	0	0	0	0	0	0	0	0	0	0	0	199450.4
71	0	0	0	0	0	0	0	0	0	0	0	0	0	0
72	0	450.5	3090.1	1429.7	17126.6	0	644.2	5520.2	4699.3	25388.1	10360.7	26871.9	30559.8	2568.4
73	6241.1	514.6	6647.7	11011.3	112288.6	7457.2	63085.2	936764.3	37929.4	615842.9	80004.5	118207.2	38815.6	1309.6
74	35894	333.2	1700.6	610.2	14777.9	103026.6	1352.9	20328.8	2468.8	78795.2	18444.1	77119.8	13221.8	1068.2
75	5462.9	12405.7	60535.4	21687.4	157027.3	12899.5	401412.5	521024.9	172036.7	365099.3	122934.5	311514.7	114619.1	12408.9
76	396802.6	303.3	2570.8	1096.2	8402.6	153929.9	5478.6	6847.1	1774.4	28347.7	999.1	16981.8	787.5	1934.2
77	10723.3	1389.1	4441.7	2769.4	32391.7	2588.7	3967.4	22623.3	5382.5	15024.2	1442.9	29641.7	2661.6	979.9
78	21594.7	1831.2	9031.6	5340.2	24600	5962.1	64440	663689.3	3755.6	552985.2	11872.8	30705.8	2320.2	7092.7
79	22226.1	788.3	5927.3	2393	35672	11958.8	3885	14845	3044	120682.4	1506.2	11003.6	69281.5	8874.2
80	30983.6	3434.3	18016.8	9197	75843	3865.4	38073.1	71686.9	18044.7	80315.6	10854.4	115749	32895.3	18135.7
81	79798.9	414.1	2922.1	1296	12781.1	42370.8	1436.6	9284.2	3339.8	8411.7	1530.1	5074.5	697.3	1048.2
82	151036.1	3623.8	37498.4	12058.5	62355	3473.3	5918.3	37869	7412.3	40075.2	7574.4	36317.6	6331	1095.2
83	77769.8	5945.5	23267.7	16897.3	106311	14856.8	10509.4	62884.9	16410.4	44638.4	4808.9	59441.2	6426.1	1770.1
84	505290.4	18912.3	74965.9	31970.1	383620.5	25978.8	66432.1	288571.9	29250.3	101925.8	23906.9	167635.1	17564.1	3683.6

Continued

Input\Output	57	58	59	60	61	62	63	64	65	66	67	68	69	70
85	867766.3	14920.9	1098750.7	26769.4	72696.4	7830.4	6129.6	264518.5	7282.7	33473.8	6835.1	11179.4	115.7	73
86	1512685.5	351382.4	931194.8	382954.1	1700155.3	95661.8	75606.1	582360.5	72678.5	409797.7	101364.7	300710.3	41875.7	12585.3
87	19638	1233.1	7953.6	1881.4	17324	2623	2699.4	7087.1	2150.5	11045.8	2668.5	9267.5	1512.1	1213.3
88	60404.8	2859.2	3241.4	4040.8	9708.6	2641	797.7	7484.8	1210.2	6613.2	1442.6	4743.2	425	206.4
89	63481.8	3752.9	14715.4	8777.4	62264.2	9679.1	4417.3	35294.6	4449.3	26092	3390.6	25428.4	3800.4	2899.8
90	35533.9	1516.8	16660.2	5181.4	43830.9	11938.8	9040.7	36130.2	10323.8	66206.5	5491.1	32300	2176	2134.5
91	413099.5	41928.7	95571.2	36746.9	265560.8	25989.4	21996	141139.5	27250.3	114375.7	14802.3	85221.1	10719.1	2996.2
92	315630.3	50749.2	158091.6	52844.5	557733.3	45039.6	33247.3	271254.6	57235.2	208576.9	13679.9	214658.5	14226.8	4670.4
93	7237.4	182	969.8	405.9	1503.3	388.6	241.8	1198.8	262.6	926.7	146.2	1141	124	49.1
94	116541.1	15106.4	11031.3	2446.1	64628.5	4278.4	2594.1	19341.8	3830.6	15173.6	1409.2	20412.6	1393.3	330.4
95	3547.6	151	650.6	502.6	7317	1094.7	1032.6	5786.7	1452.9	4727.4	392.2	15842.1	882.3	231.1
96	150189.6	15850.3	45482	20039.1	204633.4	13324.1	10333.1	55898.3	12380.2	44385.7	4179.9	35258.1	5769.7	1174.5
97	1943.5	101.3	231.7	35.9	685.9	3274.7	183.3	1101.3	228.1	1222.6	197.9	2970.6	9.4	107.1
98	10.3	3.4	10.2	12.3	139.4	119.6	4.7	73.4	2.9	157.6	4.6	50.9	1.7	2.7
99	217894.9	29097	326959.6	16672.3	915477.2	129680.8	28797	263555.9	172822	213369.2	38357.9	263066.5	26452.7	42187.6
100	1253000.7	94691.8	337769.5	202632	1551041.2	175415.9	129973.3	854205.5	227754.1	634873.6	67478	864446.9	88153.2	22106.4
101	120363.9	15623.4	35122	48393.2	377530.4	54918.7	43301.6	311425	333267.5	177246.9	10479.7	80005.7	27491.3	3333.1
102	229.1	30.6	10621.4	17.5	1436.2	136.4	30.3	277.2	452.8	224.5	40.3	276.7	27.8	44.4
103	0	0	17740.4	0	1173.2	0	0	0	669.3	0	0	0	0	0
104	216.8	29.3	3548.5	16.6	930.4	129.1	28.6	262.8	182.1	212.8	38.2	262.1	26.3	42.1
105	3587.2	479.5	8213	274.2	12044.2	2135.1	474	4339.7	1118.6	3514.1	631.6	4331	435.5	694.5
106	475868.3	34696.9	169918.5	128389	1763997.4	88044.6	99953.2	437188.4	50320.1	337608.4	15298.2	267327.1	35157.4	15668.5
107	50180.9	3993.1	18380	14363.9	66321.9	10163.4	9478.2	48547.7	8882.6	42035.2	3112.4	26723	4747.6	2039.4
108	8105.6	829.6	3633.1	5768.6	83846.7	4886.4	3379	26964.5	2584	28611.1	299.9	17196.2	514.7	625.3
109	169.6	22.7	2806	13	729.8	100.9	22.4	205.1	144.3	166.2	29.9	204.7	20.6	32.8
110	303230	14000.6	44584	17350.6	251674.9	30685.4	14704.2	75712.2	15617.4	78540.6	19428.1	71935.3	9826.2	5858.3
111	15406.8	1145.7	41993.1	985.8	38850.7	3988.6	748.6	37563.3	3135	30942.8	855	7152.9	681.4	983.8
112	0	0	553.9	35.9	35.9	0	0	0	21.5	0	0	0	0	0
113	14157.9	1797.5	4099.6	5720.4	44108.3	6458.3	5129.3	36538.4	3849.6	20867.3	1226	9236.3	3259.4	375.9
114	65510.4	1808.3	14417.2	22828.9	300666.6	37823	23050.7	218357.7	31435.4	218187.7	7799.2	138647.8	4343.8	3313.1
115	18156.6	1349.8	22424.7	1950.8	10424.5	3047.9	1710.4	26171.6	1018.3	50951.6	5756.4	1914.5	7775.4	62.1
116	0	0	0	0	0	0	0	0	0	0	0	0	0	0
117	0	0	0	0	0	0	0	0	0	0	0	0	0	0
118	29275	1858.9	7466.7	5928.9	82362.1	12490.1	6123.2	40478.9	7503.6	29746.3	5481.7	22287.8	5257.1	1879.2
119	70.7	23.6	4083.9	84.5	1224.6	823.3	32.6	504.9	171.5	1085.1	31.9	350.4	12.1	18.4
120	7864	70.3	1034.8	506.7	7221.9	2932	794.9	25009.7	1794.2	9781.6	1178.5	22537.1	217.6	580
121	48512.9	2333.7	22482.4	11891.5	89681.2	12309.6	7501	44135.3	5944.6	43363.3	3141.1	28419.7	9918	2346.8
122	0	0	0	0	0	0	0	0	0	0	0	0	0	0
123	0	0	0	0	0	0	0	0	0	0	0	0	0	0
124	0	0	0	0	0	0	0	0	0	0	0	0	0	0
Total	29895888.5	2904686.1	11399284.1	8065219.3	38202786.2	5468275.8	4371667.9	22967222.9	9499336.1	16254912.9	1967452.6	22451801	2945383.9	662709.9

Continued

Input \ Output	71	72	73	74	75	76	77	78	79	80	81	82	83	84
1	0	0	0	0	0	0	0	0	0	0	0	0	6710.6	15395.8
2	0	0	0	0	0	0	0	0	0	0	0	26295	329281.8	102.5
3	0	0	0	0	0	0	0	0	0	0	0	0	779513.6	752566.4
4	0	0	4610.7	0	0	0	0	0	0	0	0	0	17599.8	388.7
5	0	0	0	0	0	0	0	0	0	0	0	0	23915.4	1019.8
6	5794.2	10056.4	30970	4195.9	49558.4	381.9	3989.6	14472.8	3630.9	7560.2	0	41455.5	39676.9	192078.7
7	33.2	128.6	241.1	328.4	1415.6	12.5	23.8	99.1	16.3	78.6	0	4411.6	0	0
8	25.6	749.3	4115	353	6699.9	0	0	12244.4	224.1	72.6	3062.4	433.2	0	308.9
9	939.5	8055.2	28790.8	15541.2	220527.4	0	0	4904.4	207.3	2726.6	0	6115.4	0	2777.5
10	36712.1	16270.6	10439.8	86356.9	83375.2	0	0	10004.8	798.4	1495.2	0	1139.5	0	438.3
11	0	0	0	0	0	0	0	0	0	0	0	0	0	0
12	23.6	1690.9	3681.9	308.1	14257.2	0	0	24140.1	178.8	4056.7	69.8	3143.6	51145.6	3258.6
13	115.7	793	6768	1062.3	7899.8	0	0	1076.9	1243.9	1145.5	85.2	8919.4	14877.2	4598.8
14	0	0	0	0	0	0	0	0	0	0	0	0	0	0
15	0	0	0	0	0	0	0	0	0	0	0	0	0	0
16	0	0	0	0	0	0	0	0	0	0	0	0	0	0
17	0	0	0	0	0	0	0	0	0	0	0	0	0	83426.5
18	0	0	0	0	0	0	0	0	0	0	0	0	0	0
19	0	0	0	0	0	0	0	0	0	0	0	0	0	0
20	0	0	0	0	0	0	0	0	0	3894.8	0	0	0	3649.6
21	0	0	0	0	0	0	0	0	0	0	0	0	0	0
22	229.7	11837.3	14819	39647.7	35536.4	349.7	763.4	5492.3	5625.5	28358.3	361.5	106745.2	607380.6	682302.8
23	78.3	412.1	258.8	434.7	1652.1	124.7	253	1298.8	139.4	2070.3	128.1	18470.6	130460.9	18838.7
24	172.9	83.4	23.9	89.8	551	27.4	52	239.9	29.7	164	27.8	513	15707.2	6135
25	52.8	185.1	74.5	281.7	3688.9	83.9	187.5	841.4	77.5	2108.7	95.9	11722.7	344375.3	122614.4
26	47.4	2778.6	324.5	695.3	2222.7	35.5	234.2	563.8	136.1	775.1	48.7	1670.1	9103.2	8748.2
27	10321.1	0	0	0	8046.3	0.5	5622.6	2380.3	0	0	0	15663.1	80609.9	111957.4
28	2789.9	10030.3	10380.9	17971.8	59616	2659.7	1862.8	11672	8449.8	8020.2	2164.9	20378.5	23190.9	20726
29	1845.6	6728.4	2144.4	9924.3	24455.1	1172.3	7006.6	3469.3	3956.2	18409.6	2693.1	8129.6	138506.7	6152.3
30	1914	12754.3	4009.5	4684	40426.7	2268	6771.7	35475.8	4475.3	11962.1	234.5	17409.4	72819.1	16030.8
31	774	22417.8	14317.2	84547.4	146506.7	3510.9	139016.2	7441.2	9768.6	23375.1	1215.3	3397.8	42160.3	63680.6
32	22766.8	33748.1	10363.1	463847.1	548940.2	40453.2	35986.4	176727	44875.5	56417.6	43782.7	2016.6	857658.9	103330.7
33	1485.1	6128.4	1997.2	59194.9	27394.8	5399.5	1191.1	64996.7	7480.9	0	139.5	1406.2	15525	1536.2
34	585.4	2555	771.3	2703	11773.3	26303.5	58.5	12911.1	1337.6	2860	1365.3	1282.6	2062.1	3541.5
35	468	0	0	18.6	0	0	0	849.6	0	0	0	112.3	835.1	10.1
36	7906.8	28397.5	35182.5	83754.9	255839.3	7800.7	14935.2	86669.5	18676.6	34612.6	3650.7	81804.5	101258.2	67765.2
37	226.7	796	5447.1	451.9	1399.3	0.7	1.6	351	13.8	1055.8	0	2686.2	0	11892.6
38	5814	7389.9	5053.1	258570.4	107888.1	507.3	5792	127214	44144.1	5207.4	1150.5	13845.7	24382.1	68353.7
39	0	0	0	0	0	0	0	0	0	0	0	0	0	0
40	0	0	0	0	0	0	0	0	0	0	0	0	0	0
41	15230.2	30563	28800.3	319327.2	854688.5	705.2	10424.9	203860.5	31980.9	15575.2	284398.7	70338.1	221353.1	105857.1
42	1847.5	3031.9	2773.9	11573.8	30707.9	1364.1	4822.3	6599.7	7648.3	37510	475.7	5593.5	60183	20017.3

Continued

Output / Input	71	72	73	74	75	76	77	78	79	80	81	82	83	84
43	3087.8	20169.6	17548.7	220530.9	1038574.2	138108.3	302537.1	1052187.2	43886	22377.3	55318.1	44197.7	220787.6	3387817
44	0.2	4.1	52.7	93.1	117.1	29.5	7.9	229.7	144.9	609.6	5.3	36.5	44.2	10.7
45	371.4	1336.3	2401.5	6248.7	244710	236.4	58622.9	7137.5	9594.7	689.7	452.2	2159.6	77696.9	501657.3
46	62859.4	661287.6	38252.7	161048.7	188008.6	38324.6	56501.1	115683	227710	58995	14249.5	46182.2	16802.6	184511.6
47	24420.3	111491.6	59971.2	1453334.2	2557427.8	279840.6	656296.4	439497.3	243745.6	139769.4	139571.5	20538.4	227660.2	287885.4
48	471.2	3765.8	993	4006.1	19026.6	473.3	753.5	1923.8	1538	1079.2	192.3	9086.6	4164.9	29354.2
49	283.5	1953.5	317.1	1377.6	9125.5	22.3	17.1	690.3	151.1	521.8	169.9	3235.6	10608.4	7464.5
50	560.5	4006	1696.7	3983	58403.4	416.1	1002.5	2389.6	1833.4	1984.4	165.9	14361	4915.8	19850.4
51	1792.1	11691.8	6158.7	33975	406204.5	69300.7	72645.2	1411845.5	98614	124669.7	93976.6	33175.2	5760.9	63885.6
52	2719.5	11114	12650.4	24732.6	227075.1	9312.7	5122.3	45064.1	10543.5	10558.2	1088.4	10602.9	7994.6	67928.3
53	55.4	1844	6309.7	7579.7	185213	0.4	772.2	23146.3	201.7	3429.3	3.7	7476.4	423.7	22914.5
54	512.6	2923.7	18935	14590.7	716215.6	0.9	6441.3	167386.9	58834.4	5356.9	1341.8	8177.9	4231.6	78046.1
55	186.3	3544.4	33592.8	3188.9	89202.3	2.3	46.3	5996.7	2873.6	5846.6	31.2	6754.2	0	25609.9
56	5359.7	31322.3	38140.2	51219.4	288331.3	187.5	201.2	2937.2	2432.8	12951.9	3	85852.1	0	33006.3
57	559123.4	492965.8	439758.1	608386.6	800221	48309.3	14299.4	78316.6	45585.6	271190	107447.5	193155.4	1913.9	97125.4
58	1366.8	2244.2	28115.7	9467.2	27096.6	23351.2	5823.6	26968.3	899.8	2176.3	6359.1	9598.7	0	42725.6
59	3807.3	42954.8	106208	95380.4	3095080.2	598.8	3350.4	68234.3	36254.5	37601.2	2529.8	9494.1	514075.1	39722.1
60	19193.6	68875	145195.1	376190.4	4650529.5	12655.8	5592.9	224343.4	110729.9	74150.2	4151.1	18485.6	248393.7	71411
61	360507.4	366971	208028	1324586.7	1175765.4	137776.8	203281.2	511136.1	236935.1	210959.1	105946	408524.7	698901	370029.8
62	1640	1415492.5	261360.2	175.5	2335.2	295.3	286.2	120	158.8	209.6	11117	26435.4	394.9	899.1
63	2750.8	48268.7	88452	35104.2	87634	11833.1	18778.7	314114	23066.2	19965.8	76179.9	41423	5370.7	8627.6
64	19977.4	1465596.9	416367.5	759666	556364	6999.1	66061.1	48541.9	62960.4	1933111	7339.9	273064.7	26451.8	36777.2
65	0	8682.5	0	0	0	0	0	0	0	0	0	9048.2	0	0
66	3471.3	7311.7	14691.7	84924.8	44081.9	6975.3	3013.3	19273.1	13747.3	10130	650.1	37259.8	20827.6	14431.6
67	0.1	0.1	5074.1	0	0	0.1	0.1	0	0.9	0	0.4	28171.0	0	0
68	5555.5	26280.3	7573.4	28115.3	94494.4	26720.6	8932.4	28656.2	22079.6	8222.8	6874.7	895244.9	10306.8	23342.7
69	0	5075.8	0	0	0	0	0	0	0	0	0	12012.5	0	0
70	0	3808.5	0	0	0	0	0	0	0	0	0	125681.4	0	0
71	966857.6	3173.8	0	0	0	0	0	0	0	0	0	0	0	0
72	224518.6	2645272.8	778.3	1409.3	2791.9	54.3	242.9	1048.1	739.9	697.7	501.3	5083.8	1379	14825.6
73	2172.3	55119.6	323870.7	589521	76841.4	690.3	13193.8	55709.6	40597.4	16417	2591.8	39968.5	7961.6	5418.9
74	59.6	651.7	4778.7	1429400.8	96792.9	119246.4	51387.5	790	854.5	1161.4	324.4	5841.4	1040.2	6061
75	3047.1	280920.9	503001.9	827082.5	2552810.8	472947.6	1669349.7	83693.7	1280036.7	237132.5	200920.9	133549.3	10552.3	240410
76	145.2	666.9	790.4	2366.1	11575	1797149	4454.5	154360.1	103229	15361.2	843.8	27034.3	1929.2	751.7
77	1261.7	3222.1	5508.8	35968.2	100137.7	3177.2	407077.9	15950.3	5210.8	5504.9	670.6	5797	1851.9	6717.5
78	274.3	63158	87960.6	1122456.8	522475	3560161.2	4663560.9	2865301.3	2581358.8	650710.1	467748.4	15996	944.6	24365.2
79	382.3	1723.1	1758.2	6841.4	69916.8	208090.5	100788.8	191582.5	2734053	58055.3	10079.7	7500.1	3667.3	6731
80	2146.4	30337.3	20063.6	23958.3	261401.9	18940.8	9919.8	25802	21867.9	455579.1	4005.5	41494.7	6703.3	13697.4
81	208	2429.4	1047.8	3237.3	7492.9	667.6	396.3	2222.7	1023.8	1357.7	79705.2	7221.4	1721.2	859.3
82	2095.4	4304.7	4051	12846.5	58111.5	2961.1	3972.4	14756.4	9403.8	6866.1	2302.3	369386.6	4203.7	8214
83	4561.7	15997	9005.3	28119.9	70533	15786.9	18823.6	23076.8	19892.4	11010.2	4165.9	11582.8	853554.2	15576.9
84	19849.5	38546.9	23912.1	104227.2	334886.2	17315.3	28084.9	87051.2	56325.3	81739.3	79858.6	34200	75717.2	893765

Continued

Input \ Output	71	72	73	74	75	76	77	78	79	80	81	82	83	84
85	168.3	309.7	10876	1475.4	5341.9	0.4	154	43817.2	175.5	1888.9	74.9	371.4	2496.5	25959.2
86	30282.9	54045.9	63721.3	85513.7	368758.8	13579.4	23579.8	202226.8	45589	61792.5	8036	86428.1	75195.9	184647.6
87	1377.5	1634.1	1749.8	1099.1	6209	554.8	353.4	4251.5	879.3	1249.4	77.3	3304	1492	826.4
88	1489.9	656.7	389.1	494.5	14843.8	0.1	100.1	5824.3	59	1024.5	53	1315.4	227.1	324.1
89	5571.9	3987.9	4279.2	14876.8	32804.9	3056.7	2816.6	16268.7	4362.2	5971.4	1627.5	11198.1	8876.1	6929.6
90	2545	4016.4	6223.3	26863.6	30423.5	1881.7	3710	9283.5	5448.8	6475.5	2095	24875.5	11090.5	9640.6
91	11068.4	20597.1	32486.9	33358	106328.4	20561.5	16481.4	50989.5	26990.1	13757.9	7124.4	27151.7	117339.5	59394.9
92	31291.2	53251.9	38638.6	101156	267929.6	33674.7	63950.8	82978.5	40731.6	37944.9	20038.5	39139.1	73077.4	69523.4
93	31.9	128.9	155.6	369.7	1294.7	28.6	56.3	645.8	79.3	131	44.3	361.4	396.8	275
94	1334.6	3161.3	3890.1	5839.4	23020.3	3713.8	3764.6	6325.6	2882.1	2185	1076.1	3632.2	7053	5679
95	301.2	1798.3	947.7	7392.9	10341.5	15260.4	9004.4	17647.5	14743.6	3190	1755.2	2016.2	995.1	1376.2
96	5304.8	10000.9	7567.5	14798.2	48308.1	3351.7	6499.6	12548.6	6326	8013.6	2045.3	8522.8	109330.1	17709.6
97	90	290.9	227.9	2303.3	2058.1	69	1211.9	962.5	326.1	171.2	21.8	199.1	245.5	235.4
98	1.9	9.1	3.9	18.6	120.2	17.5	1.3	15.6	15	23.1	0.5	16.4	4.8	0.9
99	4087.8	66667.7	31474.3	128843	169814.8	20165.7	31262.9	51709.8	43839.6	101722.9	4974.5	29899.2	124879.9	70684.2
100	167668.1	264441	129364.6	585148.5	1246198.4	371407.7	453585	663248	540104.6	199305.9	108290.8	155054.2	618154	461372.1
101	8573.2	22616.7	38328.7	74155.2	408846.8	13446.3	12993	38322.7	55017.2	39184.5	4082.3	51172	53913.7	63449.7
102	4.3	70.1	33.1	135.5	178.6	21.2	32.9	54.3	46.2	107.1	5.2	31.4	131.4	74.1
103	0	0	0	0	0	0	0	0	0	0	0	0	0	0
104	4.1	66.4	31.4	128.3	169.2	20.1	31.2	51.5	43.5	101.5	5	30.9	125	70.4
105	67.4	1097.3	518	2121.3	2796.6	332	514.6	851.2	722.5	1675.9	81.9	491	2054.7	1162.4
106	48102.2	54298.6	59272.4	119560.1	333807.4	42869	62006.5	119427.5	67447.8	97570	13535.5	65533.6	84510.7	87434.8
107	1989	6464.3	7162.3	16001.7	69564.3	5201.7	10219.9	17713	6980.3	8734.4	4263.7	14749.6	20236.4	14818.7
108	12359.2	3864	4287.6	40823.9	50260.6	20342.5	18402.8	40954.2	20778.6	13646.1	4005.6	5282.1	17582.9	24212.1
109	3.2	51.9	24.5	100.3	132.2	15.7	24.3	40.2	34.1	79.3	3.9	23.1	97	55
110	4498.9	11316.7	13114.5	56014.7	153146.5	12538.4	15210	27854.1	16263.2	16752.3	3872.1	32016	9008.2	15719.3
111	101.6	1516	1507.5	14131.7	9989	1144.5	1202.8	3038.9	5666.4	2956.4	254.3	2829.7	3993.7	5655.4
112	0	0	0	0	0	0	0	0	0	0	0	0	0	0
113	1017.3	2630	4503.4	8719.2	48400.4	1590	1529.7	4520.7	6350.3	4577.7	483.4	6056.2	6223.3	7502.1
114	15351.9	57772.5	14287.4	442991.5	196015.4	43126.1	128333.6	40180.1	101610.3	29046.1	8894.5	38004.9	46267.4	149436.8
115	162.5	312.7	2616.6	1033.4	8460.5	86.3	196.6	284.1	310.3	961.5	65.7	1481.4	2266.9	11105.5
116	0	0	0	0	0	0	0	0	0	0	0	0	0	0
117	0	0	0	0	0	0	0	0	0	0	0	0	0	0
118	1579.4	4333	10734.7	4927	28260.2	1391.7	6835.5	6184.4	9138.5	7078.5	1676.2	11499.5	6842.7	6316.4
119	13.3	62.7	26.9	128.3	827.2	120.3	9	107.6	103.3	159.1	3.4	113	32.7	5.8
120	270.8	3153.1	1906	18158.3	10844	8389.7	3351	10012.5	10596.2	1845.2	1920.6	919.7	701.8	3161.9
121	1499.1	6242.4	5249.1	12939	38525.1	2936.2	8398.5	12466.8	11804.2	7559.4	3384.5	8086.3	5771.9	19277.8
122	0	0	0	0	0	0	0	0	0	0	0	0	0	0
123	0	0	0	0	0	0	0	0	0	0	0	0	0	0
124	0	0	0	0	0	0	0	0	0	0	0	0	0	0
Total	2778406.1	8844953.8	3666330	12782825.4	26966037.2	7763015.8	9586458.7	10423967.3	9349695.3	5450747.3	2037906.9	4190995.9	8499859.7	10475964.5

Continued

Input＼Output	85	86	87	88	89	90	91	92	93	94	95	96	97	98
1	0	0	0	0	0	381494.5	0	0	0	136.1	0	0	111508.7	0
2	0	3931.5	0	0	0	0	0	0	0	0	0	0	36.3	0
3	0	0	0	0	0	0	0	0	0	0	0	0	5	0
4	0	0	0	0	0	0	0	0	0	0	0	0	10.7	0
5	0	1199	0	0	0	339097.1	0	0	0	0	1317	0	87.7	0
6	0	6582482.9	340512.1	336926.2	8302.9	105071.6	160413.7	38037.6	1057.7	356.8	0	19834	3160.8	0
7	0	731521.2	52711.7	41305.3	1043.6	0	0	0	0	0	0	36121.1	484.4	0
8	0	25570.3	238.9	1218.2	147.5	0	0	0	0	0	0	0	0	0
9	0	4629.1	409.4	0	0	0	0	0	0	0	0	0	0	0
10	0	73313.4	132.1	0	144.6	3261566.7	36200.6	0	0	0	0	0	0	0
11	0	1282.1	498.4	0	726.5	1190263.9	2152.2	51.5	60.3	291.7	0.4	54050.9	297.7	0
12	0	1431.5	290.3	375.2	1495.8	0	0	403	115.7	14268.5	6.9	4120.2	186.1	0
13	0	8398	395.9	0	0	0	0	0	0	417.4	0	809.2	137351.7	0
14	0	0	0	0	0	0	1437.3	0	0	26.3	0	0	1.6	0
15	0	0	0	0	0	0	0	0	0	725.4	0	1621	92.7	0
16	0	0	0	0	0	0	0	0	0	356.3	0	728.7	68.5	0
17	0	0	0	0	0	0	0	0	0	871.9	0	11054.3	30.4	0
18	0	0	0	0	0	0	921.8	0	0	968.7	0	3619.3	75.8	0
19	0	0	0	0	0	111106	0	0	0	501.9	0	11298.5	303.6	0
20	0	0	0	0	0	0	6053.8	0	0	0	0	4706.3	142.4	0
21	0	0	0	0	0	0	0	0	0	0	0	0	0	0
22	0	13247.3	462.9	7057.2	462.9	322162.4	17821.7	3827	66	427.9	269.3	9146.2	2936.5	906.3
23	0	49518	74.3	276.4	115	5582.1	3193.9	400.7	17.1	16.8	124.7	1435.9	2	0.9
24	0	796.2	35.6	23	31.4	10205.5	808.8	223.1	0	0	0	1178.7	3584.6	0.7
25	0	2408	39.2	302.8	74.3	12371	261.4	0	0	0	0	54.5	0	0
26	0	1963.3	37.7	28.6	116.8	0	687.6	4763.7	0	0	0	3658.9	3332.1	0
27	0	0	0	0	0	9428.2	1067.4	0	0	0	0	0	0	0
28	0	99769.7	3169.8	4447.8	13962.3	210891.5	18302.1	31082.9	639.4	7971.5	1534.3	52014.8	1720.5	43879.4
29	0	23170.6	462.5	871.1	2349.8	1329.1	12913.4	17190.4	266.4	282.3	217.7	11067	136.9	11
30	0	22267.6	999.4	774.9	3384.7	2744646.7	34650.6	7267.9	32	191.2	8.2	4549	374.1	347.8
31	0	13173.2	253	560.8	2456.9	927195.3	9387.8	12823.7	245.5	156	218.2	19751.9	3324.6	14548
32	0	7769.2	77.7	704.3	155	70403.2	5215.7	17249.5	242.3	243.2	2125.9	14878.9	472.1	39868.5
33	0	9871.2	225.1	697.7	1922.6	34373.8	7327	22343.4	389.6	1709.5	2094.2	13250	834.7	254450.2
34	0	7069.8	176	353.6	675.7	16606.6	7909.8	26023.1	500.9	2688	1881.6	12189.7	2798	77858
35	0	428.1	0	0	0	0	1365.1	1897.8	108.6	232.9	68.8	1700.3	0	0
36	0	1984682	65773.2	123322.9	22285.4	4903074	705877.4	2150991.2	37949.3	513529.3	161628.9	202224.6	23747.8	29765.2
37	0	1021.5	21.3	4643.6	0	58215.3	209.8	36105.3	370.4	97.4	0	0	0	0
38	0	69412.1	8484.6	12525.9	108917.3	284882.5	3192.8	908.5	0	0	0	1509.3	190.5	0
39	0	0	0	0	0	0	0	0	0	0	0	0	0.4	0
40	0	0	0	0	0	0	0	0	0	0	0	0	480.3	0
41	0	30506.9	2088.8	4793	5447.7	1001861.1	12942	4381.3	157.8	8918.8	4.6	19209.8	325.4	374.7
42	0	28090.7	538	1243.4	35232.1	19546.1	6024	11640.4	159.7	1176.2	756.7	18811.3	525.3	0

Continued

Input\Output	85	86	87	88	89	90	91	92	93	94	95	96	97	98
43	0	46338.3	2300	3840.8	13403.7	805030.9	2521	3985.3	619.1	1097.3	44.6	3717.8	345.7	161.8
44	0	2658.2	72	5.2	170.4	62862.6	1570.6	7638.9	302.8	4271.3	134.7	8186.9	568.7	410
45	0	0	0	0	0	428517.3	279.5	829	95.1	196	11.5	7205.4	135.6	0
46	0	57028.7	2715.6	2007.1	3729.7	514807.5	13557	648198.2	1227.2	1788.9	598.4	33873.8	892.4	2860.9
47	0	38584	1432	1758.6	8409.1	519065.9	3171	6057.8	260.3	2017.2	202.4	11543.6	1524.2	3505.1
48	0	57330.6	3685.8	1743.7	8292.4	14401283.8	8806.9	5533.7	832.8	545.6	33.7	38387.2	259.2	1209.5
49	0	38282.2	809.7	497.4	4116.1	9208559.4	60191.1	5862.9	393	20.6	0.1	10870.3	301.5	0
50	0	65805.1	2984.2	1693.4	11419.6	15705935.1	6311.7	2719.6	350	211.2	0.6	27122.8	303.5	31.3
51	0	31699.8	1020.8	1165.8	5338.6	1037248.7	2076.3	5821.7	828.9	141.9	34.4	1313.6	83.8	1442.6
52	0	54219.1	1371.5	1771.9	7826.9	3378623.8	1399	350	43.8	1318	27.6	1143.8	1.2	2893.3
53	0	23179.9	1109.6	505.7	276.8	1932432.4	3299.1	1215.8	185.4	72.1	10.5	7498.2	172.9	0
54	0	23518.6	198.5	19.8	266.2	1391716.4	5735.7	1072.3	23.5	33	0	6842.4	6	0
55	0	0	0	0	0	0	0	0	0	0	0	0	0	0
56	0	0	0	0	0	0	0	0	0	0	0	0	0	0
57	0	69176.1	3008.5	4141.3	5470.3	9842701.4	98878.7	5241.4	2047.5	328.2	6.5	14645.6	9377.2	1583.7
58	0	0	0	0	0	27387.5	0	0	0	0	0	0	0	0
59	0	0	0	0	0	0	0	0	0	0	0	0	0	0
60	0	17313.2	421	364.8	1172.2	930839.6	1316.9	2687.1	113.7	98.3	0	140.8	0	0
61	0	174385	6358.7	5370.6	108470.1	10493494.4	25473.6	50870.6	905.6	3798.5	860.3	44409.5	1310	10969.7
62	0	224157.9	6457.3	502.8	247.9	0	11583.7	4376.9	172.6	1709.2	8.1	5379.5	151.7	0
63	0	46019.8	991.5	1632	2191.5	507719.4	1465.6	49512	1.5	941.3	2393.8	7379.6	0	0
64	0	1128308.9	24978.8	36079.5	56688.6	2843689.4	142097.2	271052	9738.8	88478.9	43471.3	246314.9	7543.8	22796.2
65	0	0	0	0	0	0	0	0	0	0	0	0	0	0
66	0	203304.9	6535.9	9948.9	3934.6	1624202	3691.6	5324.4	1715.9	989.1	504.8	15742.6	205.2	21406.5
67	0	0	0	0	0	0	495039.8	1135083.5	0	0	0	1009	138.8	0
68	0	184085.1	19956.5	11223	25584.8	61228.2	14027.5	0	808	2505.9	413.2	45488.3	2805.7	26533.9
69	0	0	0	0	0	0	0	0	0	143432.7	0	6304.6	0	0
70	0	0	0	0	0	0	0	0	0	0	66926.3	383.1	0	0
71	0	0	0	0	0	0	230.2	0	0	0	0	339.9	0.9	11129.7
72	0	15300.4	1039.4	351.5	1637.3	19133.3	3720.4	57904.1	599.1	9558.2	4337.5	38070.5	414.4	337.6
73	0	47520.3	1084.6	375.1	3829.8	14200.8	2593.3	3047	227	1341.4	51.6	9015.4	75.5	606.2
74	0	15158.7	176.6	490.9	1088.9	809102	3371.5	2669.2	111.1	252.5	2317.3	6736.1	66.5	2273.1
75	0	1541830.5	12961.7	5841.5	35038.5	6994314.6	43457.3	9894.2	875.6	752.6	611.9	31093.3	490.7	33229.3
76	0	32876.3	402.4	525.4	2809.8	18900.2	4694.4	4728.6	251.8	552.4	1679.9	4987.6	278.6	333.5
77	0	15252.1	479.5	432.4	2233.4	192030.8	1516.9	6148.9	0.4	591	3311.6	3382.1	35.1	6142
78	0	43904.2	557.1	468.2	2825.4	25864.8	5902.9	2643.6	66	81.5	349.2	2981.9	16.5	0
79	0	94092.9	2453.8	942.4	21420.3	1028974.1	11402.8	3507.7	137.7	951.3	268.2	34443.8	42.7	4083.3
80	0	592604.4	7404	5981.2	3982.9	17650.5	4728.8	4644.8	1081.5	418.2	590.6	16040	102.2	8423.6
81	0	30195.6	488.9	764.1	22790.6	369469.8	2323.7	12055	9.2	390.8	219.4	3614.9	1720.8	311.5
82	0	264837	7876.9	13019.4	6067.6	507719.4	96853	267078.8	7803.2	83694.6	50435.2	162080.8	6717	37694.6
83	0	63163.5	2258.2	2332.1			15226	26094.9	570.6	3448.2	2563.9	12010.5	1018.2	2882.9
84	0	241997.9	8422.3	5828.2	44275.2	53383.9	9696.5	33488.2	777.9	10484.9	98.9	8386.7	452	312.5

Continued

Output \\ Input	85	86	87	88	89	90	91	92	93	94	95	96	97	98
85	0	0	0	0	0	0	0	0	0	0	0	0	0	0
86	0	1264401.1	43151.2	73875.3	765326.5	1142954.2	234898.1	94502.1	69744.8	1711.8	770.1	124299.7	7456.6	36436.8
87	0	21998.6	30259.9	4211.9	1427	76677.2	5738.3	17253.8	375.9	148.4	172.2	3250.2	676.1	6118.9
88	0	3216.8	1022.7	68609.4	60.4	15104.2	184.4	1543	0	396.8	3.7	2428	290	454.1
89	0	110059	12863.4	3946.2	224208.3	181249.7	8251.1	33562.1	383.1	2052.6	546.1	24334	1327.7	6182.3
90	0	104976.8	2593.6	6557.5	10404.5	100902.4	304835.1	48773.5	6809.3	2389.3	1676.7	136500.1	11579	156275
91	0	532932.6	40795.9	26339.4	5894.1	824110.7	25910.4	29691.8	625.1	3471.1	1058.9	28067.1	9432.8	40946.4
92	0	237403.5	11451.6	12822.2	7558.5	1654223.8	35085.7	155419	1000.1	4292.3	4823	34288.3	10844.8	11964
93	0	15360.8	787.6	902	101.5	18138	2541.9	7714.5	1545.4	1842.4	579.5	1111.9	90.3	106.8
94	0	59955	3487.4	3152.5	634.5	93916.7	4312.1	10344.6	162.5	12782.8	635.9	2737.3	340.5	789.3
95	0	3448.8	85.5	78.7	249.1	24798.6	388.9	2525.9	25	118.5	5027	8496.7	38.6	49044
96	0	265975.3	13546.2	14737.2	2881	326321.2	23206.2	589254.1	227.8	172367.3	55263.5	246042.3	7821.6	5649.3
97	0	175	77.7	71.2	57.1	183833.2	0	12254.8	260.4	9353.7	126.3	63177.4	34372.5	59.7
98	0	50.2	11	1.1	12.6	475.5	7610.7	16477.6	75	532	263.4	3944.2	323.2	0
99	0	80170	3076.9	15629.8	55530.4	3204103.4	123070.8	126588.7	936.8	14374.3	3098.8	67256.5	4580.1	10030.5
100	0	1889763.7	91996.2	89989.3	37064.4	7726505.6		200294	3382.7	112124.4	11981.8	99121.8	12332.5	87694.3
101	0	75327.6	12076.7	4568.3	16027.5	588579.4	10650.3	137118.5	838.9	6112.1	12271	73666	8848.7	8739.6
102	0	84.2	3.2	16.4	58.4	10637.2		32903.4	948.1	1780	903.1	40075.2	3411.1	16.7
103	0	0	0	0	0	8939		41398	175.1	1607.9	734.1	11599.5	1187.3	0
104	0	80.1	3	15.6	55	7355.4		10485.7	193.2	3534.9	33.2	46933.3	142.5	15.8
105	0	1318.6	50.7	257.3	914.2	16315.2		15529.2	112	2826.2	3779.2	46267	592.4	261.1
106	0	629836.4	12628	10352.2	18314.2	649191.6	86200.6	141363.4	3554.9	45325.9	23319.8	83020.9	27097.6	6937.2
107	0	144777.7	4886.3	4116.6	7127.6	422163.1		187619.1	1086.8	22085.4	9265.4	41506	5693	7401.8
108	0	4287.8	331.3	820.6	280.2	20015.7	452.8	18218.8	58.1	256.5	2012.3	32799.2	2779.7	5606.1
109	0	62.3	2.4	12.2	43.3	323.8	7246.1	100933.5	490.6	4956.4	1348.6	68671.4	2621.7	12.3
110	0	345050.1	18662.6	14748.6	69909.5	98717.9		16386.1	143.9	154.9	405.2	4646.4	274.1	10094.9
111	0	2854.8	112	325.1	1152.4	55829.5	37095.2	164336.5	1030	4148.5	5943.5	57424.7	2443.4	736.8
112	0	0	0	0	0	0		2324.3	283.1	13.2	18.5	721.2	38.2	0
113	0	8917.9	1434	534.1	1882.1	68204.3		884.5	0	29	724.1	1322.5	0.9	1032.9
114	0	56999.1	2850.8	2847	5830.1	3259819.5	1313.1	83045.5	18077.7	23210.5	61289	113668.6	8097.4	33511.6
115	0	30117.2	1560	340.9	979.3	20599.4		13975.4	344	775	111.2	2597.4	342.7	60.5
116	0	0	0	0	0	0	15550.7	0	0	0	0	0	0	0
117	0	0	0	0	0	0		0	0	0	0	0	0	0
118	0	31343.4	1577.4	2313.7	6573.3	250342.7	20423.8	58741.8	2166.4	3375	6064.6	26395.1	1087.6	7035.8
119	0	345.8	14.3	7.3	87	27599.5	4561.2	5379.9	30.4	185.6	828.7	4879.7	295.4	11574.2
120	0	3921.9	79.8	102.8	740.1	7174.5	176.5	141.5	42.9	1.8	7.3	151.8	2.6	725.7
121	0	116176.1	4471.8	1127.2	2774.4	97118.6	693.3	11761.8	611.3	176.4	96.1	4949.8	277.6	44.1
122	0	0	0	0	0	0	0	0	0	0	0	0	0	0
123	0	0	0	0	40006	1413185.8	0	0	0	0	0	0	0	0
124	0	0	0	0	0	0	0	0	0	0	0	0	0	0
Total	0	21324441.8	926072.3	1027406.5	1912091.2	123880171.2	3124946.2	7423696.5	191296.7	1381587.7	569957.2	2919032.5	507502.5	1171133.5

Continued

Input\Output	99	100	101	102	103	104	105	106	107	108	109	110	111	112
1	0	50223.8	655061.3	0	0	110.4	0	0	0	0	26276.6	12370.9	60779	299.7
2	0	66214.4	50645.8	0	0	0	0	0	0	0	21071.2	9368.9	2288.6	102.7
3	0	567906	1806961.9	0	0	0	0	0	0	0	6784.5	9540.1	83091.2	74.7
4	0	87043.9	2512168.5	0	0	0	0	0	0	0	0	1506.1	55030.2	318.8
5	0	39870.1	0	37921.2	4970.2	992.6	641.1	145617	4069.6	5860	5333.6	237.5	3728.3	3.5
6	0	100687.3	62476.2	0	0	0	0.8	0	0	100888.2	45668.5	109713.9	51576.6	230
7	0	0	0	0	0	0	0	0	0	0	8364.9	1149	0	0
8	0	0	53519	36.6	100.4	0	0	0	0	0	39	754.3	366	0
9	0	0	0	0	0	0	0	0	0	0	0	0	0	0
10	0	0	0	0	0	0	0	0	0	0	0	0	0	0
11	0	0	0	155114	0	0	0	0	0	0	21581	460.6	0	0
12	0	0	0	749.5	0	0	0	0	0	26731.6	3255.5	5780.2	0	0
13	0	16875.2	475.9	0	80.2	874	34.2	0	0	0	15239.5	1568	2461	7.7
14	0	1061159	848922.4	5918	0	566.3	3990.5	0	0	0	1704	5809.7	20127.4	191.3
15	0	15643.1	88624.2	0	0	59.1	508.6	0	0	0	5652.3	531.4	752.4	13.2
16	0	33110.7	1529519.6	0	0	3419.5	7705.4	0	0	0	1034.2	6875.5	87105.8	420.3
17	0	92441.9	682143.8	583	0	872.9	933.4	47413	1582.7	0	7040.2	1587.4	43285.8	493.5
18	0	105176.5	638588.8	0	0	82.3	93105.1	0	0	6271.7	7650	10618	40133.5	2027
19	0	286016.7	1365872.8	0	0	455.6	8251.1	0	0	0	17330.9	510351.9	167587.2	901
20	0	663824.5	638578.6	31719	0	1121	97211.5	0	0	0	1730.9	27762.1	89038.4	1483.1
21	7015	902296.8	732242	0	0	0	733.4	3072.3	1771.3	6061.8	10749.6	17498	51091.5	297.2
22	10.5	282906.8	25813.8	18167.1	698.2	1551.4	736.8	2516.4	929	2636.6	1388.4	752651.6	64864	698.1
23	8.7	130314.1	0	7312	530.2	3	257.3	0	0	0	0	113247	10074.1	32
24	0	2996.9	0	288.9	0.9	0	0	0	0	2139.3	1964.7	4769.4	420.1	0
25	0	279977.7	0	914	2.7	0	1.4	0	0	0	4582.3	149513.6	8676.5	538.7
26	0	46809.6	0	1463.1	185	0	0	0	0	0	0	5553.5	26440.3	26.7
27	0	0	0	12619	995	0	0	0	0	0	0	8713.4	0	0
28	118813.4	1057142.8	59610.1	7408.2	9456.1	7883.4	8053.8	65705.5	11115.6	27979	33218.1	105344.8	59298.5	2010.1
29	25.6	19527.3	0	4042.9	51.2	259.3	3620	913.5	2229	1410.6	7656.3	6149.4	5046.4	131.4
30	5566	72389.5	4381.4	15713.6	142.6	28.3	106.4	7517.9	334	11316.2	8824.9	110238.3	3466.7	151.5
31	62596	841113.1	140914.4	4869.9	1687.5	1411.8	737.5	142534.2	31528	17747.1	11811.2	434696.4	27655.6	277.6
32	44309.5	766792.2	34850	2236.8	1791.5	2131.3	2646.7	200280.7	42689.4	39502	18974	175877.9	34665.4	621.4
33	105946.5	2592557	11756.1	3805.5	7652.2	2919	4802.1	243258.1	113814.8	27867.9	34761.8	59645.9	30899.7	1239.3
34	45368.6	17294.1	12712.6	3570.6	5629.2	2461	6305.8	233457	59219.9	56267.6	30716.8	70628	25719.1	2499.3
35	0	82310.4	437.8	507.9	507.9	96.9	271	10497.5	980.1	4840.4	23383.5	32068.2	3480.8	362.7
36	98667.7	1656160	51706.3	213110.9	16948.4	547112	375917.5	97406.6	23592.6	45870.2	1242080	78446.7	55057.7	757.7
37	0	0	0	89.3	476132.7	741.1	0	4.6	0	0	12096.1	0	0	0
38	0	0	0	1427.7	59.7	0	0	0	0	0	4828.4	2308.7	245.6	0
39	0	0	0	0	0.3	0	0.5	0	0	0	2082.9	0	0	0
40	0	835417.1	850.8	0	0	0	0	0	0	0	2371.7	0	0	0
41	5556	158672.1	32189.3	6236.7	1076.5	2848.9	12.5	2706.4	1060	5057.1	14085.5	201025.8	3534.6	91
42	0	60392.2	40219.9	4196.2	1367.6	1620.8	1903.2	27913	16825.7	14903.5	17513.9	723157.1	97650.8	464.2

Continued

Input \\ Output	99	100	101	102	103	104	105	106	107	108	109	110	111	112
43	3080.8	1920.3	0	1104.3	112.6	300.7	439.6	13992.3	4472.4	5645.3	11813.8	443490.4	23080.9	367.3
44	2239.1	464328.8	8871.4	652.9	4255.8	765.7	344.9	5769.4	932.7	2332.2	42502.4	8151.6	7351.2	282.8
45	0	0	0	592.5	3.2	74.6	259.6	0	0	2	0	0	0	0
46	14591.9	132532.4	7365.2	6131.9	96247.9	6444.1	6380.3	3945.6	670.6	1177.7	151250	26454.1	10275.7	58.2
47	30675.6	671675.2	29247.9	4071.3	945.7	1003	4475.1	7170.3	3009.6	3016.6	9612.7	53019.7	23752	63.4
48	17897.7	376507.2	3095	3258.3	1018.5	4.7	128.8	6590.3	446.4	260884.5	42203.7	9008.5	1907.6	224.5
49	0	0	0	25234.9	661.4	0	30.6	2240.9	1777.1	28801.1	45560.2	4634.8	8985.9	99.3
50	364.2	459648.9	3113.5	2805.3	1158.8	0	38.6	14867.7	1730.1	411407.7	36495.6	111097.5	7263	173.2
51	5724.4	55487.7	7590.4	2987.7	1041.5	22	285.3	11546.8	3219.3	8209.8	7151.9	6223.4	3566	20.6
52	10819	108650.2	56459.4	974.3	178	201.5	144.5	5295.5	1400.9	8526.3	3286.5	4299.2	23467.4	65.4
53	0	0	0	1035	30.1	149.6	78	663.9	426	3589.8	1817.2	21278.1	662.5	0
54	0	0	0	1673.3	606.1	37.8	1	1785.1	168.1	359.1	4704.1	4626.3	467.3	171.9
55	0	0	0	0	0	0	0	0	0	0	819.9	0	0	0
56	0	0	0	0	0	0	0	0	0	0	1164.6	0	0	0
57	27657.7	48640	5825.6	36905.6	2456	95.2	62.4	0	0	17705.5	20102.4	50.3	20.2	116.3
58	0	0	0	0	0	0	0	0	0	0	5685	453.7	0	0
59	0	0	0	0	0	0	0	0	0	0	821.8	1708.6	13.6	0
60	0	0	1891.3	403.7	42.3	9	0	0	4517.6	11169.6	2368.6	31467.7	21036.9	353.3
61	60229.9	530609.2	53359	10823.6	5943.6	1305.7	2226.6	57520	0	56555.5	70155.8	2690.8	5915.2	0
62	0	126759.4	3956.5	6589.5	1035	247.9	66.1	0	0	7948.2	3267.3	1003.5	0	0
63	0	0	0	631.2	5345.8	84.9	1318.3	0	0	0	25899.8	0	0	0
64	408271.7	504532.1	13156.6	44138.2	34926.3	85770.8	157124.7	184939.4	43297.1	75125.9	382324.7	44674.9	40281.7	1847.1
65	7.2	0	0	0	0	0	0	0	0	0	1199.3	0	0	0
66	14461.9	1231336.3	7536.4	2102.1	215.7	839.2	670.9	109347.7	7323.4	5703.9	11154.1	87688.3	6429.7	25.8
67	0	0	0	229409.8	0	0	0	0	0	0	27755.8	0	0	0
68	103778.4	2818439.8	5709.9	6244.5	235747.1	656.3	25315.8	33614.5	10104.7	33456	411319.8	43369.5	9318.2	3379.2
69	0	0	0	0	0	205884.3	481182.2	0	0	0	61.4	0	0	0
70	0	0	0	0	0	0	0	0	0	0	0	0	0	0
71	39651.2	112481.5	1258.1	89.4	19.8	0.6	6089.4		6554.8		722.6	150	8	0.7
72	3501.9	860883.3	0	1315.2	5576.9	599.8	403.4	12908.5	20556.4	6093.2	39579.2	9592.6	504.1	21.9
73	8372.2	0	0	2078	209.3	2049.6	3327	1866.4	3276	1509.1	13002.5	490.2	637	38.3
74	16493	19868.8		2354.7	645.5	1736	2951.6	81518.6	4494.3	11919	5647.4	57430.4	63297.4	248.5
75	1832330.9	2539166.2	11775.2	22901.5	2216.8	1747.5	4155.8	65701.7	2867.2	153311.3	62367.4	12859.1	32207.6	736.7
76	3459.2	211932.6	6524.7	894.5	2027.8	3371.2	7252.8	361614.7	21293.9	16079.4	8960.8	5171.7	15460.1	752.3
77	35407.3	1637443.3	10996	362.1	494.6	89.1	129.7	41832.8	14429.2	2533	2283.1	13415.9	16639.2	150.3
78		47752		4430.1	103.4	525.9	1087.3	40603.7	859.4	1024.1	2230.3	20632.2	1794.2	18.2
79	438258.4	138358.4	566.9	3102.4	320.6	944.7	1325.8	117606.4	8565.5	19213.2	11409	2418.6	9244.5	309.9
80	288968.7	78537.2	0	2308.9	167.5	212.4	398.2	6729.6	404.8	669.4	11679	2648.8	7184.4	42
81	3230.5	221775.1	1160.6	957.3	1562.5	191.6	181510.2	321215	13719.5	22562.6	4354.4	225914.9	18806.7	652.1
82	718962	431031.2	2606.8	55480.2	64218.4	133825	27338.9	172375.9	39875.6	59995.3	327252.6	38563.8	36239.8	1693.1
83	27102.5	506809.3	43792.4	5414.7	6623.4	3925.7	266.9	150538.5	28668.7	28260.3	26783.2	92507.9	72699.1	6257.6
84	2189.4	196154.1	18264.3	4791.4	1607.5	2590.9		83945.5	42357.8	120371.7	44045.3	159098.9	33360.1	532.7

Continued

Input \ Output	99	100	101	102	103	104	105	106	107	108	109	110	111	112
85	0	0	0	0	0	0	0	0	0	0	0	0	0	0
86	422635.1	852853.7	122449.4	66318.2	21658.1	3084.8	6363	139970.7	21415.3	75210.5	228844.1	121063.4	182976.3	8814.6
87	24544.4	130744.8	29945.7	3047.8	2260.9	668.1	383.7	26604.4	4313.4	12760.2	13584.4	6050	10066.2	277.3
88	5276.4	29693.1	59865.3	149.6	1558.4	211.2	24.2	5286.6	123.4	5638.5	17243	15167.3	9265.6	80.4
89	45041.7	164541	47462.5	4119.4	10177.5	4042.8	2025.3	38209.9	8025.4	13481.8	29672.1	41924.9	37104.2	1475.6
90	500658.5	563088.8	15171.4	162396.6	23982.2	709.6	6134.2	603112.1	61872.6	722604.1	357510	280136	218152.7	5887.3
91	26952.1	77135.8	31530.3	8723.8	3472.7	3637.4	5376.9	10990.8	3511.3	19612.1	17631.4	27548.3	8671.3	850
92	17897.3	185151.4	57078.6	16072.6	13619.6	4450.1	13795.9	19322.1	5087.9	33026.4	32747.4	46055.9	12032	1896.9
93	354.5	5982.5	434.3	768.8	1709.9	1962.1	1348.6	350.4	357.6	168.9	4580.8	2835	225.9	2.8
94	1264.1	14942.5	4554.2	1300.8	2128.2	2359.7	1608	1314.1	825.9	1691.7	6247.5	3870.9	966.8	30.7
95	29141.7	11830.3	740.1	170.8	479.3	121.9	1025.1	13267.6	11935.9	1863.7	1455.4	28567.5	1675	489.6
96	4293.8	699508.8	14942.9	8990.1	81783	235468.9	195852.2	16397.3	9221.4	9795.3	17189.9	844.9	4965.8	6302.2
97	931.1	217765.9	513	0	1821.6	1805.6	280.1	3805.4	15581.8	1481.1	517.8	7519.5	486.8	91.8
98	98.2	149877	4141	0	1300.8	205.5	336.6	111716.8	86015.9	3545.2	6142.7	62479.9	9451.5	3059.5
99	0	697293.5	39181.9	3955.9	29194.2	7008.5	12149.9	382260.3	67671.3	47908.3	67638.1	459364.1	101234.5	35349
100	220666.8	9063035.9	1002111	44353	45231.6	35494.9	63324.9	289521.3	140198.4	160665	176110.8	89804.7	149980.9	62210.5
101	55077.9	1822221.1	26295.2	5807.3	33584.3	7300.5	54469.6	287504.2	35729.9	124752.7	83023.6	24860.5	38324.8	311950.3
102	172.9	363579.1	34.5	0	5498.4	1070.3	866.4	135626.5	18762.3	10641.3	17676.4	8822.1	7829.1	401779.7
103	29.6	159745.8	0	0	37448.7	1359.9	924.5	60327.8	24627.4	6964.9	24969.7	10693.8	6367.2	186308.9
104	154.8	226618.9	32.6	0	3968.3	23972.8	149	82198	57328.9	8906.6	18398.4	34092.5	2276.8	82713
105	2210.4	324427.6	535.4	0	3766	2886	33330.4	213910	4117	28434.2	22910	9549.2	16682.4	694290.8
106	76328.1	3613660.3	53227.5	30232	28748.8	45310.2	78517	2852930.4	141243.1	813220.6	32472.3	21192.6	99015	66664.3
107	45741.8	582704.4	16868.4	0	57335.2	17694.2	35552.8	122345.2	319914.2	41610.9	109402.2	197268.9	38721.4	11842.5
108	57414.8	1446493.8	75314.6	23.7	10432.8	1338.9	14174.1	1038237.9	81727.8	111168.7	29052.8	42992	31865.4	21033.8
109	103	104574.5	25.5	4552.4	25703.8	4332.3	10716.8	164554.5	18898.5	60327.1	423149.9	213757.8	26915.6	153994.4
110	92807.4	1150170.6	13533.7	0	4603.7	1267.4	224	39836.7	155944.8	19157.3	32801.7	48280.2	42880.9	171132.6
111	11421.4	397486	19375.4	24216	22851.2	4807.9	43433.8	441042.5	1560.2	44488.4	51492.1	5894.6	122247	280200.8
112	0	0	0	0	286	452.1	158.1	8171.5		3797	5373		15534.3	488083.5
113	6489.4	212262.7	3114.2	0	182.6	27.1	1872	20782.8	1827.4	2074.6	1053.3	1596.5	4009.1	7064.2
114	713734.5	2874555	421381.1	2970	15465.8	7652.9	176356.9	1444768.1	341516	262100.8	114497	140246.4	83520.5	60505.1
115	882	3360	20377.6	0	2346.1	211.3	333.4	7563.1	2433.1	2110.3	14968.3	1379.4	2260.3	2299
116	0	0	0	0	0	0	0	0	0	0	0	0	0	0
117	0	0	0	9179	18847.8	0	0	0	0	0	0	0	0	0
118	29697.7	264889.8	13366.9	10908.1	6915.5	5405.5	18466.3	87723.2	16426.7	13015.5	45575.1	25794.9	9123.4	1943.7
119	151782.4	0	0	3139.3	1745.8	588.2	2230.5	51844.3	20756.2	6232.9	18625.1	7039	11836.3	17611.7
120	4963.8	13699.5	4.3	136.4	734.2	0	22.5	2797.4	44.7	813.8	480.9	37.6	4.1	61.1
121	211.3	247707.9	0	369.5	0	589	90.2	22014.7	3758	12400	15904.5	1249.9	5816.4	829
122	0	0	0	0	0	0	0	0	0	0	1079.8	0	0	0
123	0	0	0	0	0	0	0	0	0	0	2645	0	0	0
124	0	0	0	0	0	0	0	0	0	0	0	0	0	0
Total	7160595.2	54136385.7	14538713.8	1446173.5	1528172.2	1462826.5	2317214.8	11894869.6	2286559.5	4470180.1	5602828.2	7302264.7	3030001.5	3122594.6

Continued

Input \ Output	113	114	115	116	117	118	119	120	121	122	123	124	Total
1	44565.5	23969.7	26641.6	1799.5	10642.1	31116.7	980	5851.8	2086.8	710324	16105.8	0	81760389
2	9231.2	5155.6	1203.4	435.1	165.6	15515.2	155.8	6509.3	1226.9	23859.7	1317.6	0	6916372.3
3	18332.6	13785.6	1017.3	439.5	1911	33161.4	18.4	3475.6	0	62395.3	176.6	0	28711889
4	28327.7	18355.7	729.3	10.3	749	6334.3	188	388.8	0	63099.3	13896.7	0	10196782
5	142.8	53.2	42499.4	114.6	187.2	5744.1	35.8	52	408.2	783.7	1536.1	0	6565181.3
6	8868.4	79888.5	96613.2	4793.7	14851.7	383519.7	8955.2	21407.2	27275.6	9131.1	11156.4	155206.9	21521983
7	0	0	1321.1	0	0	0	0	44.9	410.1	0	0	1537.8	18889019
8	579.4	0	1479.8	0	0	0	0.3	363.8	0	0	0	72.8	816689.7
9	0	0	0	0	0	0	0	186.2	0	0	4711	0	4878398.2
10	0	0	0	0	0.2	0	0	0.2	0.2	0	27276	0	8761439.6
11	0	0	0	0	0	0	0	28.1	0	0	0	0	1016779.8
12	0	0	0	114.3	43.9	0	0	2791.8	2329.7	449.7	33332	0	13068093
13	43.3	4262.8	2475.4	54.4	14.9	6313.5	385.7	205.7	2053.1	192.5	1784.4	232872.3	4844272.4
14	23158.1	3391.5	4533.4	1284.9	14857.2	85148.7	425.3	4893.9	575	190483.7	489.5	10442.6	27923327
15	1439.9	498.4	1070.6	102.3	519.9	3217.8	0	594.1	51	0	0	465.1	2152286.2
16	29471.5	5480	3047.1	6489.1	11013.5	73867.4	1032.3	539.2	225.5	1928.5	2446.7	307.1	5475252.2
17	24924.9	4186	1002.3	887.4	2308.6	4351	476.1	908.8	171.6	1878.1	1424	8296.9	1833356.5
18	61708.6	38906.6	2189.4	1452.8	7786.6	35708.7	1230.2	1990.3	4703.9	1544.3	13856.9	0	2895299.4
19	128764.5	30853	65956.2	455.3	1752.6	15058.6	2159.6	1052.1	1630.3	1182.9	1573.2	0	5177577.9
20	140066.2	36170.9	2068.5	3645	1702.2	19959.4	3804.9	1198.7	4948.3	1401.3	4084.3	0	2240955.5
21	140975.5	27236	1054.6	150.7	2141	28124	4364.9	2539.9	4425.3	445.6	4154.8	0	3647138.6
22	5782.2	29254.1	100578.2	459.4	2991.8	8246.3	1325.8	975.9	5379.8	231.6	160.5	0	35182988
23	399.8	6776.5	1668.5	121.8	685.7	782.8	3090.8	32241.1	639.4	448.4	626.5	0	11498530
24	0.1	7603.8	98.9	0.7	20.2	3945.2	592.2	95.5	64.8	74.4	15.8	0	1291035.4
25	242.1	4525	273.9	40.8	193.8	8591.3	1603.2	373.4	789.6	610.1	409.9	0	12943830
26	2374.2	5418.1	3627.7	831.9	3363.4	3178.5	601	685.3	1558.5	0	0	0	1472468.4
27	5927.2	20316.2	4726.5	1056.7	133.1	58145.6	880.8	123.1	647	8298	16876.9	567742	10688027
28	16285.7	102652.8	33056.7	6612	5677.7	4736.7	5914.5	3709.8	12764.9	625.8	2064.5	82460.8	5967176.6
29	2231.4	7175.4	5939.5	980.8	347.3	34500.2	975.2	564.5	3836.3	609.2	992.2	90606.5	8084647.3
30	1022.7	69554	6925.6	428.2	1620	182955.4	3201.3	1509.9	4342.9	4450.5	8108.1	237746.4	9030424.7
31	7152.7	209061.1	59165	862.7	2131.7	319369.8	5357.8	2457.3	14382.2	12681	13710.3	345994.8	6018908.4
32	3176.3	528540	45851	438.5	3228.9	567389.8	901931	9861	57828.4	9221.2	14232.1	1671317.2	25296562
33	9630.2	874401.6	85922.7	1163.5	12079.1	216705.9	375845.8	20015	41407.4	9777.1	20460.1	337980.6	9908527.4
34	8386.6	175558.6	22248.7	1654.7	13357.2	53733.2	24596.8	11523	41514.1	841.4	1280.8	16646.7	1983477
35	25095.9	14608	3536.3	13162.8	1269.8	91669.8	6604.1	979.3	2270.6	16115.4	65369.3	651637.5	573670.4
36	9356.1	150830.2	29369.1	1232.3	10375.8	2232.6	8107.5	14861	31583.7	0	130.9	0	30899361
37	0	0	0	0	0	26720.6	0	207	238.1	205.1	3515.1	0	2483072.8
38	200.2	4354.9	20900.1	70.9	40.8	2057.3	2712.1	10838.9	4491.2	3021.6	380.1	2134.8	12047987
39	0	1	0	0	39.4	1756.3	0	6898.7	44.8	44.8	238.1	0	15980432
40	0	0	0	320.9	171.5	62044.4	0	5226.5	949.3	14048.8	1857.1	0	4103383.8
41	2933.8	120422.7	19431.9	421.9	2038.1	31060.8	18828.9	18496.2	4161	2138.6	1597	122921.9	25193870.2
42	11076.9	62893.1	62506.1	1901.4	2780.7		5054.3	4322.9	8473.8	1885.6		87568.9	4969175.2

Continued

Input \ Output	113	114	115	116	117	118	119	120	121	122	123	124	Total
43	5979.4	137256.9	190406.3	504	861.6	87987.4	24635.1	16185	6744.1	784.6	7734	14466.4	27862015.1
44	694.7	12469.1	6638468	3051.5	20026.3	224371	8530.3	11560.2	13498	41383.3	8156	142004.7	11438025.6
45	3850.1	0	0	0	0	0	0	872.3	0	1256.1	0	0	14085016.3
46	1444.5	11651.6	104132.6	375.6	531.3	8120.6	1144.6	4062.7	2720.8	352.4	3969.1	29735.8	10146817
47	2145.2	29514	53815.5	296.6	497.1	18623.8	3496.8	6230.6	4475.8	13855.5	15148.8	15465.1	25459442.6
48	786	56883.8	28475.6	1521.1	4464.7	92193.4	9369.6	4179	15597.9	5925.7	76285.6	407110.9	20872785.3
49	620.4	39595	11745.8	188.1	834.9	45811.6	657.4	3685.4	4598	769.8	102782.4	31923	10732095.6
50	1090.3	26252.6	10995.4	911.6	2101.8	87304.6	6550.2	2283.9	5123.8	1566.2	12884.6	151175	21326008.1
51	1784.8	20934.4	9521.7	189.1	354	23868.1	4318.5	4568.3	2820.6	156.2	979.8	3052.4	6267227.5
52	2798.3	8208.7	2967.9	77.8	737.8	14498.1	503.4	1186.5	715.6	334.9	4090	11126.7	5709429.8
53	679.5	10748.9	287.7	799.3	57.4	3157.8	114.5	4364.5	1300.9	32.8	401	0	5465183.1
54	209	17140.3	819.1	414.3	10.8	7631.2	78.2	4368.9	1021.9	220.8	587.1	0	7626504.9
55	0	0	0	0	0	1590.2	0	1186.7	135.9	0	0	0	11211518
56	0	0	0	0	0	3474	0	3131.4	111.8	0	0	0	7428124.2
57	89.8	17212.2	1058.1	91.1	243.6	5400.5	1053.4	6411.4	2980.6	287.2	40002.8	0	45695948.3
58	0	5820.6	0	0	0	927.7	0.8	2632.6	313.8	0	0	0	28961192.5
59	0	13.9	0	0	0.2	755.5	0	10761.7	111.5	0	0	0	16552856.5
60	0	3202.4	75.1	2.9	1.6	1272	70	5072.5	695.4	469.8	3742	0	10668752.2
61	7280.5	411879	57746.7	2287	6435.7	127071.2	10791.6	64926.3	28101.1	3744.2	212162.5	624776.6	42287925.6
62	0	41207.8	9072.7	1453.7	1210.6	20191.4	3328.5	1407.8	6965	1251.8	334	16101.9	5874379.9
63	1339.6	10858.5	880.1	11.8	0	15615.6	163.4	32344.4	3294.3	1571.7	2585	11263.6	3698911.5
64	17973	273194.2	117308.5	3500.7	8371.7	125315.6	28642.1	41529.8	68552.5	13501.1	121116.5	917734.1	33505442.9
65	0	0	0	211.6	0	0	0	658.1	1.4	4713.1	2045.7	61090.4	3683226.2
66	4115.1	29620.4	736376	969.9	1607.9	36472.4	3213.2	16415.1	33840.7	3501.8	15742.1	100134.6	12556511.3
67	0	0	0	0	0	780.8	0	199.9	0	0	363.3	0	1176074.9
68	3670.3	413910.2	16378.9	2260.4	4348.7	68224.6	7965.1	6379	42109.8	8689.6	12985.7	751812	22175362.3
69	99.3	0	0	0	0	556.8	0	232.1	0	0	3303.2	0	1124637.3
70	0	0	0	0	0	11.8	0	23632.4	6389.1	0	0	0	907465.2
71	35.9	3337.3	22.4	102.4	22.6	413.5	96	62.5	289.3	58.3	53.9	0	1039387.2
72	49.4	29034.7	3617	514.5	498.7	12477.5	3850.8	394.1	1410.4	2575.1	444.3	42337.4	3715532.4
73	2511	17022	1589	96.1	124.4	24188.8	748.5	6138.4	1351.6	997.4	2517.7	15610.3	4168782.2
74	20077.5	370241.1	19795.4	2123.3	3407.2	79429.4	7525.2	73289.3	6168.5	1365	2206.6	67935.5	4709507.3
75	11985	411034.2	32803.6	1502.5	2421	66973.5	23127.8	93741.2	34591.1	883	32698.2	102620.7	28075746.5
76	4413.8	2724192.7	19531.1	311.8	5169.6	186078.9	11245.8	39127.3	106994	2481.8	4984.3	254751.4	6408729
77	3327	20978.9	4990.9	503.8	523.8	46360.2	9756	17705.2	3410.4	760.3	1136.1	24016.8	3139322.7
78	2018.1	288725.8	5417.7	85	100.7	91485.9	8534.8	110394.3	6314.8	34.9	2876.3	89552.7	19375386.9
79	7930.5	448738.4	9336.9	536.3	225.6	65264.7	36834.4	260055.8	16881	490.5	3800.4	48992.5	5837427.2
80	1240.7	65402	10425.6	121.9	32.7	38817.9	1934.5	54579.9	39720.4	3671.8	4266.8	32474.9	4871072
81	6185.6	229139.7	9849.6	82	1308.4	65295.2	13395.2	5889.3	16056.7	2290	1864.3	74730.1	1665275
82	9215.1	192557.5	96881.5	2799.6	7454.3	91270.7	25107.7	14324.2	32113.3	11633.9	92241	562597.6	7367029.1
83	6711.4	157183	25653.7	878.7	2144.2	48740.8	27228.3	11635.3	10939.1	4302.7	8376.7	140459.2	4969077.3
84	9723.6	35363	94208.7	1485.6	2869.8	60694.6	11680.2	20695.1	5530.9	416.9	18456.6	24470.1	9198139.2

Continued

Input	113	114	115	116	117	118	119	120	121	122	123	124	Total
85	0	0	0	0	0	0	0	0	0	0	0	0	5146206.4
86	34688.5	73364.1	232182.1	10865.3	16032.5	693652.4	42246.5	35667.9	46791.5	21692.8	62137.4	522080.6	35014944.8
87	2032	8194.9	7782	1133.5	2307	47069.6	5530.8	2518.3	5481.8	3528	6542.5	114771.2	1256581.3
88	1734.9	6334.9	11689.9	530.6	1624.3	37064.2	2292.7	1119.6	2576.5	61.2	231.2	22635.1	605383.9
89	9481.1	45916.8	43600.4	3487.7	3934.7	265255.5	9519.6	6776.4	14110.1	5236.9	11743.9	146490.3	2874388.1
90	50203.1	329694.6	353056.4	21122.2	68973.5	1576539.7	61713.5	30898.2	133422.1	34591.1	176263.1	1505197.4	10285093.2
91	4980.9	33884.6	31633.5	952.2	1511.7	51705.6	21303.8	9331.8	5894.1	12588.3	21491.4	43321.5	8442804.5
92	8178.8	47885.6	42207.4	1046.4	1513.8	53647.3	17330.8	7456.8	18096.4	18408.8	44434.2	112902.8	14255697
93	58.9	565	323.7	41.9	231.8	397.4	50.9	77	121.1	58.5	261	2371.9	330618.5
94	425.4	4228.8	1874.4	70.6	200.3	4127.5	635.1	641.8	568.6	902.5	3394.2	5927.6	1336360.2
95	371.6	22408.5	10946.7	126.2	219.7	5099.9	1176.8	2489.3	5962.4	322.7	570.2	9936.2	428760.7
96	1626.5	18669.1	15810.1	440.9	1668.2	31836.9	6704.5	3052.1	10946.8	5987.2	19597.1	132284	5984743.7
97	323.2	2956.2	116.2	4		179.4	7640.9	328	15940.5	256.4	0	0	763363.3
98	1801.3	26459.8	39824.6	577.8	1306.2	73242	83529.4	5656.9	73523.5	4240.3	3357.3	292431.1	938686.1
99	12637	211564.9	109307.6	4390.3	12985.3	285676	343575.6	21503.3	75607.9	22928	41861.2	1691203	14801710.6
100	81316	761232.4	1076568	8745.7	19342.3	512090.2	224530	90361	95743.3	59826.4	103189.3	952305.2	72995384.1
101	37942	269011.7	63349.2	7210.2	31747.5	237779.5	61619.1	25941.4	64494	51702.1	94736.1	1360135.5	11907620
102	4346.7	123062.9	27784.7	5931.9	2741.6	265248.4	28024.3	39620	32521	13467.1	41887.7	668341.3	2503710.4
103	2542.2	58779.8	7577.1	845.1	1762.3	103537.4	14411.2	7027.9	26196.7	9826.3	23149.3	277871	1284732.2
104	3144.8	88992.4	11124.8	463	1802.6	175763.9	32279.4	10753.8	47001.8	11160.7	34975.3	990023.6	1418161.3
105	5914	141675.9	23422.1	3847	5523.1	166432	35716.5	18918.3	11114.1	13263.7	12366.7	1054350.4	3105238.1
106	53128	113822	19142.5	687.3	1235.8	40093.9	89587.9	8958.2	33697.8	27373.3	61245.9	277554.3	22744712.7
107	16258.9	48969.3	12801.9	1356.6	5339.4	34793.1	13869.9	4491.7	30916.6	9218.8	27437.4	508621.4	4196837.2
108	79522.4	298895	23914.2	422.4	31893.7	63789.6	29234.7	5012.3	49444.9	4240.4	2396.7	502437.5	5612860.1
109	11237	208954.9	56892.8	2256.3	9508.6	284444.3	45809.8	23250.5	21438.9	13808.4	41054.5	161449.8	2757970
110	5296.5	66272.2	19873.5	4304.1	5234	117932.6	21827.7	9730.4	168352.2	5890.5	7488.1	2139354.9	6301313.1
111	9740.9	263937.6	41797.7	7832.6	15376.2	379629.8	67049.9	37334.5	2406.3	30747.4	40331.1	52477.2	5911221.1
112	5390.3	6075.9	1254.4	104.9	1302.3	20216.3	1114.3	306.5	1343.5	1376.1	1690.4	33563.5	628429.2
113	680.9	841.4	195.2	97.4	112.8	1905.4	3194.9	78.5	67349.8	161.9	474.6	432175	1025314.8
114	24148.9	537123.2	17509	3214.6	7962.5	222414.1	120717.1	19068.4	5929.8	5965.3	98576.6	101198.1	18554747.1
115	390.3	5622.4	32813	258.9	1329.8	58075.7	3033.9	3049.3	0	1269.6	4112.4	0	727298.7
116	0	0	0	0	0	0	0	0	0	0	0	0	0
117	0	0	0	0	0	0	0	0	0	0	0	0	43577.5
118	1737.8	42386	34693.1	2430.2	13591.2	70301.7	14016.2	9499	29384.3	7304.2	17374.9	254571.6	2404554.1
119	15437.1	166119.9	11957.6	1893.8	5076.5	241350.6	557045.2	7183.6	23743.7	5202.6	5127	276759.7	1745316.9
120	206	253637.6	2558.3	34.7	146.5	85109.9	234.8	27704.8	11606.9	2814.7	748.2	12538.3	694708.1
121	4585.4	13396.6	10953.1	220.1	0	36120.9	3737.6	13342.8	157117.5	82223.8	26597	59996.3	2166573.3
122	3983.7	0	0	0	0	0	0	0	0	0	252088.4	0	2772687.7
123	0	0	0	0	0	0	0	0	0	0	0	0	2369279.1
124	0	0	0	0	0	0	0	0	0	0	0	0	0
Total	1409441.6	13557602.3	11481921	181883.2	485731.6	10568205	3657434.8	1699707	2112727.5	1778921.6	2403604	24339302.6	

APPENDIX 3

Input Coefficients among Industries in China in 1997

a_{ij} ╲ b_{ij}	1	2	3	4	5	6	7	8	9	10	11	12	13	14
1	0.25592	0.03941	0.41670	0.13538	0.29757	0	0	0	0	0	0	0	0	0.72326
2	0.00177	0.26117	0.00138	0.00071	0.06685	0.00837	0	0	0.00037	0.00017	0.00013	0.00009	0.34592	0.00017
3	0.01896	0.00333	0.09913	0	0	0	0	0	0	0	0	0	0	0.01212
4	0	0	0	0.15529	0.00668	0.01358	0	0	0	0	0	0	0	0.01204
5	0.01792	0.01449	0.01247	0.00558	0.15788	0.05749	0	0	0.00597	0.00598	0.05740	0.00060	0.00315	0.00017
6	0	0.00333	0.00369	0.00301	0.00914	0	0.02335	0.00233	0.00329	0.00782	0	0.00337	0.01460	0.00206
7	0	0	0.00249	0	0	0	0.01622	0.01165	0.00001	0.00023	0	0.00841	0.00060	0.00004
8	0	0	0	0	0	0	0.00006	0.06028	0	0.00002	0	0	0	0
9	0	0	0	0	0	0.00003	0	0	0.15574	0.00059	0	0	0	0
10	0	0	0	0	0	0	0	0	0	0.31027	0	0	0	0
11	0	0	0.00025	0.00081	0.00549	0	0.00029	0.00009	0.00001	0.00003	0.02190	0	0	0.00076
12	0	0	0	0	0.01132	0.00264	0.00113	0.00273	0.00077	0.00068	0.00716	0.12119	0.00004	0.00001
13	0.00050	0.00675	0.00080	0.00332	0.00815	0.08922	0.00164	0.00106	0.00541	0.00492	0.00104	0.00143	0.04045	0.10608
14	0.03299	0.00688	0.32571	0.25205	0	0	0	0	0	0	0	0	0	0.00071
15	0	0	0	0	0	0	0	0	0	0	0	0	0	0.00233
16	0	0	0	0	0	0	0	0	0	0	0	0	0	0.01684
17	0	0	0	0	0	0	0	0	0	0	0	0	0	0.00339
18	0	0	0	0	0	0	0	0	0	0	0	0	0	0
19	0.00055	0.00070	0.00025	0.00059	0.00172	0.00036	0.00201	0	0	0	0	0	0	0
20	0	0	0	0	0	0	0	0	0	0.00139	0	0	0	0
21	0	0	0	0	0	0	0	0	0	0	0	0	0	0
22	0.00043	0.00058	0.00036	0.00023	0.01222	0.00499	0.00540	0.00458	0.00041	0.00378	0.00151	0.00086	0.00081	0.00163
23	0.00003	0.00001	0	0.00006	0.00372	0.00077	0.00230	0.00102	0.00012	0.00052	0.00018	0.00034	0.00011	0.00004
24	0.00103	0.00091	0.00299	0.00035	0.01933	0.00021	0.00049	0.00025	0.00008	0.00014	0.00103	0.00605	0.00010	0.00046
25	0	0	0	0	0.01235	0.00046	0.00108	0.00063	0.00005	0.00026	0.00023	0.00022	0.00002	0.00006
26	0.00009	0.00039	0.00002	0.00012	0.00103	0.00036	0.00057	0.00068	0.00037	0.00099	0.00020	0.00020	0.00025	0.00002
27	0	0	0	0	0.01689	0	0	0	0	0	0	0	0	0
28	0.00037	0.00239	0.00022	0.00051	0.01139	0.00883	0.01616	0.01101	0.00700	0.00857	0.00905	0.00498	0.00725	0.00105
29	0.00031	0.00039	0.00174	0.00009	0.00136	0.00276	0.00681	0.00355	0.00156	0.00467	0.00256	0.00072	0.00126	0.00011
30	0.00048	0.00400	0.00121	0.00446	0.00425	0.00556	0.00568	0.00397	0.00259	0.00477	0.00222	0.00122	0.00424	0.00005
31	0.00184	0.00124	0.00040	0.00327	0.01324	0.00183	0.00198	0.00211	0.00165	0.00110	0.00142	0.00129	0.00102	0.00005
32	0.00074	0.00224	0.00012	0.00107	0.01502	0.00101	0.00126	0.00043	0.00030	0.00026	0.00018	0.00050	0.00013	0.00082
33	0.00035	0.00167	0.00010	0.00073	0.00387	0.00082	0.00215	0.00126	0.00054	0.00070	0.00050	0.00023	0.00046	0.00012
34	0.00103	0.00267	0.00031	0.00201	0.00147	0.00126	0.00269	0.00167	0.00051	0.00053	0.00074	0.00031	0.00017	0.00004
35	0.00010	0.00052	0.00005	0.00042	0.00026	0.00011	0	0	0	0.00039	0	0.00056	0	0
36	0.02503	0.02413	0.00416	0.07378	0.01811	0.02829	0.07486	0.09123	0.03884	0.02414	0.02555	0.07066	0.05661	0.00256
37	0	0	0	0	0	0	0	0	0.00537	0.00027	0	0	0	0
38	0.00004	0.00009	0.00005	0.00012	0.00031	0.00259	0.02237	0.02022	0.00197	0.02340	0.00931	0.00213	0.00027	0.00331
39	0.30529	0.12526	0.00009	0	0	0	0	0	0	0	0	0	0	0.00042

Continued

a_{ij} \ b_{ij}	1	2	3	4	5	6	7	8	9	10	11	12	13	14
40	0.05308	0.03714	0	0	0	0	0	0	0	0	0	0	0	0
41	0.00035	0.00071	0.00006	0.00030	0.00151	0.00386	0.02525	0.08441	0.00108	0.00719	0.00235	0.00408	0.00147	0.00126
42	0.00020	0.00067	0.00014	0.00026	0.00058	0.00268	0.00694	0.00315	0.00192	0.00398	0.00328	0.00245	0.00115	0.00091
43	0.00012	0.00020	0.00065	0.00040	0.00078	0.02309	0.04122	0.00252	0.03595	0.03643	0.01102	0.06885	0.00266	0.00844
44	0.00038	0.00445	0.00804	0.00586	0.00065	0.00056	0.00001	0.00001	0.00007	0.00029	0.00125	0.00002	0.00036	0.00418
45	0	0	0	0.00657	0.00072	0	0	0	0	0	0	0	0	0
46	0.00148	0.00220	0.00050	0.00151	0.00290	0.03197	0.01276	0.00974	0.01169	0.02049	0.08637	0.00614	0.05721	0.00018
47	0.02114	0.00968	0.00067	0.01111	0.00831	0.00499	0.00594	0.00365	0.00374	0.01206	0.16450	0.06109	0.00292	0.00639
48	0.00166	0.00466	0.00117	0.00243	0.00378	0.00772	0.00945	0.01517	0.00384	0.00535	0.00511	0.00343	0.00217	0.00006
49	0.00168	0.00316	0.00055	0.00291	0.00680	0.00564	0.00352	0.00566	0.00264	0.00323	0.00118	0.00262	0.00015	0.00002
50	0.00204	0.00395	0.00219	0.00330	0.00217	0.00676	0.01925	0.02352	0.00317	0.00709	0.00371	0.00270	0.00313	0.00010
51	0.00016	0.00060	0.00009	0.00031	0.00263	0.00306	0.00320	0.00185	0.00236	0.00142	0.00228	0.00236	0.00175	0.00013
52	0.00004	0.00016	0.00004	0.00006	0.00650	0.00442	0.00396	0.00309	0.00352	0.00217	0.01106	0.00395	0.00258	0.00041
53	0.00003	0.00014	0.00001	0.00004	0.00040	0.00057	0.00087	0.00056	0.00025	0.00010	0.00034	0.00044	0.00006	0.00258
54	0.00002	0.00012	0.00001	0.00004	0.00023	0.00087	0.00079	0.00038	0.00070	0.00068	0.00015	0.00288	0.00015	0.00041
55	0	0	0	0	0	0.00055	0.00197	0.00776	0.00610	0.00042	0.00003	0.00020	0	0.00001
56	0	0	0	0	0	0.00913	0.00537	0.00687	0.00321	0.00200	0.00109	0.00445	0.00013	0.00008
57	0.00030	0.00054	0.00004	0.00031	0.00348	0.08978	0.02148	0.03022	0.01417	0.00793	0.00652	0.00082	0.00175	0
58	0	0	0	0	0	0.00129	0.00355	0.00244	0.00476	0.00164	0.00064	0.00019	0.00006	0.00001
59	0	0	0	0	0	0.00099	0.00156	0.00051	0.00022	0.00344	0.00003	0.00004	0.00001	0.00001
60	0.00002	0.00001	0.00089	0.00004	0.00006	0.00864	0.00360	0.00160	0.00048	0.00100	0.00038	0.00037	0.00008	0.00127
61	0.01064	0.01719	0	0.00487	0.02852	0.04252	0.02979	0.01918	0.04739	0.04220	0.01637	0.01440	0.04056	0.00014
62	0	0	0	0	0	0.00196	0.00350	0.00336	0.00005	0.00080	0.00069	0.00062	0.00169	0.00011
63	0	0	0	0	0	0.00662	0.00271	0.00213	0.00408	0.00633	0.00362	0.00356	0.00053	0.00106
64	0.00646	0.01241	0.00117	0.01905	0.01136	0.04460	0.05706	0.07093	0.04464	0.03729	0.03690	0.02487	0.04078	0.00011
65	0.02825	0.02222	0.00423	0.03535	0.01632	0.00062	0.00024	0.00015	0.00027	0.00052	0.00479	0.00208	0.04594	0.00127
66	0.00059	0.00142	0.00011	0.00135	0.00162	0.03477	0.08005	0.08966	0.13202	0.01750	0.02201	0.07746	0.01568	0.00082
67	0	0	0	0	0	0.00072	0.00094	0.00100	0.00022	0.00010	0.00002	0	0.00017	0
68	0.00059	0.00426	0.00065	0.00150	0.00171	0.01155	0.03351	0.03231	0.05686	0.01063	0.01735	0.03940	0.04746	0.00246
69	0.00011	0.00010	0	0.05868	0.00034	0	0	0	0	0	0	0	0	0
70	0	0	0	0	0	0	0	0	0	0	0	0	0	0
71	0.00007	0.00017	0.00001	0.00011	0.00023	0.00086	0.00125	0.00150	0.00087	0.00189	0.00117	0.00116	0.00099	0.00005
72	0.00044	0.00558	0.00050	0.00346	0.00264	0.00350	0.00981	0.00731	0.00604	0.00304	0.00447	0.00107	0.00118	0.00022
73	0.00064	0.00055	0.00040	0.00080	0.00021	0.00061	0.00079	0.00149	0.00064	0.00122	0.00048	0.00026	0.00033	0.00003
74	0.00023	0.00011	0.00002	0.00135	0.00248	0.04725	0.03803	0.02788	0.01450	0.00736	0.01213	0.00750	0.00835	0.00064
75	0.00081	0.00304	0.00037	0.00065	0.00060	0.00071	0.00196	0.00293	0.00088	0.00046	0.00094	0.00055	0.00083	0.00004
76	0.00002	0.00011	0	0.00005	0.00017	0.00256	0.00058	0.00128	0.00112	0.00039	0.00093	0.00212	0.00059	0.00005
77	0.00006	0.00013	0.00002	0.00022	0.00004	0.00099	0.00882	0.00446	0.00053	0.00473	0.00066	0.00079	0.00003	0.00007
78	0	0.00001	0	0.00001	0	0.00146	0.00760	0.01067	0.00133	0.00126	0.00158	0.00125	0.00080	0.00010
79	0.00008	0.00014	0.00002	0.00052	0.00029	0.00439	0.02197	0.03250	0.00800	0.00573	0.00404	0.00310	0.00142	0.00035
80	0.00005	0.00013	0.00008	0.00017	0.00009	0.00150	0.00124	0.00159	0.00139	0.00059	0.00142	0.00081	0.00131	0.00005
81	0.00011	0.00028	0.00001	0.00008	0.00002	0.01049	0.00805	0.00409	0.00548	0.00656	0.00835	0.00797	0.00296	0.00038
82	0.00580	0.00846	0.00103	0.01831	0.00756	0.01049	0.00805	0.00409	0.00548	0.00656	0.00835	0.00797	0.00296	0.00038
83	0.00359	0.00517	0.00218	0.00326	0.00307	0.00428	0.00752	0.00378	0.00249	0.00246	0.00407	0.00258	0.00393	0.00133

Continued

a_{ij} \ b_{ij}	1	2	3	4	5	6	7	8	9	10	11	12	13	14
84	0.00140	0.00039	0.00057	0.00108	0.00065	0.03030	0.01932	0.01749	0.01058	0.00807	0.01626	0.00985	0.00193	0.00085
85	0	0	0	0	0	0.00023	0.00004	0.00002	0.00002	0.00005	0.00015	0.00001	0	0.00673
86	0.02866	0.01876	0.00374	0.01451	0.03844	0.09715	0.11588	0.11620	0.15088	0.11586	0.15657	0.08628	0.02504	0.00667
87	0	0.00002	0.00041	0.00012	0	0.00091	0.00050	0.00069	0.00007	0.00017	0.00187	0.00006	0.00065	0.00007
88	0	0	0	0	0	0.00026	0.00004	0.00002	0.00006	0.00001	0.00009	0.00001	0.00001	0.00001
89	0.00057	0.00040	0.00012	0.00042	0.00045	0.00276	0.00333	0.00240	0.00388	0.00123	0.00333	0.00318	0.00083	0.00034
90	0.00653	0.00929	0.00345	0.00305	0.00325	0.00504	0.00709	0.00423	0.00525	0.00312	0.00534	0.00412	0.00297	0.00034
91	0.00603	0.00722	0.00255	0.00384	0.00584	0.01222	0.01162	0.00726	0.01704	0.01635	0.00883	0.03964	0.01114	0.00273
92	0.01718	0.05454	0.00860	0.02402	0.01775	0.01767	0.00674	0.00545	0.03288	0.05496	0.05596	0.04304	0.02155	0.00855
93	0.00009	0.00019	0.00004	0.00026	0.00007	0.00010	0.00044	0.00191	0.00024	0.00009	0.00010	0.00035	0.00021	0.00001
94	0.00073	0.00063	0.00053	0.00101	0.00061	0.00133	0.00112	0.00090	0.00878	0.00240	0.00264	0.00362	0.00134	0.00051
95	0.00005	0.00010	0.00005	0.00036	0.00015	0.00017	0.00028	0.00027	0.00022	0.00012	0.00023	0.00019	0.00014	0.00002
96	0.00218	0.00212	0.00158	0.00228	0.00182	0.00696	0.00310	0.00248	0.00767	0.01742	0.01132	0.00591	0.00413	0.00246
97	0.00171	0.00126	0.00090	0.00320	0.00096	0.00001	0	0	0	0.00009	0.00007	0	0	0.00002
98	0.00049	0.00112	0.00045	0.00095	0.00051	0.00001	0.00001	0	0	0	0	0	0	0
99	0.00076	0.00259	0.00051	0.00155	0.00067	0.01279	0.01926	0.00913	0.00843	0.00247	0.00333	0.07535	0.00447	0.00097
100	0.03970	0.05098	0.04804	0.03830	0.05384	0.05818	0.04114	0.03534	0.03788	0.04347	0.05797	0.05214	0.05774	0.03404
101	0.00158	0.00445	0.00058	0.00157	0.00247	0.01352	0.00433	0.00357	0.01721	0.00885	0.01762	0.02027	0.01438	0.00197
102	0.00095	0.00143	0.00045	0.00137	0.00080	0.00001	0.00002	0.00001	0.00001	0	0	0.00249	0	0
103	0.00177	0.00626	0.00057	0.00195	0.00409	0	0	0	0	0	0	0.00419	0	0
104	0.00220	0.00220	0.00042	0.00480	0.00192	0.00001	0.00002	0.00001	0.00001	0	0	0.00082	0	0.00002
105	0.00011	0.00025	0.00003	0.00029	0.00011	0.00021	0.00032	0.00015	0.00014	0.00004	0.00005	0.00163	0.00007	0
106	0.01160	0.02485	0.00646	0.01048	0.01370	0.01635	0.03621	0.01843	0.00979	0.01913	0.02539	0.02301	0.02034	0.00642
107	0.00196	0.00426	0.00059	0.00508	0.00141	0.00452	0.01368	0.00753	0.00302	0.00260	0.00464	0.00347	0.00725	0.00052
108	0.00072	0.00339	0.00025	0.00075	0.00064	0.00172	0.00015	0.00025	0.00024	0.00129	0.00047	0.00163	0.00052	0.00028
109	0.00261	0.00956	0.00085	0.00189	0.00344	0.00001	0.00001	0.00001	0.00001	0	0	0.00065	0.03294	0.00147
110	0.00529	0.01275	0.00079	0.00332	0.00105	0.01273	0.01473	0.00830	0.00928	0.00582	0.02570	0.00885	0.00013	0.00003
111	0.00214	0.00388	0.00091	0.00191	0.00582	0.00047	0.00041	0.00019	0.00020	0.00081	0.00008	0.00936	0	0
112	0	0	0	0	0	0	0	0	0	0	0	0.00013	0	0
113	0.00001	0.00005	0.00001	0.00002	0.00003	0.00143	0.00051	0.00042	0.00188	0.00104	0.00206	0.00240	0.00167	0.00023
114	0.00487	0.01082	0.00039	0.00334	0.00331	0.01319	0.00204	0.00143	0.00181	0.00615	0.00892	0.00249	0.00326	0.00212
115	0.00022	0.00114	0.00010	0.00038	0	0.00638	0.00078	0.00041	0.00091	0.00061	0.00043	0.00053	0.00148	0.00004
116	0	0	0	0	0	0	0	0	0	0	0	0	0	0
117	0	0	0	0	0	0	0	0	0	0	0	0	0	0
118	0.00214	0.00482	0.00063	0.00253	0.00134	0.00641	0.00093	0.00087	0.00127	0.00211	0.00383	0.00135	0.00361	0.00020
119	0.00065	0.00147	0.00023	0.00046	0.00062	0.00005	0.00004	0.00002	0	0.00001	0.00002	0.00095	0	0.00001
120	0.00005	0.00033	0.00001	0.00012	0.00004	0.00040	0.00047	0.00048	0.00013	0.00039	0.00007	0.00024	0.00003	0.00006
121	0.00210	0.00401	0.00077	0.00425	0.00142	0.00305	0.00150	0.00182	0.00249	0.00352	0.00072	0.00314	0.01141	0.00026
122	0.03601	0.09610	0.01365	0.02975	0.01327	0	0.01932	0.01045	0.06606	0	0	0.00466	0	0
123	0.00692	0.00591	0.00021	0.00155	0.00111	0.01338	0	0	0	0.00648	0	0	0	0
124	0	0	0	0	0	0	0	0	0	0	0	0	0	0

Appendix 3   Input Coefficients among Industries in China in 1997

Continued

b_{ij} \ a_{ij}	15	16	17	18	19	20	21	22	23	24	25	26	27	28
1	0.68966	0.00140	0.00034	0.22605	0.39259	0.32700	0.33646	0.08348	0.00029	0.50132	0.01167	0.00103	0.49605	0.00060
2	0.00021	0.00008	0	0.00312	0.00015	0.00050	0.00002	0	0	0.00004	0.00019	0	0.00001	0.00009
3	0	0.80735	0.80295	0.10040	0	0.01097	0	0	0.28701	0.00128	0.11219	0.00739	0.09773	0.00023
4	0	0	0.00190	0.02209	0	0	0.00001	0	0	0	0	0	0	0
5	0.00046	0.00007	0.00060	0.03647	0.00067	0.00959	0.00308	0.00002	0.00492	0.00003	0.00680	0.00002	0.00001	0.00001
6	0.01774	0.00135	0	0.01056	0.01213	0.00288	0	0.00518	0	0.00423	0.00481	0.00777	0.00266	0.00158
7	0.00041	0	0	0	0	0	0.00004	0	0.00002	0	0	0	0	0
8	0.00003	0	0	0.00002	0.00011	0.00001	0	0.00005	0	0	0	0	0	0
9	0	0	0	0	0	0	0	0	0	0	0	0	0	0
10	0	0	0	0	0	0	0	0	0	0	0	0	0	0
11	0	0	0	0.00332	0	0	0	0	0	0	0	0	0	0
12	0	0	0	0	0	0	0	0	0	0	0	0	0	0
13	0.00019	0.00002	0.00137	0.00019	0.00023	0.00010	0.00012	0.00005	0.00005	0.00008	0.00003	0.00004	0.00005	0.00005
14	0.00349	0.00208	0.01882	0.16801	0.09022	0.04494	0	0	0	0	0	0	0	0
15	0.09825	0.00048	0.00018	0.05091	0.00760	0.09016	0	0	0	0	0	0	0	0
16	0	0.03813	0.00051	0.02215	0.00006	0.00011	0	0	0	0	0	0	0	0
17	0	0	0.02620	0.00507	0.00005	0	0	0	0	0	0	0	0	0
18	0.00283	0.00804	0.00204	0.05074	0.01340	0.02957	0.00190	0.00045	0	0.00008	0.00001	0.00001	0	0
19	0	0.00030	0.00007	0.00184	0.11954	0.03561	0.00283	0	0	0	0	0	0	0
20	0	0.00230	0	0.00315	0.00251	0.04924	0	0	0	0	0	0	0	0
21	0	0	0	0	0	0	0.26556	0	0	0	0	0	0	0
22	0.00063	0.00005	0.00012	0.00020	0.00022	0.00008	0.01026	0.42206	0.01950	0.01795	0.01584	0.30951	0.04932	0.27422
23	0.00016	0.00002	0.00005	0.00005	0.00008	0.00005	0.00005	0.00367	0.34682	0.02243	0.00148	0.18449	0.00114	0.16784
24	0.00083	0.00001	0.00001	0.00027	0.00031	0.00014	0.00068	0.00116	0.00100	0.05899	0.00032	0.00227	0.00145	0.02628
25	0.00016	0.00001	0.00003	0.00017	0.00021	0.00001	0.00003	0.00188	0.00952	0.00235	0.63365	0.06421	0.00337	0.12634
26	0.00029	0.00001	0.00001	0.00003	0.00007	0.00010	0	0.00103	0.00174	0.00057	0.00032	0.04581	0.00091	0.00675
27	0	0	0	0	0	0	0	0.13666	0.12026	0.16466	0.00532	0.00629	0.15703	0.08226
28	0.00120	0.00046	0.00046	0.00115	0.00141	0.00125	0.00259	0.00633	0.00071	0.00574	0.00276	0.00255	0.00079	0.02062
29	0.00034	0.00031	0.00011	0.00026	0.00049	0.00045	0.00017	0.00181	0.00552	0.00013	0.00010	0.00092	0.00017	0.00670
30	0.00024	0.00004	0.00013	0.00051	0.00056	0.00115	0.00016	0.00025	0.00031	0.00022	0.00006	0.00016	0.00040	0.00013
31	0.00012	0.00012	0.00039	0.00043	0.00078	0.00063	0.00021	0.00019	0.00012	0.00014	0.00006	0.00006	0.00014	0.00021
32	0.00071	0.00624	0.00664	0.03158	0.04979	0.04439	0.09071	0.00320	0.00194	0.00264	0.00052	0.00794	0.00469	0.00934
33	0.00011	0.00041	0.00006	0.01011	0.01467	0.01602	0.05314	0.00015	0.00045	0.00021	0.00019	0.00042	0.00022	0.00038
34	0.00011	0.00004	0.00052	0.00009	0.00030	0.00005	0	0.00009	0.00027	0.00025	0.00007	0.00020	0.00008	0.00018
35	0	0	0	0	0	0	0	0	0.00010	0	0	0	0	0
36	0.00525	0.00129	0.00361	0.00453	0.00304	0.00486	0.00117	0.00205	0.00310	0.00672	0.00072	0.00334	0.00259	0.00207
37	0.00078	0	0	0	0	0	0	0	0	0	0	0	0	0
38	0.00384	0.00011	0.00025	0.00685	0.00288	0.00197	0.00468	0.00187	0.00103	0.00263	0.00090	0.00296	0.00274	0.00163
39	0	0	0	0.00114	0.00015	0.00003	0	0	0	0	0	0	0	0
40	0	0	0	0	0	0	0	0	0	0	0	0	0	0
41	0.00045	0.00014	0.00349	0.00079	0.00092	0.00193	0.00064	0.02154	0.01083	0.00105	0.00566	0.02455	0.00515	0.00234
42	0.00044	0.00017	0.00025	0.00294	0.00208	0.00488	0.01446	0.00322	0.00065	0.00151	0.00071	0.00136	0.00032	0.00062

Continued

a_{ij} \\ b_{ij}	15	16	17	18	19	20	21	22	23	24	25	26	27	28
43	0.00080	0.00045	0.00248	0.00615	0.00251	0.01242	0.00420	0.00174	0.00017	0.00097	0.00406	0.00056	0.00070	0.00124
44	0.00006	0.00020	0	0.00012	0.00038	0.00149	0.00002	0.00002	0.00001	0.00001	0	0	0.00009	0
45	0	0	0	0	0.00025	0.00090	0.05535	0.10323	0.04495	0.05784	0.09154	0.12474	0.04564	0.10154
46	0.00093	0.00020	0.00046	0.00435	0.00108	0.00148	0.00091	0.00164	0.00050	0.00093	0.00027	0.00090	0.00057	0.00661
47	0.01001	0.00770	0.00273	0.03660	0.00669	0.04870	0.00658	0.00338	0.00298	0.00309	0.00043	0.00652	0.00727	0.01099
48	0.00047	0.00009	0.00010	0.00028	0.00566	0.00019	0.00028	0.00019	0.00026	0.00025	0.00007	0.00019	0.00016	0.00016
49	0.00031	0.00004	0.00006	0.00017	0.00037	0.00008	0.00004	0.00006	0.00013	0.00011	0.00003	0.00006	0.00003	0.00004
50	0.00158	0.00011	0.00010	0.00034	0.00079	0.00024	0.00038	0.00024	0.00027	0.00029	0.00008	0.00024	0.00020	0.00023
51	0.00034	0.00003	0.00013	0.00279	0.05526	0.00498	0.00031	0.00043	0.00016	0.00017	0.00020	0.00033	0.00018	0.00031
52	0.00074	0.00024	0.00026	0.00090	0.00340	0.00483	0.00068	0.00155	0.00057	0.00042	0.00032	0.00125	0.00088	0.00106
53	0.00021	0	0	0.00004	0.00004	0.00002	0	0.00001	0	0	0	0.00012	0	0
54	0.00011	0	0	0.00006	0.00057	0.00003	0.00001	0.00001	0.00006	0	0.00003	0	0.00130	0.00004
55	0	0	0	0	0	0	0	0	0	0	0	0	0	0
56	0	0	0	0	0	0	0	0	0	0	0	0	0	0
57	0.00134	0.00002	0.00008	0.00059	0.00053	0.00042	0.00011	0.00020	0.00019	0.00039	0.00008	0.00009	0.00039	0.00035
58	0	0	0	0	0	0	0	0	0	0	0	0	0	0
59	0	0	0	0	0	0	0	0	0	0	0	0	0	0
60	0.00021	0.00001	0.00007	0.00014	0.00041	0.00238	0.00047	0.00005	0.00019	0.00004	0.00160	0.00001	0.00007	0.00004
61	0.00244	0.00081	0.00452	0.01390	0.02333	0.05815	0.00440	0.00275	0.00329	0.00312	0.00004	0.00222	0.00329	0.00474
62	0.00023	0.00002	0.00002	0.00013	0.00014	0.00002	0.00006	0.00002	0.00004	0.00006	0.00008	0.00001	0	0.00012
63	0.00060	0.00007	0.00012	0.00034	0.00049	0.00081	0.00041	0.00046	0.00041	0.00092	0.00024	0.00036	0.00226	0.00029
64	0.00767	0.00088	0.00138	0.00342	0.00499	0.00332	0.00949	0.00387	0.00247	0.00391	0.00138	0.00237	0.00437	0.00124
65	0	0	0	0	0	0	0	0	0	0	0	0	0	0
66	0.00576	0.00052	0.00076	0.00234	0.00440	0.00164	0.00652	0.02691	0.00292	0.00413	0.00071	0.00156	0.00205	0.00115
67	0	0	0	0	0	0	0	0	0	0	0	0	0	0
68	0.00502	0.00084	0.00050	0.00382	0.00272	0.00213	0.00180	0.00125	0.00070	0.00155	0.00089	0.00128	0.00116	0.00115
69	0	0	0	0	0	0	0	0	0	0	0	0	0	0
70	0	0	0	0	0	0	0	0	0	0	0	0	0	0
71	0	0	0	0	0	0	0	0	0	0	0	0	0	0
72	0.00011	0.00004	0.00004	0.00007	0.00026	0.00008	0.00014	0.00005	0.00003	0.00006	0.00002	0.00007	0.00003	0.00005
73	0.00038	0.00011	0.00030	0.00036	0.00044	0.00018	0.00008	0.00218	0.00148	0.00016	0.00008	0.00026	0.00029	0.00041
74	0.00008	0.00008	0.00004	0.00007	0.00010	0.00012	0.00012	0.00006	0.00011	0.00009	0.00009	0.00008	0.00053	0.00007
75	0.00374	0.00035	0.00059	0.00140	0.00230	0.00135	0.00284	0.00600	0.00172	0.00124	0.00056	0.00102	0.00124	0.00090
76	0.00017	0.00005	0.00001	0.00010	0.00015	0.00011	0.00018	0.00009	0.00004	0.00008	0.00003	0.00011	0.00005	0.00007
77	0.00024	0.00011	0.00008	0.00018	0.00021	0.00031	0.00013	0.00013	0.00013	0.00010	0.00015	0.00022	0.00018	0.00026
78	0.00025	0.00002	0.00008	0.00008	0.00014	0.00008	0.00032	0.00032	0.00033	0.00016	0.00004	0.00015	0.00020	0.00005
79	0.00039	0.00006	0.00009	0.00015	0.00026	0.00017	0.00082	0.00035	0.00036	0.00022	0.00006	0.00024	0.00018	0.00011
80	0.00170	0.00025	0.00053	0.00079	0.00168	0.00122	0.00121	0.00086	0.00047	0.00033	0.00027	0.00075	0.00044	0.00056
81	0.00023	0.00007	0.00002	0.00014	0.00022	0.00017	0.00026	0.00012	0.00007	0.00014	0.00004	0.00017	0.00007	0.00010
82	0.00194	0.00032	0.00079	0.00124	0.00155	0.00109	0.00457	0.00103	0.00041	0.00092	0.00040	0.00063	0.00151	0.00042
83	0.00141	0.00104	0.00130	0.00169	0.00161	0.00167	0.00364	0.00189	0.00145	0.00152	0.00108	0.00159	0.00142	0.00170
84	0.00226	0.00194	0.00656	0.00574	0.00689	0.00412	0.00182	0.01044	0.00454	0.00898	0.00267	0.00605	0.00264	0.01521

Continued

a_{ij} \ b_{ij}	15	16	17	18	19	20	21	22	23	24	25	26	27	28
85	0.00032	0.00002	0	0.00016	0.00052	0.00002	0.00004	0.00007	0.00001	0.00001	0.00011	0.00001	0.00010	0.00002
86	0.02212	0.00454	0.00853	0.01614	0.02161	0.00887	0.00544	0.01990	0.00335	0.02310	0.00066	0.00623	0.00314	0.00395
87	0.00107	0.00006	0.00002	0.00117	0.00130	0.00046	0.00028	0.00125	0.00064	0.00028	0.00037	0.00065	0.00025	0.00020
88	0.00002	0	0.00015	0.00012	0.00018	0.00012	0.00020	0.00014	0.00005	0.00002	0.00005	0.00003	0.00016	0.00005
89	0.00091	0.00055	0.00105	0.00164	0.00146	0.00173	0.00047	0.00168	0.00101	0.00185	0.00093	0.00130	0.00049	0.00097
90	0.00110	0.00080	0.00040	0.00101	0.00110	0.00120	0.00114	0.00158	0.00036	0.00057	0.00033	0.00022	0.00058	0.00067
91	0.00446	0.00071	0.00213	0.00544	0.00440	0.00475	0.00198	0.00207	0.00185	0.00228	0.00157	0.00196	0.00306	0.00209
92	0.01469	0.00353	0.00785	0.01036	0.00896	0.01979	0.00607	0.00634	0.00448	0.01035	0.00319	0.04009	0.00770	0.00708
93	0.00002	0	0.00001	0.00002	0.00002	0.00002	0.00001	0.00001	0.00001	0.00002	0	0.00002	0.00001	0.00001
94	0.00074	0.00020	0.00044	0.00076	0.00059	0.00082	0.00038	0.00028	0.00028	0.00047	0.00016	0.00136	0.00039	0.00030
95	0.00003	0.00010	0.00003	0.00008	0.00048	0.00012	0.00004	0.00006	0.00010	0.00003	0.00008	0.00007	0.00004	0.00008
96	0.00344	0.00059	0.00047	0.00196	0.00249	0.00195	0.00250	0.00100	0.00070	0.00235	0.00073	0.00220	0.00209	0.00081
97	0.00006	0.00001	0.00005	0.00022	0.00007	0.00004	0.00013	0.00003	0.00003	0	0.00001	0.00006	0.00001	0.00014
98	0	0	0	0	0	0.00002	0	0	0	0	0	0	0.00001	0
99	0.00810	0.00125	0.00203	0.00365	0.00325	0.00133	0.00368	0.00242	0.00403	0.00135	0.00329	0.01784	0.00081	0.00629
100	0.03673	0.08995	0.07070	0.05569	0.05393	0.05579	0.06197	0.05870	0.07608	0.04746	0.06207	0.06356	0.05040	0.06485
101	0.00276	0.00137	0.00178	0.00398	0.00504	0.00619	0.00238	0.01425	0.00212	0.00586	0.00236	0.00548	0.00539	0.00420
102	0.00001	0	0	0	0	0	0	0	0	0	0	0.00059	0	0.00001
103	0	0	0	0	0	0	0	0	0	0	0	0.00099	0	0
104	0.00001	0	0	0	0	0	0	0	0	0	0	0.00019	0	0.00001
105	0.00013	0.00002	0.00003	0.00006	0.00005	0.00002	0.00006	0.00004	0.00007	0	0.00005	0.00040	0.00001	0.00010
106	0.01262	0.00344	0.00550	0.00855	0.00810	0.00549	0.00586	0.00953	0.00934	0.01058	0.00845	0.01420	0.00928	0.00657
107	0.00184	0.00029	0.00044	0.00126	0.00161	0.00157	0.00141	0.00170	0.00096	0.00168	0.00052	0.00155	0.00166	0.00185
108	0.00035	0.00141	0.00023	0.00106	0.00065	0.00287	0.00047	0.00037	0.00032	0.00080	0.00038	0.00078	0.00415	0.00193
109	0.00001	0	0	0	0	0	0	0	0	0	0	0	0	0
110	0.00344	0.00082	0.00064	0.00235	0.00253	0.00159	0.00267	0.00313	0.00098	0.00102	0.00055	0.00102	0.00109	0.00196
111	0.00026	0.00017	0.00005	0.00058	0.00024	0.00008	0.00015	0.00006	0.00013	0.00002	0.00008	0.00222	0.00006	0.00026
112	0	0	0	0	0	0	0	0	0	0	0	0.00003	0	0
113	0.00032	0.00016	0.00021	0.00047	0.00059	0.00066	0.00028	0.00169	0.00025	0.00067	0.00028	0.00065	0.00064	0.00049
114	0.00231	0.00199	0.00134	0.02787	0.03117	0.04369	0.00362	0.00491	0.00448	0.00098	0.00116	0.00669	0.00123	0.01217
115	0.00031	0.00036	0.00007	0.00004	0.00045	0.00018	0.00015	0.00012	0.00005	0.00009	0.00002	0.00003	0.00005	0.00009
116	0	0	0	0	0	0	0	0	0	0	0	0	0	0
117	0	0	0	0	0	0	0	0	0	0	0	0	0	0
118	0.00074	0.00017	0.00015	0.00048	0.00077	0.00041	0.00054	0.00042	0.00053	0.00060	0.00056	0.00046	0.00044	0.00065
119	0.00001	0	0.00002	0.00001	0.00001	0.00017	0.00002	0	0.00003	0	0.00001	0.00023	0.00004	0.00004
120	0.00008	0.00001	0.00002	0.00011	0.00026	0.00028	0.00024	0.00002	0.00003	0.00002	0.00004	0.00002	0.00004	0.00006
121	0.00030	0.00021	0.00039	0.00115	0.00108	0.00623	0.00034	0.00030	0.00022	0.00006	0.00015	0.00132	0.00025	0.00043
122	0	0	0	0	0	0	0	0	0	0	0	0	0	0
123	0	0	0	0	0	0	0	0	0	0	0	0	0	0
124	0	0	0	0	0	0	0	0	0	0	0	0	0	0

Continued

b_{ij} \ a_{ij}	29	30	31	32	33	34	35	36	37	38	39	40	41	42
1	0.00003	0.00070	0.00959	0.05462	0	0	0	0	0	0	0	0	0.00338	0.00538
2	0.00004	0.10660	0.00605	0.01031	0	0.00031	0.01731	0	0.00011	0.00001	0.00002	0.00013	0.00031	0.00074
3	0.12486	0	0	0	0	0.00303	0.00021	0	0	0	0	0	0.03003	0.00262
4	0	0.00001	0	0.03383	0.00262	0.01244	0	0	0	0	0.00003	0	0	0.00002
5	0.00049	0.03601	0.01832	0.01333	0	0.00269	0.00734	0.00831	0.46067	0.04951	0.08565	0.00083	0.00073	0.00576
6	0.00102	0.01700	0.00252	0	0	0.00003	0.00287	0.70873	0.04929	0.00505	0.03537	0.02153	0.02290	0.01095
7	0	0	0	0.00003	0	0	0.00002	0.00082	0.00097	0.00156	0.02959	0.00102	0.02996	0.00140
8	0	0	0	0	0.00004	0	0	0	0	0	0	0.00033	0.00761	0.00005
9	0	0	0	0	0	0	0	0	0	0	0	0	0	0
10	0	0	0	0	0	0	0	0	0	0.03687	0.00346	0.00102	0.00177	0.00043
11	0	0	0	0	0	0	0	0	0	0.08060	0.01116	0.00282	0.00125	0.01193
12	0	0	0	0.07530	0	0	0.00106	0.00017	0.00124	0.04228	0.08210	0.00445	0.00118	0.00049
13	0.00002	0.11174	0.00717	0.00033	0	0.00015	0.00006	0	0	0.00018	0.00023	0.00018	0.01038	0.04419
14	0	0	0	0	0	0	0	0	0	0	0	0.00580	0.00016	0.00128
15	0	0	0	0	0	0	0	0	0	0.00006	0.00001	0.00007	0.00030	0.03943
16	0.14023	0	0	0	0	0	0	0	0	0	0	0	0	0
17	0	0	0	0	0	0	0	0	0	0	0	0	0	0
18	0.00003	0.00006	0	0	0	0	0	0	0	0	0	0.01767	0.00012	0.00064
19	0	0	0	0	0	0	0	0	0	0	0.00092	0.00456	0.01373	0.01001
20	0	0	0	0	0	0	0	0	0	0	0	0	0	0
21	0	0	0	0	0	0	0	0	0	0	0	0	0	0
22	0.08748	0.03369	0.08838	0.01554	0.00320	0.10149	0.05275	0.00023	0.00030	0.00694	0.00792	0.00049	0.00488	0.00346
23	0.00538	0.00262	0.01223	0.02758	0.00029	0.00008	0.07532	0.00005	0.00007	0.00011	0.00624	0.00005	0.00019	0.00007
24	0.00088	0.00001	0.00204	0.00015	0.00004	0.00001	0.00011	0.00002	0.00002	0.00005	0.00010	0.00001	0.00004	0.00002
25	0.03048	0.00003	0.00222	0.01529	0.00120	0.00100	0.00414	0	0.00005	0.00015	0.00007	0.00003	0.00012	0.00004
26	0.00899	0.00001	0.01415	0.00032	0.00047	0.00038	0.00673	0.00004	0.00002	0.00013	0.00009	0.00005	0.00005	0.00003
27	0.01010	0	0.01543	0	0	0.00008	0.02989	0	0	0	0	0	0	0
28	0.01174	0.00757	0.00238	0.00355	0.00255	0.00179	0.04407	0.00192	0.00260	0.00382	0.00328	0.00187	0.00201	0.00142
29	0.33983	0.01583	0.03983	0.00041	0.00056	0.00117	0.03051	0.00095	0.00081	0.00058	0.00066	0.00042	0.00041	0.00093
30	0.00024	0.26584	0.27860	0.00384	0.00030	0.04460	0.01803	0.00039	0.00048	0.00054	0.00045	0.00064	0.00041	0.00147
31	0.00014	0.00558	0.08189	0.00486	0.00089	0.00481	0.00322	0.00034	0.00062	0.00042	0.00024	0.00037	0.00044	0.00026
32	0.00796	0.02315	0.01731	0.30165	0.50734	0.25285	0.03102	0.00027	0.00017	0.00140	0.00038	0.01723	0.00269	0.08337
33	0.00056	0.00013	0.00048	0.01873	0.09685	0.02945	0.00482	0.00022	0.00019	0.00020	0.00161	0.00285	0.00042	0.00860
34	0.00009	0.00037	0.00009	0.00028	0.00142	0.03019	0.00037	0.00012	0	0.00014	0.00012	0.00014	0.00009	0.00026
35	0	0	0	0.00001	0	0.00004	0.03968	0	0	0.00001	0.02102	0	0.00001	0
36	0.00225	0.00663	0.00458	0.00811	0.00694	0.00350	0.00683	0.06204	0.00946	0.01471	0.01282	0.00920	0.05572	0.03404
37	0	0	0	0	0	0	0	0	0.01008	0.00616	0.04158	0.00050	0.00161	0
38	0.00680	0.00605	0.00071	0.03930	0.00194	0.00914	0.00428	0.00438	0.00162	0.18334	0.14192	0.16729	0.08357	0.09577
39	0	0.00734	0	0.00132	0	0	0	0	0	0.01299	0.00061	0.00332	0.00352	0.00021
40	0	0	0	0	0	0	0	0	0	0	0	0.21327	0.00069	0.01044
41	0.01096	0.06207	0.02089	0.01427	0.04932	0.02835	0.01794	0.00887	0.00057	0.05422	0.00436	0.18577	0.39619	0.08892
42	0.00084	0.00219	0.00100	0.00106	0.00524	0.01986	0.00541	0.00078	0.00090	0.00208	0.00885	0.00766	0.01688	0.18846

Continued

b_{ij} / a_{ij}	29	30	31	32	33	34	35	36	37	38	39	40	41	42
43	0.00661	0.00845	0.00446	0.00297	0.02188	0.06786	0.02494	0.01089	0.00329	0.00698	0.00206	0.01540	0.04512	0.00673
44	0	0.00001	0.00002	0.00030	0.00012	0	0	0.00001	0.00001	0.00004	0.00118	0.00150	0.00237	0.00153
45	0.00299	0.00029	0.00427	0.00937	0.00094	0.00173	0.07001	0	0	0.00154	0.00161	0.00039	0.00043	0.00155
46	0.02010	0.00169	0.00204	0.00367	0.00710	0.01038	0.05268	0.00243	0.00648	0.00231	0.00421	0.00486	0.00106	0.01787
47	0.01655	0.00671	0.01727	0.01041	0.02927	0.06615	0.09524	0.00152	0.00136	0.02055	0.14798	0.03228	0.00443	0.07462
48	0.00008	0.00585	0.00039	0.00083	0.00035	0.00018	0.00057	0.00167	0.00102	0.00223	0.00100	0.00131	0.00252	0.00054
49	0.00001	0.00344	0.00057	0.00038	0.00007	0.00003	0.00048	0.00121	0.00053	0.00120	0.00034	0.00074	0.00189	0.00149
50	0.00011	0.00185	0.00053	0.00223	0.00048	0.00031	0.00054	0.00208	0.00086	0.00956	0.00390	0.00422	0.00609	0.00319
51	0.00033	0.00048	0.00211	0.00133	0.00172	0.01060	0.00158	0.00048	0.00082	0.00130	0.00076	0.01444	0.00212	0.00572
52	0.00060	0.00125	0.00214	0.00141	0.00564	0.00173	0.00659	0.00108	0.00175	0.00193	0.00115	0.00125	0.00158	0.00083
53	0.00002	0.00185	0.00237	0.00015	0.00010	0.00011	0.00025	0.00056	0.02445	0.00457	0.00039	0.00173	0.00113	0.00040
54	0.00017	0.00329	0.00028	0.00044	0.00005	0.00256	0.00209	0.00250	0.00003	0.01591	0.00080	0.00194	0.00579	0.00063
55	0	0	0	0	0	0	0.00125	0	0	0.00005	0.00002	0.00007	0.00030	0.00004
56	0	0	0	0	0	0	0.00223	0	0	0.00016	0.00018	0.00003	0.00006	0.00001
57	0.00012	0.00077	0.04432	0.00099	0.00075	0.00323	0.01011	0.00443	0.00186	0.00170	0.00219	0.00122	0.00141	0.00029
58	0	0	0	0	0	0	0.00399	0	0	0.00031	0.00047	0.00051	0.00018	0.00008
59	0	0	0	0	0	0	0.01854	0	0	0.00397	0.00131	0.00310	0.00164	0.00140
60	0.00009	0.00012	0.00520	0.00188	0.00153	0.01938	0.00227	0.00027	0.00022	0.00104	0.00109	0.00022	0.00046	0.00197
61	0.00931	0.03319	0.06460	0.00953	0.00954	0.08107	0.06307	0.00370	0.00268	0.00784	0.00530	0.01494	0.02918	0.00877
62	0.00044	0.00019	0.00001	0.00014	0.00003	0	0	0.00023	0.00007	0.00026	0.00022	0.00033	0.00009	0.00008
63	0.00077	0.00053	0.00148	0.00086	0.00169	0.00037	0.00054	0.00021	0.00097	0.00157	0.00101	0.00083	0.00101	0.00035
64	0.00208	0.00537	0.00708	0.01238	0.00926	0.00250	0.00974	0.01149	0.03149	0.01318	0.02780	0.00718	0.00904	0.00332
65	0	0	0	0	0	0	0	0	0	0	0	0	0	0
66	0.00156	0.00658	0.00165	0.00944	0.00758	0.00143	0.00197	0.00914	0.02925	0.01373	0.01466	0.00644	0.00692	0.00211
67	0	0	0	0	0	0	0	0	0	0	0	0	0	0
68	0.00120	0.00472	0.00402	0.00892	0.00690	0.00389	0.00282	0.00471	0.00320	0.00639	0.00527	0.00711	0.00425	0.00662
69	0	0	0	0	0	0	0	0	0	0	0	0	0	0
70	0	0	0	0	0	0	0	0	0	0	0	0	0	0
71	0	0	0	0	0	0	0	0	0	0	0	0	0	0
72	0.00002	0.00009	0.00016	0.00014	0.00042	0.00007	0.00003	0.00023	0.00111	0.00020	0.00023	0.00022	0.00039	0.00012
73	0.00021	0.00028	0.00026	0.00134	0.00107	0.00018	0.00389	0.00056	0.00030	0.00129	0.00074	0.00052	0.00053	0.00024
74	0.00002	0.00006	0.00011	0.00032	0.00017	0.00006	0.00036	0.00017	0.00012	0.00019	0.00021	0.00025	0.00010	0.00006
75	0.00089	0.00337	0.00131	0.00505	0.00479	0.00106	0.00472	0.00833	0.00299	0.00817	0.00803	0.00427	0.00358	0.00154
76	0.00004	0.00012	0.00024	0.00020	0.00042	0.00010	0.00005	0.00041	0.00029	0.00024	0.00034	0.00025	0.00018	0.00014
77	0.00024	0.00032	0.00053	0.00061	0.00159	0.00031	0.00156	0.00014	0.00064	0.00078	0.00032	0.00040	0.00033	0.00023
78	0.00010	0.00017	0.00026	0.00024	0.00016	0.00015	0.05988	0.00042	0.00018	0.00034	0.00065	0.00029	0.00026	0.00013
79	0.00008	0.00026	0.00031	0.00144	0.00119	0.00037	0.00044	0.00150	0.00140	0.00051	0.00065	0.00065	0.00033	0.00024
80	0.00048	0.00076	0.00073	0.00180	0.00249	0.00089	0.00061	0.00243	0.00159	0.00518	0.00500	0.00296	0.00280	0.00114
81	0.00004	0.00018	0.00037	0.00029	0.00245	0.00016	0.00009	0.00035	0.00062	0.00035	0.00048	0.00037	0.00022	0.00020
82	0.00060	0.00107	0.00115	0.00143	0.00442	0.00113	0.00058	0.00551	0.00388	0.00324	0.00403	0.00164	0.00377	0.00169
83	0.00115	0.00216	0.00208	0.00195	0.00230	0.00252	0.00185	0.00255	0.00217	0.00244	0.00200	0.00197	0.00202	0.00235
84	0.00563	0.00377	0.01118	0.02640	0.01445	0.00519	0.01338	0.00269	0.00440	0.00622	0.00551	0.00660	0.00431	0.00382

Continued

a_{ij} \ b_{ij}	29	30	31	32	33	34	35	36	37	38	39	40	41	42
85	0.00006	0.00080	0.00003	0.04747	0.00004	0.00002	0.00013	0.00074	0.00001	0.00081	0.00083	0.00083	0.00008	0.00467
86	0.00287	0.02516	0.01542	0.04474	0.01541	0.01109	0.01333	0.01756	0.08790	0.19859	0.08818	0.03227	0.03804	0.00977
87	0.00010	0.00072	0.00021	0.00304	0.00035	0.00047	0.00010	0.00137	0.00529	0.00502	0.00234	0.00218	0.00300	0.00068
88	0.00001	0.00011	0	0.00006	0.00005	0.00007	0.00001	0.00002	0.00422	0.00102	0.00087	0.00035	0.00048	0.00351
89	0.00044	0.00156	0.00145	0.00330	0.00139	0.00100	0.00079	0.00195	0.00159	0.00428	0.00286	0.00821	0.00288	0.00116
90	0.00102	0.00059	0.00081	0.00085	0.00336	0.00178	0.00089	0.00142	0.00061	0.00155	0.00194	0.00123	0.00130	0.00181
91	0.00222	0.00824	0.00438	0.00963	0.00503	0.00288	0.00371	0.00454	0.03749	0.01788	0.01628	0.01634	0.01064	0.00646
92	0.00629	0.01636	0.01246	0.01040	0.00968	0.01390	0.01397	0.00338	0.01994	0.01880	0.02030	0.01306	0.01140	0.01488
93	0.00001	0.00003	0.00002	0.00003	0.00003	0.00002	0.00003	0.00744	0.00057	0.00017	0.00116	0.00008	0.00074	0.00016
94	0.00037	0.00111	0.00083	0.00078	0.00054	0.00063	0.00068	0.00463	0.00380	0.00248	0.00233	0.00109	0.00127	0.00103
95	0.00032	0.00007	0.00013	0.00009	0.00013	0.00044	0.00033	0.00007	0.00005	0.00008	0.00016	0.00019	0.00009	0.00020
96	0.00097	0.00255	0.00215	0.00205	0.00142	0.00161	0.00110	0.00522	0.01879	0.00576	0.00663	0.00283	0.00287	0.00243
97	0	0.00001	0.00011	0.00005	0.00007	0.00020	0.00006	0.00004	0.00002	0.00006	0.00004	0	0.00019	0.00014
98	0	0	0.00001	0.00001	0	0	0	0	0	0	0	0	0	0
99	0.00884	0.00639	0.01513	0.00217	0.00298	0.00453	0.00853	0.00512	0.00120	0.00499	0.00290	0.00714	0.00389	0.00514
100	0.09308	0.09563	0.09161	0.06891	0.09809	0.08571	0.06641	0.03730	0.11521	0.04957	0.05971	0.05041	0.04573	0.05185
101	0.00233	0.00541	0.00689	0.01180	0.00939	0.01425	0.00453	0.00152	0.00943	0.00482	0.00540	0.01104	0.00613	0.00698
102	0.00001	0.00001	0.00005	0.00001	0	0	0.00001	0.00001	0	0.00001	0	0.00001	0	0.00001
103	0	0	0.00008	0	0	0	0	0	0	0	0	0	0	0
104	0.00001	0.00001	0.00002	0.00004	0.00005	0.00007	0.00001	0.00001	0.00002	0.00008	0.00005	0.00001	0.00006	0.00001
105	0.00015	0.00011	0.00005				0.00014	0.00008				0.00012		0.00008
106	0.00529	0.01185	0.00925	0.01167	0.01159	0.00953	0.00734	0.01049	0.00967	0.01503	0.01900	0.00998	0.01381	0.00726
107	0.00065	0.00204	0.00183	0.00294	0.00277	0.00226	0.00121	0.00241	0.00249	0.00196	0.00243	0.00263	0.00235	0.00190
108	0.00110	0.00123	0.00504	0.00065	0.00249	0.00154	0.00466	0.00021	0.00003	0.00201	0.00019	0.00050	0.00050	0.00296
109	0.00001	0	0.00001	0.00001	0	0	0.00001	0	0	0	0	0.00001	0	0
110	0.00087	0.00275	0.00223	0.00365	0.00428	0.00301	0.00202	0.00645	0.00752	0.00716	0.00677	0.00536	0.00510	0.00308
111	0.00026	0.00015	0.00030	0.00018	0.00024	0.00014	0.00023	0.00016	0.00003	0.00012	0.00007	0.00017	0.00019	0.00105
112	0	0	0	0	0	0	0	0	0	0	0	0	0	0
113	0.00027	0.00064	0.00082	0.00140	0.00110	0.00167	0.00053	0.00018	0.00112	0.00057	0.00064	0.00130	0.00072	0.00083
114	0.00598	0.00360	0.01744	0.00454	0.00468	0.00785	0.00445	0.00223	0.00120	0.00226	0.00282	0.01323	0.00323	0.06320
115	0.00014	0.00033	0.00008	0.00130	0.00011	0.00003	0.00003	0.00019	0.00025	0.00028	0.00024	0.00027	0.00016	0.00005
116	0	0	0	0	0	0	0	0	0	0	0	0	0	0
117	0	0	0	0	0	0	0	0	0	0	0	0	0	0
118	0.00032	0.00133	0.00044	0.00143	0.00141	0.00112	0.00087	0.00049	0.00067	0.00109	0.00103	0.00174	0.00069	0.00088
119	0	0	0.00003	0.00005	0.00002	0	0	0.00001	0	0.00001	0.00003	0.00001	0.00001	0.00001
120	0.00011	0.00013	0.00004	0.00033	0.00013	0.00009	0.00005	0.00072	0.00012	0.00029	0.00012	0.00046	0.00042	0.00097
121	0.00184	0.00062	0.00133	0.00358	0.00101	0.00090	0.00089	0.00152	0.00049	0.00164	0.00089	0.00294	0.00223	0.00170
122	0	0	0	0	0	0	0	0	0	0	0	0	0	0
123	0	0	0	0	0	0	0	0	0	0	0	0	0	0
124	0	0	0	0	0	0	0	0	0	0	0	0	0	0

Continued

b_{ij} / a_{ij}	43	44	45	46	47	48	49	50	51	52	53	54	55	56
1	0.00358	0.06160	0.00662	0.00047	0	0	0	0.01003	0	0	0	0	0	0
2	0.08172	0.00035	0	0.20782	0	0	0.00125	0.00119	0.00074	0.00026	0.00084	0.00118	0	0
3	0	0.06296	0	0	0	0	0	0	0	0	0	0	0	0
4	0.00084	0.00204	0	0.00446	0	0	0.00162	0.00024	0.00332	0.00085	0.00057	0	0	0
5	0.09094	0.02823	0.00754	0.00925	0.00257	0.09325	0.01343	0.07228	0.04644	0.02801	0.06123	0.00011	0.03882	0.01104
6	0.01801	0.00316	0.00457	0.00082	0.00041	0.00019	0.00001	0.00027	0.03153	0.00259	0.00051	0.01267	0.00577	0.00187
7	0.04506	0.00051	0.00047	0.00003	0.00001	0.00004	0	0.00008	0.00505	0.00529	0.00011	0.00149	0.00009	0.00012
8	0.00296	0.00010	0	0	0	0	0	0	0	0	0.00126	0.00243	0.17242	0.02917
9	0	0	0	0	0	0	0	0	0	0	0.00670	0.00066	0.00140	0.00823
10	0.00560	0	0	0	0	0	0	0.00109	0	0	0.00019	0.00484	0	0
11	0.00229	0.00023	0	0	0	0.00041	0.00005	0.00392	0.00046	0.00013	0	0.00011	0.04721	0.00498
12	0.00426	0.00020	0	0	0	0.07389	0.06564	0.09924	0.08649	0.26638	0.12051	0.05321	0.00045	0.00014
13	0.00018	0.00010	0	0	0	0.00021	0.00028	0.00016	0.00087	0	0	0	0	0
14	0.00182	0.01543	0	0	0	0	0	0	0	0	0	0	0	0
15	0.00021	0.00794	0	0	0	0	0	0	0	0	0	0	0	0
16	0.00004	0.00265	0	0	0	0	0	0	0	0	0	0	0	0
17	0.00001	0.00316	0	0	0	0	0	0	0	0	0	0	0	0
18	0.00048	0.01296	0.00060	0.00003	0.00006	0	0	0	0	0	0	0	0	0
19	0.00649	0.02351	0.00038	0.00060	0.00011	0.00112	0	0	0.00080	0.00576	0.00324	0.00108	0	0
20	0	0	0	0	0	0	0	0	0	0	0	0	0	0
21	0	0	0	0	0	0	0	0	0	0	0	0	0	0
22	0.01854	0.00621	0.01986	0.22274	0.05103	0.00793	0.01571	0.00173	0.00896	0.02381	0.00332	0.01611	0.00061	0.00014
23	0.00008	0.00096	0.00258	0.00089	0.01017	0.00259	0.01296	0.00104	0.00030	0.00223	0.00204	0.00031	0.00023	0.00005
24	0.00003	0.00003	0.00023	0.00009	0.00005	0.00021	0.00007	0.00004	0.00043	0.00032	0.00002	0.00004	0.00006	0.00001
25	0.00005	0.00008	0.00228	0.00077	0.00006	0.00008	0.00015	0.00005	0.00027	0.00250	0.00016	0.00005	0.00015	0.00002
26	0.00005	0.00028	0.00018	0.01340	0.00523	0.00008	0.00005	0.00003	0.00010	0.00008	0.00186	0.00003	0.00012	0.00003
27	0	0.00501	0.07221	0.01375	0.00578	0	0	0.00145	0	0	0.00211	0.00154	0	0
28	0.00201	0.00209	0.00101	0.00272	0.00163	0.00480	0.00176	0.00442	0.00374	0.00216	0.00658	0.00117	0.00256	0.00138
29	0.00058	0.00080	0.00023	0.01261	0.00185	0.00205	0.00046	0.00177	0.00117	0.00078	0.01009	0.00048	0.00052	0.00025
30	0.00079	0.00031	0.00023	0.00028	0.00023	0.00049	0.00108	0.00127	0.00778	0.00247	0.00189	0.00044	0.00106	0.00032
31	0.00043	0.00053	0.00018	0.00029	0.00062	0.00031	0.00034	0.00294	0.00584	0.00457	0.00250	0.00550	0.00033	0.00009
32	0.00920	0.04241	0.02030	0.01012	0.01725	0.10380	0.00082	0.08301	0.03757	0.05186	0.06035	0.00340	0.00018	0.00032
33	0.00050	0.01949	0.00018	0.00032	0.00052	0.00041	0.00027	0.00018	0.00029	0.00238	0.00027	0.00124	0.00040	0.00005
34	0.00009	0.00071	0.00008	0.00030	0.00016	0.00017	0.00020	0.00019	0.00050	0.00209	0.00006	0.00011	0.00024	0.00007
35	0	0	0	0	0.00005	0.00001	0	0	0	0.00006	0	0.00010	0	0
36	0.04324	0.00183	0.01408	0.01054	0.00807	0.01128	0.00700	0.06283	0.07433	0.06994	0.01573	0.01909	0.01267	0.00822
37	0.00143	0	0	0.00065	0	0	0	0.00174	0	0	0.00239	0.02598	0.19059	0.01350
38	0.07932	0.03741	0.01566	0.02607	0.00588	0.00172	0.00091	0.03280	0.09667	0.02170	0.00411	0.01042	0.00086	0.00193
39	0.01333	0.00105	0	0	0	0	0	0	0	0	0	0	0	0
40	0.00026	0.00026	0	0	0	0	0	0	0	0	0	0	0	0
41	0.11836	0.06627	0.13482	0.02403	0.09825	0.00266	0.00236	0.00549	0.01354	0.03519	0.05656	0.00437	0.00062	0.00047
42	0.01485	0.00550	0.00386	0.00568	0.00372	0.00227	0.00055	0.02105	0.00219	0.00109	0.00305	0.01301	0.00644	0.00027

Continued

a_{ij}＼b_{ij}	43	44	45	46	47	48	49	50	51	52	53	54	55	56
43	0.13702	0.01099	0.29812	0.08435	0.35332	0.00400	0.00093	0.05780	0.02276	0.00681	0.00634	0.01455	0.01965	0.00096
44	0.00027	0.27716	0.00003	0.00002	0.00002	0.00002	0	0.00002	0.00006	0.00006	0	0	0.00001	0.00001
45	0.00135	0.00138	0.22470	0.03394	0.00523	0.00030	0.00050	0.00359	0.00030	0.00030	0.00025	0.00405	0.00365	0
46	0.00626	0.01005	0.00174	0.09046	0.00371	0.00676	0.00243	0.00386	0.00619	0.00320	0.00789	0.00208	0.00120	0.00065
47	0.01912	0.01673	0.01004	0.00608	0.24627	0.05286	0.00296	0.01423	0.01680	0.01230	0.01321	0.00683	0.00138	0.00056
48	0.00142	0.00071	0.00054	0.00060	0.00035	0.03454	0.29059	0.03266	0.00165	0.01109	0.01483	0.00508	0.00103	0.00040
49	0.00079	0.00033	0.00053	0.00038	0.00013	0.00721	0.02282	0.00426	0.00082	0.00618	0.00451	0.00197	0.00341	0.00295
50	0.00228	0.00198	0.00173	0.00054	0.00043	0.01063	0.01146	0.12500	0.00866	0.00829	0.01696	0.00667	0.00064	0.00036
51	0.00343	0.01578	0.00036	0.00060	0.00072	0.00100	0.00079	0.00287	0.10280	0.00162	0.00300	0.00118	0.00975	0.00065
52	0.00143	0.00196	0.00077	0.00145	0.00118	0.00374	0.00115	0.00272	0.00162	0.03391	0.02170	0.00182	0.03001	0.02362
53	0.00051	0.00011	0.00036	0.00001	0.00002	0.01925	0.00916	0.01107	0.00561	0.01638	0.06660	0.00480	0.00242	0.00643
54	0.00297	0.00058	0.00085	0.00033	0.00168	0.01078	0.09902	0.01617	0.02971	0.01391	0.13153	0.20738	0.01539	0.54961
55	0.00002	0.00001		0.00100	0.00015	0.00613	0.00008	0.00011	0.02157	0.00011	0.00005	0.00009	0.02400	0.02479
56	0.00005	0.00006	0.00005	0.00006	0.00001	0.00185	0.00338	0.00070	0.00013	0.00019	0.00002	0.02936	0.00427	0.00578
57	0.00104	0.00369	0.00082	0.00919	0.00066	0.02307	0.09979	0.03550	0.00015	0.00249	0.00334	0.00113	0.00099	0.01929
58	0.00011	0.00011	0.00036	0.00024	0.00010	0.00124	0.00059	0.00027	0.00161	0.00073	0.00033	0.00072	0.00009	0.03706
59	0.00877	0.00013	0.00013	0.00033	0.00006	0.00018	0.00009	0.00062	0.00021	0.00072	0.00021	0.14421	0.00168	0.00060
60	0.00222	0.00052	0.00027	0.00131	0.00039	0.00108	0.00155	0.00104	0.00441	0.00088	0.00188	0.00187	0.00399	0.04910
61	0.01049	0.00862	0.00265	0.02825	0.00746	0.07144	0.09014	0.01336	0.04958	0.01799	0.01190	0.00853	0.00004	0.00005
62	0.00041	0.00062	0.00004	0.00004	0	0.00007	0.00001	0.00003	0.00048	0.00008	0.00003	0.00006	0.00380	0.00535
63	0.00087	0.00053	0.00064	0.00662	0.00130	0.00513	0.00096	0.00256	0.00218	0.00555	0.00123	0.00162	0.01228	0.00940
64	0.01143	0.00543	0.00528	0.00444	0.00487	0.09108	0.00901	0.00550	0.01039	0.01811	0.00996	0.15916		
65	0	0	0	0	0	0	0	0	0	0	0	0	0	0
66	0.00976	0.00325	0.00420	0.00363	0.00266	0.01424	0.00347	0.00251	0.01012	0.00997	0.00886	0.00705	0.01077	0.00780
67	0	0	0	0	0	0	0	0	0	0	0	0	0.00014	0.00018
68	0.00470	0.00311	0.00310	0.00322	0.00380	0.00638	0.00273	0.00297	0.00673	0.00761	0.00795	0.00256	0.00749	0.00419
69	0	0	0	0	0	0	0	0	0	0	0	0	0	0
70	0	0	0	0	0	0	0	0	0	0	0	0	0	0
71	0	0	0	0	0	0	0	0	0	0	0	0	0	0
72	0.00022	0.00013	0.00010	0.00013	0.00011	0.00030	0.00009	0.00036	0.00025	0.00022	0.00012	0.00012	0.00080	0.00011
73	0.00133	0.00074	0.00054	0.00048	0.00232	0.00156	0.00052	0.00036	0.00113	0.00036	0.00073	0.00596	0.00091	0.00066
74	0.00012	0.00011	0.00008	0.00007	0.00009	0.00035	0.00009	0.00011	0.00009	0.00050	0.00009	0.00010	0.00011	0.00009
75	0.00560	0.00231	0.00227	0.00245	0.00248	0.00719	0.00280	0.00280	0.00600	0.00741	0.00678	0.00738	0.00692	0.00368
76	0.00023	0.00041	0.00016	0.00013	0.00016	0.00042	0.00013	0.00011	0.00039	0.00021	0.00018	0.00011	0.00018	0.00016
77	0.00035	0.00035	0.00022	0.00053	0.00061	0.00065	0.00023	0.00107	0.00169	0.00122	0.00105	0.00073	0.00046	0.00026
78	0.00045	0.00035	0.00031	0.00026	0.00034	0.00131	0.00042	0.00013	0.00111	0.00100	0.00025	0.00040	0.00148	0.00022
79	0.00038	0.00057	0.00044	0.00030	0.00034	0.00086	0.00052	0.00020	0.00060	0.00102	0.00032	0.00033	0.00163	0.00037
80	0.00353	0.00253	0.00179	0.00189	0.00188	0.00346	0.00081	0.00167	0.00335	0.00287	0.00255	0.00163	0.00238	0.00097
81	0.00030	0.00029	0.00022	0.00030	0.00026	0.00059	0.00021	0.00016	0.00058	0.00045	0.00028	0.00017	0.00020	0.00014
82	0.00428	0.00262	0.00203	0.00148	0.00163	0.00213	0.00340	0.00167	0.00224		0.00168	0.00169	0.00209	0.00167
83	0.00209	0.00207	0.00181	0.00190	0.00194	0.00102	0.00225	0.00241		0.00296	0.00218	0.00172	0.00216	0.00102
84	0.00537	0.00732	0.00355	0.00697	0.00358	0.00805	0.01037	0.01295	0.01553	0.01608	0.01114	0.00494	0.00649	0.00912

Appendix 3   Input Coefficients among Industries in China in 1997

Continued

a_{ij} \ b_{ij}	43	44	45	46	47	48	49	50	51	52	53	54	55	56
85	0.00022	0.00001	0.00012	0.00064	0.00170	0.00961	0.00058	0.00330	0.00711	0.00004	0.00087	0.00265	0.10774	0.04770
86	0.05265	0.02930	0.03302	0.02429	0.02734	0.10974	0.02944	0.04434	0.04444	0.03921	0.06311	0.04239	0.04919	0.04983
87	0.00288	0.00174	0.00124	0.00251	0.00023	0.00011	0.00027	0.00010	0.00042	0.00141	0.00043	0.00037	0.00052	0.00015
88	0.00067	0.00013	0.00004	0.00058	0.00020	0.00004	0.00003	0.00008	0.00365	0.00291	0.00016	0.00009	0.00033	0.00050
89	0.00230	0.00241	0.00140	0.00138	0.00125	0.00100	0.00246	0.00105	0.00196	0.00270	0.00097	0.00121	0.00161	0.00196
90	0.00168	0.00129	0.00070	0.00065	0.00060	0.00111	0.00042	0.00130	0.00098	0.00137	0.00138	0.00098	0.00065	0.00033
91	0.00762	0.00375	0.00397	0.00500	0.00626	0.01666	0.02138	0.02411	0.01366	0.02715	0.01746	0.00958	0.03089	0.00491
92	0.01149	0.00989	0.00699	0.01126	0.01052	0.02842	0.02631	0.02181	0.02400	0.03235	0.02330	0.01884	0.03937	0.00664
93	0.00070	0.00003	0.00012	0.00005	0.00005	0.00006	0.00003	0.00025	0.00044	0.00044	0.00009	0.00009	0.00019	0.00006
94	0.00123	0.00047	0.00050	0.00062	0.00068	0.00217	0.00182	0.00242	0.00213	0.00355	0.00219	0.00015	0.01006	0.00179
95	0.00011	0.00058	0.00007	0.00008	0.00024	0.00016	0.00013	0.00009	0.00091	0.00011	0.00016	0.00016	0.00005	0.00003
96	0.00287	0.00153	0.00168	0.00158	0.00177	0.00793	0.00416	0.00659	0.00558	0.00657	0.00658	0.00351	0.00725	0.00271
97	0.00006	0.00009	0.00004	0.00009	0.00003	0.00005	0	0.00008	0.00019	0.00002	0.00251	0.00002	0.00002	0.00002
98	0	0.00002	0	0	0	0	0	0	0.00004	0	0.00001	0	0.00001	0
99	0.01332	0.00387	0.00274	0.00452	0.00775	0.00118	0.00316	0.00333	0.00528	0.00370	0.02816	0.01123	0.00403	0.00098
100	0.04837	0.07383	0.04592	0.05721	0.04792	0.07237	0.07183	0.07384	0.07428	0.07435	0.07563	0.04660	0.05839	0.01673
101	0.00529	0.00812	0.00325	0.00507	0.00573	0.00708	0.01333	0.00697	0.00823	0.00863	0.01204	0.01334	0.00307	0.00244
102	0.00003	0	0	0	0.00001	0	0	0	0.00001	0	0.00003	0	0	0
103	0.00004	0	0	0	0	0	0	0	0	0	0	0	0	0
104	0.00001	0	0	0	0	0	0	0	0	0	0	0.00001	0	0
105	0.00011	0.00006	0.00005	0.00007	0.00001	0.00002	0.00005	0.00001	0.00001	0.00006	0.00046	0.00018	0.00001	0.00002
106	0.01314	0.01104	0.01222	0.00919	0.01110	0.01675	0.00847	0.01833	0.01813	0.02349	0.01493	0.01678	0.01050	0.00497
107	0.00227	0.00241	0.00137	0.00205	0.00188	0.00162	0.00151	0.00139	0.00262	0.00185	0.00184	0.00146	0.00117	0.00063
108	0.00089	0.00232	0.00048	0.00150	0.00202	0.00022	0.00033	0.00124	0.00134	0.00346	0.00145	0.00031	0.00015	0.00012
109	0.00001	0	0	0	0	0	0	0	0	0	0	0.00001	0	0
110	0.00572	0.00285	0.00248	0.00254	0.00209	0.00423	0.00310	0.00300	0.00610	0.00632	0.00238	0.00341	0.00423	0.00304
111	0.00031	0.00072	0.00007	0.00027	0.00018	0.00022	0.00047	0.00452	0.00015	0.00106	0.00092	0.00048	0.00002	0.00003
112	0	0	0	0	0	0	0	0	0	0	0	0	0	0
113	0.00062	0.00096	0.00038	0.00059	0.00067	0.00084	0.00158	0.00082	0.00097	0.00102	0.00141	0.00156	0.00036	0.00029
114	0.00351	0.03899	0.00081	0.00298	0.00360	0.00309	0.00150	0.00537	0.00594	0.00678	0.01056	0.00246	0.00062	0.00062
115	0.00015	0.00011	0.00003	0.00021	0.00017	0.00021	0.00015	0.00034	0.00022	0.00017	0.00015	0.00021	0.00045	0.00006
116	0	0	0	0	0	0	0	0	0	0	0	0	0	0
117	0	0	0	0	0	0	0	0	0	0	0	0	0	0
118	0.00080	0.00093	0.00037	0.00103	0.00050	0.00116	0.00096	0.00063	0.00094	0.00099	0.00156	0.00078	0.00071	0.00036
119	0.00002	0.00011	0.00001	0.00001	0.00003	0.00001	0.00001	0.00002	0.00026	0.00001	0.00009	0.00002	0.00008	0
120	0.00035	0.00076	0.00008	0.00015	0.00050	0.00004	0.00002	0.00003	0.00027	0.00028	0.00010	0.00020	0.00002	0.00004
121	0.00195	0.00171	0.00082	0.00117	0.00098	0.00195	0.00061	0.00060	0.00058	0.00161	0.00128	0.00049	0.00064	0.00051
122	0	0	0	0	0	0	0	0	0	0	0	0	0	0
123	0	0	0	0	0	0	0	0	0	0	0	0	0	0
124	0	0	0	0	0	0	0	0	0	0	0	0	0	0

Continued

b_{ij} \ a_{ij}	57	58	59	60	61	62	63	64	65	66	67	68	69	70
1	0	0	0	0	0	0	0	0	0	0	0	0	0	0
2	0.00105	0	0	0	0.00057	0	0	0.00013	0.00016	0.00059	0	0.00083	0.00091	0
3	0	0	0	0	0	0	0	0	0	0	0	0	0	0
4	0	0	0	0	0	0	0	0	0	0	0	0	0	0
5	0	0	0	0.00340	0.00051	0	0	0	0	0.00131	0	0	0	0.00657
6	0.03962	0.01350	0.02059	0.00022	0.00326	0.00439	0.00514	0.00847	0.00350	0.00636	0.01226	0.00389	0.00092	0.00046
7	0.00849	0.00049	0.00173	0.00009	0.00005	0.00015	0.00073	0.00016	0.00011	0.00023	0.00037	0.00012	0.00014	0.00031
8	0.00135	0.00001	0.00029	0.00022	0.00009	0.00083	0.00002	0.00027	0.00009	0.00035	0.00061	0.00042	0.00001	0.00118
9	0.05953	0.09328	0.00403	0.10791	0.01967	0.00580	0.00132	0.00282	0.00125	0.00273	0.00379	0.00128	0.00108	0.00178
10	0.00396	0.21553	0.34903	0	0.00235	0.00757	0.00098	0.01184	0.00030	0.00187	0.00139	0.00028	0.00028	0
11	0	0	0	0.00029	0	0	0	0	0	0	0	0	0	0.00037
12	0.00836	0.01367	0.00308	0.00022	0.00414	0.00203	0.00200	0.00174	0.00063	0.00061	0.00421	0.00038	0.00222	0.00075
13	0.00033	0.00049	0.00067		0.00039	0.00083	0.00179	0.00040	0.00009	0.00033	0.00312	0.00013	0.00280	
14														
15														
16														
17														
18														
19														
20	0	0	0	0	0	0	0	0	0	0	0	0	0	0
21	0.00101	0.00030	0.00107	0.00211	0.00248	0.00295	0.00365	0.00489	0.00163	0.01610	0.00191	0.00575	0.01020	0.00175
22	0.00017	0.00003	0.00015	0.00077	0.00065	0.00114	0.00095	0.00092	0.00006	0.00130	0.00046	0.00516	0.00334	0.00054
23	0.00005	0.00013	0.00004	0.00005	0.00062	0.00002	0.00002	0.00003	0.00003	0.00004	0.00008	0.00006	0.00032	0.00013
24	0.00010	0.00002	0.00009	0.00096	0.00043	0.00005	0.00006	0.00073	0.00004	0.00011	0.00015	0.00064	0.00028	0.00615
25	0.00008	0.00010	0.00009	0.00008	0.00010	0.00008	0.00005	0.00020		0.00010		0.00095	0.00011	0.00344
26	0	0	0	0	0	0	0	0	0	0	0	0	0	0.00033
27	0.00281	0.00189	0.00259	0.00170	0.00229	0.00296	0.00183	0.00353	0.00167	0.00649	0.00380	0.00165	0.00069	0.00482
28	0.00055	0.00085	0.00037	0.00048	0.00061	0.00095	0.00056	0.00155	0.00345	0.00331	0.00042	0.00222	0.00289	0.00049
29	0.00098	0.00221	0.00083	0.00066	0.00320	0.00168	0.00217	0.00181	0.00057	0.00141	0.00729	0.00452	0.00046	0.00110
30	0.00021	0.00092	0.00082	0.00152	0.01061	0.00280	0.00304	0.00278	0.00096	0.00271	0.00072	0.00301	0.00448	0.00159
31	0.00061	0.00024	0.00075	0.00402	0.00845	0.00203	0.00078	0.01037	0.00039	0.00376	0.00041	0.00189	0.00122	0.00128
32	0.00022	0.00012	0.00016	0.00048	0.00054	0.00064	0.00037	0.00095	0.00033	0.00454	0.00036	0.00427	0.00019	0.00037
33	0.00021	0.00009	0.00013	0.00065	0.00019	0.00020	0.00019	0.00027	0.00057	0.00105	0.00025	0.00076	0.00028	0.00063
34	0	0.00004	0.00002	0	0.00001	0.00001	0.00001	0		0.00034	0.00005	0.00023	0.00023	
35												0.00001	0.00001	
36	0.02666	0.00438	0.01549	0.01244	0.00731	0.01286	0.01267	0.01155	0.00634	0.01223	0.01427	0.01038	0.01034	0.01647
37	0.01518	0.01920	0.00376	0.00053	0.00184	0.00331	0.00425	0.00257	0.00121	0.00128	0.00213	0.00069	0.00072	0.00012
38	0.00354	0.05357	0.03133	0.00786	0.00386	0.00243	0.00175	0.00328	0.00100	0.00301	0.00807	0.00448	0.00534	0.06238
39	0	0	0	0	0	0	0	0	0	0	0	0	0	0
40														
41	0.00248	0.00510	0.00242	0.00453	0.01285	0.00427	0.00690	0.06385	0.00851	0.00593	0.01466	0.01298	0.02072	0.01371
42	0.00108	0.00031	0.00151	0.00059	0.00096	0.00114	0.00070	0.00169	0.00084	0.00214	0.00069	0.00077	0.00036	0.00087

Continued

a_{ij} / b_{ij}	57	58	59	60	61	62	63	64	65	66	67	68	69	70
43	0.00274	0.00222	0.00785	0.00045	0.00377	0.00201	0.00183	0.00321	0.00135	0.00678	0.00973	0.00273	0.00244	0.01479
44	0.00003	0.00001	0.00002	0	0	0.00003	0.00007	0.00003	0.00001	0.00008	0.00002	0.00004	0.00052	0.00003
45	0	0	0	0	0.00725	0.00823	0.00032	0.00021	0.00006	0.00351	0.00074	0.00014	0.00048	0.00039
46	0.00387	0.00139	0.00244	0.00112	0.00633	0.00713	0.00871	0.02060	0.04455	0.01435	0.01559	0.04272	0.00605	0.00778
47	0.00209	0.00321	0.00262	0.00229	0.00676	0.00537	0.02375	0.03309	0.00271	0.01146	0.00857	0.02357	0.00191	0.00399
48	0.00148	0.01632	0.00231	0.00085	0.00133	0.01419	0.00057	0.00068	0.00033	0.00111	0.00126	0.00082	0.00097	0.00091
49	0.00082	0.00121	0.00068	0.00060	0.00039	0.00044	0.00010	0.00029	0.00007	0.00040	0.00437	0.00021	0.00033	0.00019
50	0.00296	0.00456	0.00217	0.00072	0.00465	0.00222	0.00105	0.00104	0.00047	0.00177	0.00283	0.00111	0.00313	0.00135
51	0.00100	0.00066	0.00121	0.00128	0.00607	0.00164	0.01041	0.00186	0.00163	0.00183	0.00158	0.00843	0.00504	0.00236
52	0.00156	0.00120	0.00117	0.00098	0.00336	0.00331	0.00168	0.00284	0.00140	0.00283	0.00147	0.00148	0.00209	0.00219
53	0.02318	0.00449	0.00331	0.00102	0.00170	0.01111	0.07842	0.00252	0.00056	0.00221	0.00560	0.00309	0.00724	0.00060
54	0.00507	0.01568	0.01189	0.00106	0.00773	0.00205	0.00249	0.00310	0.00191	0.00481	0.00441	0.00471	0.00285	0.01988
55	0.04888	0.00068	0.00022	0.00009	0.03129	0.01536	0.01749	0.02063	0.02996	0.00939	0.00936	0.00345	0.02299	0.00023
56	0.09693	0.00403	0.00074	0.00021	0.02350	0.00820	0.04385	0.02219	0.00751	0.02564	0.03760	0.03634	0.00511	0.00120
57	0.24400	0.04245	0.00195	0.00271	0.26073	0.13288	0.09335	0.12298	0.08394	0.20197	0.14915	0.06061	0.13538	0.03361
58	0.03365	0.10730	0.00322	0.00109	0.00671	0.01912	0.01242	0.01413	0.00166	0.00722	0.00457	0.00141	0.00204	0.00053
59	0.00933	0.07416	0.14432	0.62578	0.05556	0.01376	0.03980	0.01698	0.01123	0.01421	0.00497	0.01110	0.00916	0.00618
60	0.00212	0.00102	0.00566	0.04397	0.03386	0.01205	0.00596	0.01076	0.01525	0.01811	0.01222	0.00756	0.13102	0.02656
61	0.01761	0.00930	0.01291	0.00793	0.17008	0.03030	0.02244	0.06863	0.24185	0.06162	0.03593	0.01656	0.02602	0.01766
62	0.00014	0.00003	0.00025	0.00001	0.00005	0.23826	0.00153	0.01420	0.08560	0.00828	0.01672	0.02708	0.09569	0.02436
63	0.00870	0.00108	0.00141	0.00124	0.00854	0.02443	0.08270	0.00894	0.03142	0.01065	0.00771	0.00589	0.00365	0.00333
64	0.02623	0.00910	0.01171	0.00519	0.00987	0.14604	0.21259	0.19067	0.10290	0.10413	0.08146	0.04442	0.10895	0.08860
65	0	0	0	0	0	0	0.00075	0.00131	0.16527	0.00138	0	0	0.00344	0
66	0.02472	0.00836	0.00926	0.00489	0.00434	0.00396	0.00306	0.00582	0.00505	0.09399	0.00194	0.00300	0.04734	0.01220
67	0.00063	0.00001	0.00020	0.00001	0.00002	0.00003	0.00158	0.00027	0.00032	0.00001	0.16835	0.00011	0.00001	0
68	0.01095	0.00548	0.00378	0.00283	0.00540	0.01750	0.00431	0.01138	0.01609	0.00802	0.01065	0.46570	0.01509	0.01155
69	0	0	0	0	0	0	0	0	0	0	0	0	0.07204	0
70	0	0	0	0	0	0	0	0	0	0	0	0	0	0.30096
71	0	0	0	0	0	0	0	0	0	0	0	0	0	0
72	0.00021	0.00016	0.00027	0.00018	0.00045	0.00136	0.00015	0.00024	0.00049	0.00156	0.00527	0.00120	0.01038	0.00388
73	0.00120	0.00011	0.00058	0.00137	0.00294	0.01884	0.01443	0.04079	0.00399	0.03789	0.04066	0.00526	0.01318	0.00198
74	0.00018	0.00011	0.00015	0.00008	0.00039	0.00236	0.00031	0.00089	0.00026	0.00485	0.00937	0.00344	0.00449	0.00161
75	0.01327	0.00427	0.00531	0.00269	0.00411	0.02815	0.09182	0.02269	0.01811	0.02246	0.06248	0.01387	0.03891	0.01872
76	0.00036	0.00010	0.00023	0.00014	0.00022	0.00047	0.00125	0.00030	0.00019	0.00174	0.00051	0.00076	0.00027	0.00292
77	0.00072	0.00048	0.00039	0.00034	0.00085	0.00109	0.00091	0.00099	0.00057	0.00092	0.00073	0.00132	0.00090	0.00148
78	0.00074	0.00063	0.00079	0.00066	0.00064	0.00219	0.01474	0.02890	0.00040	0.03402	0.00603	0.00137	0.00079	0.01070
79	0.00104	0.00027	0.00052	0.00030	0.00093	0.00071	0.00089	0.00065	0.00032	0.00742	0.00077	0.00049	0.02352	0.01339
80	0.00267	0.00118	0.00158	0.00114	0.00199	0.00775	0.00871	0.00312	0.00190	0.00494	0.00552	0.00516	0.01117	0.02737
81	0.00035	0.00014	0.00026	0.00016	0.00033	0.00064	0.00033	0.00040	0.00035	0.00052	0.00078	0.00023	0.00024	0.00158
82	0.00505	0.00125	0.00329	0.00150	0.00163	0.00195	0.00135	0.00165	0.00173	0.00247	0.00385	0.00162	0.00215	0.00165
83	0.00260	0.00205	0.00204	0.00210	0.00278	0.00272	0.00240	0.00274	0.00218	0.00275	0.00244	0.00265	0.00218	0.00267
84	0.01690	0.00651	0.00658	0.00396	0.01004	0.00475	0.01520	0.01256	0.00308	0.00627	0.01215	0.00747	0.00596	0.00556

Continued

a_{ij} \ b_{ij}	57	58	59	60	61	62	63	64	65	66	67	68	69	70
85	0.02903	0.00514	0.09644	0.00332	0.00190	0.00143	0.00140	0.01152	0.00077	0.00206	0.00347	0.00050	0.00004	0.00011
86	0.05060	0.12097	0.08174	0.04748	0.04450	0.01749	0.01729	0.02536	0.00765	0.02521	0.05152	0.01339	0.01422	0.01899
87	0.00066	0.00042	0.00070	0.00023	0.00045	0.00048	0.00062	0.00031	0.00023	0.00068	0.00136	0.00041	0.00051	0.00183
88	0.00202	0.00098	0.00028	0.00050	0.00025	0.00048	0.00018	0.00033	0.00013	0.00041	0.00073	0.00021	0.00014	0.00031
89	0.00212	0.00129	0.00129	0.00109	0.00163	0.00177	0.00101	0.00154	0.00047	0.00172	0.00172	0.00113	0.00129	0.00438
90	0.00119	0.00052	0.00146	0.00064	0.00115	0.00218	0.00207	0.00157	0.00109	0.00407	0.00279	0.00144	0.00074	0.00322
91	0.01382	0.01443	0.00839	0.00456	0.00695	0.00475	0.00503	0.00615	0.00287	0.00704	0.00752	0.00380	0.00364	0.00452
92	0.01056	0.01747	0.01388	0.00655	0.01460	0.00824	0.00761	0.01181	0.00603	0.01283	0.00695	0.00956	0.00483	0.00705
93	0.00024	0.00006	0.00009	0.00005	0.00004	0.00007	0.00006	0.00005	0.00003	0.00006	0.00007	0.00005	0.00004	0.00007
94	0.00390	0.00520	0.00097	0.00030	0.00169	0.00078	0.00059	0.00084	0.00040	0.00093	0.00072	0.00091	0.00047	0.00050
95	0.00012	0.00005	0.00006	0.00006	0.00019	0.00020	0.00024	0.00025	0.00015	0.00029	0.00020	0.00071	0.00030	0.00035
96	0.00502	0.00546	0.00399	0.00248	0.00536	0.00244	0.00236	0.00243	0.00130	0.00273	0.00212	0.00157	0.00196	0.00177
97	0.00007	0.00003	0.00002		0.00002	0.00060	0.00004	0.00005	0.00002	0.00008	0.00010	0.00013	0	0.00016
98	0	0	0	0	0.00003	0.00002			0.00007	0.00001	0	0	0	0
99	0.00729	0.01002	0.02870	0.00207	0.02396	0.02372	0.06659	0.01148	0.01819	0.01313	0.01950	0.01172	0.00898	0.06366
100	0.04191	0.03260	0.02965	0.02512	0.04060	0.03208	0.02973	0.03719	0.02398	0.03906	0.03430	0.03850	0.02993	0.03336
101	0.00403	0.00538	0.00308	0.00600	0.00988	0.01004	0.00991	0.01356	0.00350	0.01090	0.00533	0.00356	0.00933	0.00503
102	0.00001	0.00001	0.00093		0.00004	0.00002	0.00001	0.00001	0.00005	0.00001	0.00002	0.00001	0.00001	0.00007
103	0	0	0.00156	0	0.00003	0			0.00007	0	0	0	0	0
104	0.00001	0.00001	0.00031	0.00003	0.00002	0.00002	0.00001	0.00001	0.00002	0.00001	0.00002	0.00001	0.00001	0.00006
105	0.00012	0.00017	0.00072		0.00032	0.00039	0.00011	0.00019	0.00012	0.00022	0.00032	0.00019	0.00015	0.00105
106	0.01592	0.01195	0.01491	0.01592	0.04617	0.01610	0.02286	0.01904	0.00530	0.02077	0.00778	0.01191	0.01194	0.02364
107	0.00168	0.00137	0.00161	0.00178	0.00174	0.00186	0.00217	0.00211	0.00094	0.00259	0.00158	0.00119	0.00161	0.00308
108	0.00027	0.00029	0.00032	0.00072	0.00219	0.00089	0.00077	0.00117	0.00027	0.00176	0.00015	0.00077	0.00017	0.00094
109	0.00001	0.00001	0.00025		0.00002	0.00002	0.00001	0.00001	0.00002	0.00001	0.00002	0.00001	0.00001	0.00005
110	0.01014	0.00482	0.00391	0.00215	0.00659	0.00561	0.00336	0.00330	0.00164	0.00483	0.00987	0.00320	0.00334	0.00884
111	0.00052	0.00039	0.00369	0.00012	0.00102	0.00073	0.00017	0.00164	0.00033	0.00190	0.00043	0.00032	0.00023	0.00148
112	0	0	0.00005	0						0	0	0	0	0
113	0.00047	0.00062	0.00036	0.00071	0.00115	0.00118	0.00117	0.00159	0.00041	0.00128	0.00062	0.00041	0.00111	0.00057
114	0.00219	0.00062	0.00127	0.00283	0.00787	0.00692	0.00527	0.00951	0.00331	0.01342	0.00396	0.00618	0.00147	0.00500
115	0.00061	0.00046	0.00197	0.00024	0.00027	0.00056	0.00039	0.00114	0.00011	0.00313	0.00293	0.00009	0.00264	0.00009
116	0	0	0	0	0	0	0	0	0	0	0	0	0	0
117	0	0	0	0	0	0	0	0	0	0	0	0	0	0
118	0.00098	0.00064	0.00066	0.00074	0.00216	0.00228	0.00140	0.00176	0.00079	0.00183	0.00279	0.00099	0.00178	0.00284
119	0	0.00001	0.00036	0.00001	0.00003	0.00015	0.00001	0.00002	0.00002	0.00007	0.00002	0.00002	0	0.00003
120	0.00026	0.00002	0.00009	0.00006	0.00019	0.00054	0.00018	0.00109	0.00019	0.00060	0.00160	0.00100	0.00007	0.00088
121	0.00162	0.00080	0.00197	0.00147	0.00235	0.00225	0.00172	0.00192	0.00063	0.00267		0.00127	0.00337	0.00354
122	0	0	0	0	0	0	0	0	0	0	0	0	0	0
123	0	0	0	0	0	0	0	0	0	0	0	0	0	0
124	0	0	0	0	0	0	0	0	0	0	0	0	0	0

Continued

b_{ij} \ a_{ij}	71	72	73	74	75	76	77	78	79	80	81	82	83	84
1	0	0	0	0	0	0	0	0	0	0	0	0	0.00079	0.00147
2	0	0	0	0	0	0	0	0	0	0	0	0.00627	0.03874	0.00001
3	0	0	0	0	0	0	0	0	0	0	0	0	0.09171	0.07184
4	0	0	0.00126	0	0	0	0	0	0	0	0	0	0.00207	0.00004
5	0.00208	0	0.00845	0.00033	0.00184	0	0	0	0	0	0	0	0.00281	0.00010
6	0.00001	0.00114	0.00007	0.00003	0.00005	0.00005	0.00042	0.00139	0.00039	0.00139	0.00150	0.00989	0.00467	0.01834
7	0.00001	0.00001	0.00112	0.00003	0.00025	0	0	0.00001	0	0.00001	0	0.00105	0	0
8	0.00066	0.00008	0.00785	0.00122	0.00818	0	0	0.00117	0.00002	0.00001	0	0.00010	0	0.00003
9	0.00034	0.00091	0.00285	0.00676	0.00309	0	0	0.00047	0.00002	0.00050	0	0.00027	0	0.00027
10	0.01319	0.00184	0	0	0	0	0	0.00096	0.00009	0.00027	0	0	0	0.00004
11	0	0	0	0	0	0	0	0	0	0	0	0	0	0
12	0.00001	0.00019	0.00100	0.00002	0.00053	0	0	0.00232	0.00002	0.00075	0.00003	0.00075	0.00602	0.00031
13	0.00004	0.00009	0.00185	0.00008	0.00029	0	0	0.00010	0.00013	0.00021	0.00004	0.00213	0.00175	0.00044
14														
15														
16														
17														0.00796
18														
19										0.00072				0.00035
20														
21	0	0	0	0	0	0	0	0	0	0	0	0	0	0
22	0.00008	0.00134	0.00404	0.00310	0.00132	0.00005	0.00008	0.00053	0.00060	0.00521	0.00018	0.02547	0.07146	0.06513
23	0.00003	0.00005	0.00007	0.00003	0.00006	0.00002	0.00003	0.00012	0.00001	0.00038	0.00006	0.00441	0.01535	0.00180
24	0.00006	0.00001	0.00001	0.00001	0.00002		0.00001	0.00002		0.00003	0.00001	0.00012	0.00185	0.00059
25	0.00002	0.00002	0.00002	0.00002	0.00014	0.00001	0.00002	0.00008	0.00001	0.00039	0.00005	0.00280	0.04052	0.01170
26	0.00002	0.00031	0.00009	0.00005	0.00008		0.00002	0.00005	0.00001	0.00014	0.00002	0.00040	0.00107	0.00084
27	0.00371	0	0	0	0.00030	0	0	0.00023	0	0	0	0.00374	0.00948	0.01069
28	0.00100	0.00113	0.00283	0.00141	0.00221	0.00034	0.00059	0.00112	0.00090	0.00147	0.00106	0.00486	0.00273	0.00198
29	0.00066	0.00076	0.00058	0.00078	0.00091	0.00015	0.00019	0.00033	0.00042	0.00338	0.00132	0.00194	0.01630	0.00059
30	0.00069	0.00144	0.00109	0.00037	0.00150	0.00029	0.00073	0.00340	0.00048	0.00220	0.00012	0.00415	0.00857	0.00153
31	0.00028	0.00253	0.00391	0.00661	0.00543	0.00045	0.00071	0.00071	0.00104	0.00430	0.00060	0.00081	0.00496	0.00608
32	0.00818	0.00382	0.00283	0.03629	0.02036	0.00521	0.01450	0.01695	0.00480	0.01037	0.02148	0.00048	0.10090	0.00986
33	0.00053	0.00069	0.00054	0.00463	0.00102	0.00070	0.00375	0.00624	0.00183	0.00114	0.00007	0.00034	0.00183	0.00015
34	0.00021	0.00029	0.00021	0.00021	0.00044	0.00339	0.00012	0.00124	0.00014	0.00053	0.00067	0.00031	0.00024	0.00034
35	0.00017	0	0	0	0	0	0.00001	0.00008	0	0	0	0.00003	0.00010	0
36	0.00284	0.00321	0.00960	0.00655	0.00949	0.00100	0.00156	0.00831	0.00200	0.00636	0.00179	0.01952	0.01191	0.00647
37	0.00008	0.00009	0.00149	0.00004	0.00005			0.00003		0.00019	0.00064	0.00064		0.00114
38	0.00209	0.00084	0.00138	0.02023	0.00400	0.00007	0.00060	0.01220	0.00472	0.00096	0.00056	0.00330	0.00287	0.00652
39	0	0	0	0	0	0	0	0	0	0	0	0	0	0
40	0	0	0	0	0	0	0	0	0	0	0	0	0	0
41	0.00547	0.00346	0.00786	0.02498	0.03169	0.00009	0.00109	0.01956	0.00342	0.00286	0.13955	0.01678	0.02604	0.01010
42	0.00066	0.00034	0.00076	0.00091	0.00114	0.00018	0.00050	0.00063	0.00082	0.00689	0.00023	0.00133	0.00708	0.00191

Continued

a_{ij} \ b_{ij}	71	72	73	74	75	76	77	78	79	80	81	82	83	84
43	0.00111	0.00228	0.00479	0.01725	0.03851	0.01779	0.03156	0.10094	0.00469	0.00411	0.02714	0.01055	0.02598	0.32339
44	0	0	0.00001	0.00001	0	0	0	0.00002	0.00002	0.00011	0	0.00001	0.00001	0
45	0.00013	0.00015	0.00066	0.00049	0.00907	0.00003	0.00612	0.00068	0.00103	0.00013	0.00022	0.00052	0.00914	0.04789
46	0.02258	0.07476	0.01043	0.01260	0.00697	0.00494	0.00589	0.01110	0.02435	0.01084	0.00699	0.01102	0.00198	0.01761
47	0.00877	0.01261	0.01636	0.11369	0.09484	0.03605	0.06846	0.04216	0.02607	0.02568	0.06849	0.00490	0.02678	0.02748
48	0.00017	0.00043	0.00027	0.00031	0.00071	0.00006	0.00008	0.00018	0.00016	0.00020	0.00009	0.00217	0.00049	0.00280
49	0.00010	0.00022	0.00009	0.00011	0.00034	0	0	0.00007	0.00002	0.00010	0.00008	0.00077	0.00125	0.00071
50	0.00020	0.00045	0.00046	0.00031	0.00217	0.00005	0.00010	0.00023	0.00020	0.00036	0.00008	0.00343	0.00058	0.00189
51	0.00064	0.00132	0.00168	0.00266	0.01506	0.00893	0.00758	0.13544	0.01055	0.02291	0.04611	0.00792	0.00068	0.00610
52	0.00098	0.00126	0.00345	0.00193	0.00842	0.00120	0.00053	0.00432	0.00113	0.00194	0.00053	0.00253	0.00094	0.00648
53	0.00002	0.00021	0.00172	0.00059	0.00687	0	0.00008	0.00222	0.00002	0.00063	0	0.00178	0.00005	0.00219
54	0.00018	0.00033	0.00516	0.00114	0.02656	0	0.00067	0.01606	0.00629	0.00098	0.00066	0.00195	0.00050	0.00745
55	0.00007	0.00040	0.00916	0.00025	0.00331	0	0	0.00058	0.00031	0.00107	0.00002	0.00161	0	0.00244
56	0.00193	0.00354	0.01040	0.00401	0.01069	0.00002	0.00002	0.00028	0.00026	0.00238	0	0.02048	0	0.00315
57	0.20083	0.05573	0.11995	0.04759	0.02968	0.00622	0.00149	0.00751	0.00488	0.04983	0.05272	0.04609	0.00023	0.00927
58	0.00049	0.00025	0.00767	0.00074	0.00100	0.00301	0.00061	0.00259	0.00010	0.00040	0.00312	0.00229	0	0.00408
59	0.00137	0.00486	0.02897	0.00746	0.11477	0.00008	0.00035	0.00655	0.00388	0.00691	0.00124	0.00227	0.06048	0.00379
60	0.00689	0.00779	0.03960	0.02943	0.17246	0.00163	0.00058	0.02152	0.01184	0.01363	0.00204	0.00441	0.02922	0.00682
61	0.12949	0.04149	0.05674	0.10362	0.04360	0.01775	0.02121	0.04903	0.02534	0.03877	0.05199	0.09748	0.08223	0.03532
62	0.00059	0.16003	0.07129	0.00004	0.00009	0.00004	0.00003	0.00001	0.00002	0.00004	0.00546	0.00631	0.00005	0.00009
63	0.00099	0.00546	0.02413	0.00275	0.00325	0.00152	0.00196	0.03013	0.00247	0.00367	0.03738	0.00988	0.00063	0.00082
64	0.00718	0.16570	0.11357	0.05943	0.02063	0.00090	0.00689	0.00466	0.00673	0.35522	0.00360	0.06516	0.00311	0.00351
65	0	0.00098	0	0	0	0	0	0	0	0	0	0.00216	0	0
66	0.00125	0.00083	0.00401	0.00664	0.00163	0.00090	0.00031	0.00185	0.00147	0.00186	0.00032	0.00889	0.00245	0.00138
67	0	0	0.00138	0	0	0	0	0	0	0	0	0.00672	0	0
68	0.00200	0.00297	0.00207	0.00220	0.00350	0.00344	0.00093	0.00275	0.00236	0.00151	0.00337	0.21361	0.00121	0.00223
69	0	0.00057	0	0	0	0	0	0	0	0	0	0.00287	0	0
70	0	0.00043	0	0	0	0	0	0	0	0	0	0.02999	0	0
71	0.34728	0.00036	0.00021	0.00011	0.00010	0.00001	0.00003	0.00010	0.00008	0.00013	0.00025	0.00121	0.00016	0.00142
72	0.08064	0.29907	0.08834	0.04612	0.00285	0.00009	0.00138	0.00534	0.00434	0.00302	0.00127	0.00954	0.00094	0.00052
73	0.00078	0.00623	0.00130	0.11182	0.00359	0.01536	0.00536	0.00088	0.00009	0.00021	0.00016	0.00139	0.00012	0.00058
74	0.00002	0.00007	0.13719	0.06470	0.09467	0.06092	0.17414	0.00803	0.13691	0.04357	0.09859	0.03187	0.00124	0.02295
75	0.00109	0.03176	0.00022	0.00019	0.00043	0.23150	0.00046	0.01481	0.01104	0.00282	0.00041	0.00645	0.00023	0.00007
76	0.00005	0.00008	0.00150	0.00281	0.00371	0.00041	0.04246	0.00153	0.00056	0.00101	0.00033	0.00138	0.00022	0.00064
77	0.00045	0.00036	0.02399	0.08781	0.01938	0.45861	0.48647	0.27488	0.27609	0.11957	0.22952	0.00382	0.00011	0.00233
78	0.00010	0.00714	0.00048	0.00054	0.00259	0.02681	0.01051	0.01838	0.29242	0.01067	0.00495	0.00179	0.00043	0.00064
79	0.00014	0.00019	0.00054	0.00187	0.00969	0.00244	0.00103	0.00248	0.00234	0.08372	0.00197	0.00990	0.00079	0.00131
80	0.00077	0.00343	0.00547	0.00025	0.00028	0.00009	0.00004	0.00021	0.00011	0.00025	0.03911	0.00172	0.00020	0.00008
81	0.00007	0.00027	0.00029	0.00100	0.00215	0.00038	0.00041	0.00142	0.00101	0.00126	0.00113	0.08814	0.00049	0.00078
82	0.00075	0.00049	0.00110	0.00220	0.00262	0.00203	0.00196	0.00221	0.00213	0.00202	0.00204	0.00276	0.10042	0.00149
83	0.00164	0.00181	0.00246	0.00220	0.00262	0.00203	0.00196	0.00221	0.00213	0.00202	0.00204	0.00276	0.10042	0.00149
84	0.00713	0.00436	0.00652	0.00815	0.01242	0.00223	0.00293	0.00835	0.00602	0.01502	0.03919	0.00816	0.00891	0.08532

Continued

b_{ij} \ a_{ij}	71	72	73	74	75	76	77	78	79	80	81	82	83	84
85	0.00006	0.00004	0.00297	0.00012	0.00020	0	0.00002	0.00420	0.00002	0.00035	0.00004	0.00009	0.00029	0.00248
86	0.01088	0.00611	0.01738	0.00669	0.01367	0.00175	0.00246	0.01940	0.00488	0.01135	0.00394	0.02062	0.00885	0.01763
87	0.00049	0.00018	0.00048	0.00009	0.00023	0.00007	0.00004	0.00041	0.00009	0.00023	0.00004	0.00079	0.00018	0.00008
88	0.00054	0.00007	0.00011	0.00004	0.00055	0	0.00001	0.00056	0.00001	0.00019	0.00003	0.00031	0.00003	0.00003
89	0.00200	0.00045	0.00117	0.00116	0.00122	0.00039	0.00029	0.00156	0.00047	0.00110	0.00080	0.00267	0.00104	0.00066
90	0.00091	0.00045	0.00170	0.00210	0.00113	0.00024	0.00039	0.00089	0.00058	0.00119	0.00103	0.00594	0.00130	0.00092
91	0.00398	0.00233	0.00886	0.00261	0.00394	0.00265	0.00172	0.00489	0.00289	0.00253	0.00350	0.00648	0.01380	0.00567
92	0.01124	0.00602	0.01054	0.00791	0.00994	0.00434	0.00667	0.00796	0.00436	0.00697	0.00983	0.00934	0.00860	0.00664
93	0.00001	0.00001	0.00004	0.00003	0.00005	0	0.00001	0.00006	0.00001	0.00002	0.00002	0.00009	0.00005	0.00003
94	0.00048	0.00036	0.00106	0.00046	0.00085	0.00048	0.00039	0.00061	0.00031	0.00040	0.00053	0.00087	0.00083	0.00054
95	0.00011	0.00020	0.00026	0.00058	0.00038	0.00197	0.00094	0.00169	0.00158	0.00059	0.00086	0.00048	0.00012	0.00013
96	0.00191	0.00113	0.00206	0.00116	0.00179	0.00043	0.00068	0.00120	0.00068	0.00147	0.00100	0.00203	0.01286	0.00169
97	0.00003	0.00003	0.00006	0.00018	0.00008	0.00001	0.00013	0.00009	0.00003	0.00003	0.00001	0.00005	0.00003	0.00002
98	0	0	0	0	0	0	0	0	0	0	0	0	0	0
99	0.00147	0.00754	0.00858	0.01008	0.00630	0.00260	0.00326	0.00496	0.00469	0.01869	0.00244	0.00713	0.01469	0.00675
100	0.06022	0.02990	0.03528	0.04578	0.04621	0.04784	0.04732	0.06363	0.05777	0.03662	0.05314	0.03700	0.07273	0.04404
101	0.00308	0.00256	0.01045	0.00580	0.01516	0.00173	0.00136	0.00368	0.00588	0.00720	0.00200	0.01221	0.00634	0.00606
102	0	0.00001	0.00001	0.00001	0.00001	0	0	0.00001	0	0.00002	0	0.00001	0.00002	0.00001
103	0	0	0	0	0	0	0	0	0	0	0	0	0	0
104	0	0.00001	0.00001	0.00001	0.00001	0	0	0	0	0.00002	0	0.00001	0.00001	0.00001
105	0.00002	0.00012	0.00014	0.00017	0.00010	0.00004	0.00005	0.00008	0.00008	0.00031	0.00004	0.00012	0.00024	0.00011
106	0.01728	0.00614	0.01617	0.00935	0.01238	0.00552	0.00647	0.01146	0.00721	0.01793	0.00664	0.01564	0.00994	0.00835
107	0.00071	0.00073	0.00195	0.00125	0.00258	0.00067	0.00107	0.00170	0.00075	0.00161	0.00209	0.00352	0.00238	0.00141
108	0.00444	0.00044	0.00117	0.00319	0.00186	0.00262	0.00192	0.00393	0.00222	0.00251	0.00197	0.00126	0.00207	0.00231
109	0	0.00001	0.00001	0.00001	0	0	0	0	0	0.00001	0	0.00001	0.00001	0.00001
110	0.00162	0.00128	0.00358	0.00438	0.00568	0.00162	0.00159	0.00267	0.00174	0.00308	0.00190	0.00764	0.00106	0.00150
111	0.00004	0.00017	0.00041	0.00111	0.00037	0.00015	0.00013	0.00029	0.00061	0.00054	0.00012	0.00068	0.00047	0.00054
112	0	0	0	0	0	0	0	0	0	0	0	0	0	0
113	0.00037	0.00030	0.00123	0.00068	0.00179	0.00020	0.00016	0.00043	0.00068	0.00084	0.00024	0.00145	0.00073	0.00072
114	0.00551	0.00653	0.00390	0.03466	0.00727	0.00556	0.01339	0.00385	0.01087	0.00534	0.00436	0.00907	0.00544	0.01426
115	0.00006	0.00004	0.00071	0.00008	0.00031	0.00001	0.00002	0.00003	0.00003	0.00018	0.00003	0.00035	0.00027	0.00106
116	0	0	0	0	0	0	0	0	0	0	0	0	0	0
117	0	0	0	0	0	0	0	0	0	0	0	0	0	0
118	0.00057	0.00049	0.00293	0.00039	0.00105	0.00018	0.00071	0.00059	0.00098	0.00130	0.00082	0.00274	0.00081	0.00060
119	0	0.00001	0.00001	0.00001	0.00003	0.00002	0	0.00001	0.00001	0.00003	0	0.00003	0	0
120	0.00010	0.00036	0.00052	0.00142	0.00040	0.00108	0.00035	0.00096	0.00113	0.00034	0.00094	0.00022	0.00008	0.00030
121	0.00054	0.00071	0.00143	0.00101	0.00143	0.00038	0.00088	0.00120	0.00126	0.00139	0.00166	0.00193	0.00068	0.00184
122	0	0	0	0	0	0	0	0	0	0	0	0	0	0
123	0	0	0	0	0	0	0	0	0	0	0	0	0	0
124	0	0	0	0	0	0	0	0	0	0	0	0	0	0

Continued

a_{ij} \\ b_{ij}	85	86	87	88	89	90	91	92	93	94	95	96	97	98
1	0	0	0	0	0	0	0	0	0	0.00010	0	0	0.21972	0
2	0	0.00018	0	0	0	0.00308	0	0	0	0	0	0	0.00007	0
3	0	0	0	0	0	0	0	0	0	0	0	0	0.00001	0
4	0	0	0	0	0	0.00274	0	0	0	0	0	0	0.00002	0
5	0	0.00006	0.36769	0.32794	0.00434	0.00085	0.05133	0.00512	0.00553	0	0	0	0.00017	0
6	0	0.30868	0.05692	0.04020	0.00055	0	0	0	0	0.00026	0.00084	0.00679	0.00623	0
7	0	0.03430	0.00026	0.00119	0.00008	0	0	0	0	0	0	0.01237	0.00095	0
8	0	0.00120	0.00014	0	0	0	0	0	0	0	0	0	0	0
9	0	0.00022	0.00054	0	0.00008	0	0	0	0	0	0	0	0	0
10	0	0.00344	0.00031	0	0.00038	0	0	0	0	0	0	0	0	0
11	0	0.00006	0.00043	0	0.00078	0.02633	0.01158	0.00001	0.00032	0.00021	0	0	0	0
12	0	0.00007	0	0	0	0.00961	0.00069	0.00005	0.00060	0.01033	0	0.01852	0.00059	0
13	0	0.00039	0	0.00037	0	0	0	0	0	0.00030	0	0.00141	0.00037	0
14	0	0	0	0	0	0	0.00046	0	0	0.00002	0	0.00028	0.27064	0
15	0	0	0	0	0	0	0	0	0	0.00053	0	0	0	0
16	0	0	0	0	0	0	0	0	0	0.00026	0	0.00056	0	0
17	0	0	0	0	0	0	0.00029	0	0	0.00063	0	0.00025	0.00018	0
18	0	0	0	0	0	0	0	0	0	0.00070	0	0.00379	0.00013	0
19	0	0	0	0	0	0.00090	0.00194	0	0	0.00036	0	0.00124	0.00006	0
20	0	0	0	0	0	0	0	0	0	0.00031	0	0.00387	0.00015	0
21	0	0	0	0	0	0	0.00570	0	0	0.00001	0	0.00161	0.00060	0.00077
22	0	0.00062	0.00050	0.00687	0.00024	0.00260	0.00102	0.00052	0.00035	0	0	0.00313	0.00028	0
23	0	0.00232	0.00008	0.00027	0.00006	0.00005	0.00026	0.00005	0.00009	0	0	0.00049	0.00579	0
24	0	0.00004	0.00004	0.00002	0.00002	0.00008	0.00008	0.00003	0	0	0	0.00040	0	0
25	0	0.00011	0.00027	0.00029	0.00004	0.00010	0.00022	0	0	0	0	0.00002	0.00706	0
26	0	0.00009	0.00004	0.00003	0.00006	0	0.00034	0.00064	0	0	0	0.00125	0	0
27	0	0	0	0	0	0.00008	0	0	0	0	0.00098	0	0.00657	0
28	0	0.00468	0.00342	0.00433	0.00730	0.00170	0.00586	0.00419	0.00334	0.00577	0.00014	0.01782	0	0.03747
29	0	0.00109	0.00050	0.00085	0.00123	0.00001	0.00413	0.00232	0.00139	0.00020	0.00001	0.00379	0.00339	0.00001
30	0	0.00104	0.00108	0.00075	0.00177	0.02218	0.01109	0.00098	0.00017	0.00014	0.00014	0.00156	0.00027	0.00030
31	0	0.00062	0.00027	0.00055	0.00128	0.00748	0.00300	0.00173	0.00128	0.00011	0.00018	0.00677	0.00074	0.01242
32	0	0.00036	0.00008	0.00069	0.00057	0.00057	0.00167	0.00232	0.00127	0.00135	0.00135	0.00510	0.00655	0.03404
33	0	0.00046	0.00024	0.00068	0.00101	0.00028	0.00234	0.00301	0.00204	0.00124	0.00133	0.00454	0.00093	0.21727
34	0	0.00033	0.00019	0.00034	0.00035	0.00013	0.00253	0.00351	0.00262	0.00195	0.00120	0.00418	0.00164	0.06648
35	0	0.00002	0	0	0	0	0.00044	0.00026	0.00057	0.00017	0.00004	0.00058	0.00551	0
36	0	0.09307	0.07102	0.12003	0.01165	0.03958	0.22588	0.28975	0.19838	0.37170	0.10295	0.06928	0	0.02542
37	0	0.00005	0.00002	0.00452	0.05696	0.00047	0.00007	0.00486	0.00194	0.00007	0	0	0.04679	0
38	0	0.00326	0.00916	0.01219	0	0.00230	0.00102	0.00012	0	0	0	0.00052	0	0
39	0	0	0	0	0	0	0	0	0	0	0	0	0.00038	0
40	0	0	0	0	0	0	0	0	0	0	0.00048	0	0.00095	0
41	0	0.00143	0.00226	0.00467	0.00285	0.00809	0.00414	0.00059	0.00082	0.00646	0	0.00658	0.00064	0.00032
42	0	0.00132	0.00058	0.00121	0.01843	0.00016	0.00193	0.00157	0.00083	0.00085	0	0.00644	0.00104	0

Continued

b_{ij} \ a_{ij}	85	86	87	88	89	90	91	92	93	94	95	96	97	98
43	0	0.00217	0.00248	0.00374	0.00701	0.00650	0.00081	0.00054	0.00324	0.00079	0.00003	0.00127	0.00068	0.00014
44	0	0.00012	0.00008	0.00001	0.00009	0.00051	0.00050	0.00103	0.00158	0.00309	0.00009	0.00280	0.00112	0.00035
45	0	0	0	0	0	0.00346	0.00009	0.00011	0.00050	0.00014	0.00001	0.00247	0.00027	0
46	0	0.00267	0.00293	0.00195	0.00195	0.00416	0.00434	0.08731	0.00642	0.00129	0.00038	0.01160	0.00176	0.00244
47	0	0.00181	0.00155	0.00171	0.00440	0.00419	0.00101	0.00082	0.00136	0.00146	0.00013	0.00395	0.00300	0.00299
48	0	0.00269	0.00398	0.00170	0.00434	0.11625	0.00282	0.00075	0.00435	0.00039	0.00002	0.01315	0.00051	0.00103
49	0	0.00180	0.00087	0.00048	0.00215	0.07433	0.01926	0.00079	0.00205	0.00001	0	0.00372	0.00059	0
50	0	0.00309	0.00322	0.00165	0.00597	0.12678	0.02202	0.00037	0.00183	0.00015	0	0.00929	0.00060	0.00003
51	0	0.00149	0.00110	0.00113	0.00279	0.00837	0.00066	0.00078	0.00433	0.00010	0.00002	0.00045	0.00017	0.00123
52	0	0.00254	0.00148	0.00172	0.00409	0.02727	0.00045	0.00005	0.00023	0.00095	0.00002	0.00039	0	0.00247
53	0	0.00109	0.00120	0.00049	0.00014	0.01560	0.00106	0.00016	0.00097	0.00005	0.00001	0.00257	0.00034	0
54	0	0.00110	0.00021	0.00002	0.00014	0.01123	0.00184	0.00014	0.00012	0.00002	0	0.00234	0.00001	0
55	0	0	0	0	0	0	0	0	0	0	0	0	0	0
56	0	0	0	0	0	0	0	0	0	0	0	0	0	0.00135
57	0	0.00324	0.00325	0.00403	0.00286	0.07945	0.03164	0.00071	0.01070	0.00024	0	0.00502	0.01848	0
58	0	0	0	0	0	0.00022	0	0	0	0	0	0	0	0
59	0	0	0	0	0	0	0	0	0	0	0	0	0	0.00937
60	0	0.00081	0.00045	0.00036	0.00061	0.00751	0.00042	0.00036	0.00059	0.00007	0.00055	0.00005	0.00901	0
61	0	0.00818	0.00687	0.00523	0.05673	0.08471	0.00815	0.00685	0.00473	0.00275	0.00001	0.01521	0.00258	0
62	0	0.01051	0.00697	0.00049	0.00013	0	0.00371	0.00059	0.00090	0.00124	0.00152	0.00184	0.00030	0
63	0	0.00216	0.00107	0.00159	0.00115	0	0.00047	0.00667	0.00001	0.00068	0	0.00253	0	0.01947
64	0	0.05291	0.02697	0.03512	0.02965	0.02296	0.04547	0.03651	0.05091	0.06404	0.02769	0.08438	0.01486	0
65	0	0	0	0	0	0	0	0	0	0	0	0	0	0.01828
66	0	0.00953	0.00706	0.00968	0.00206	0.01311	0.00118	0.00072	0.00897	0.00072	0.00032	0.00539	0.00040	0
67	0	0	0	0	0	0	0.15842	0	0	0	0	0.00035	0.00027	0.02266
68	0	0.00863	0.02155	0.01092	0.01338	0.00049	0.00449	0.15290	0.00422	0.00181	0.00026	0.01558	0.00553	0
69	0	0	0	0	0	0	0	0	0	0.10382	0.04263	0.00216	0	0
70	0	0	0	0	0	0	0	0	0	0	0	0.00013	0	0
71	0	0	0	0	0	0	0.00007	0	0	0	0	0.00012	0	0.00950
72	0	0.00072	0.00112	0.00034	0.00086	0.00015	0.00119	0.00780	0.00313	0.00692	0.00276	0.01304	0.00082	0.00029
73	0	0.00223	0.00117	0.00037	0.00200	0.00011	0.00083	0.00041	0.00119	0.00097	0.00003	0.00309	0.00015	0.00052
74	0	0.00071	0.00019	0.00048	0.00057	0.06653	0.00108	0.00036	0.00058	0.00018	0.00148	0.00231	0.00013	0.00194
75	0	0.07230	0.01400	0.00569	0.01832	0.05646	0.01391	0.00133	0.00458	0.00054	0.00039	0.01065	0.00097	0.02837
76	0	0.00154	0.00043	0.00051	0.00147	0.00015	0.00150	0.00064	0.00132	0.00040	0.00107	0.00171	0.00055	0.00028
77	0	0.00072	0.00052	0.00042	0.00117	0.00155	0.00049	0.00083	0	0.00043	0.00211	0.00116	0.00007	0.00524
78	0	0.00206	0.00060	0.00046	0.00181	0	0.00189	0.00036	0.00035	0.00069	0.00022	0.00102	0.00003	0
79	0	0.00441	0.00265	0.00092	0.01120	0.00021	0.00365	0.00047	0.00072	0.00030	0.00017	0.01180	0.00008	0.00349
80	0	0.02779	0.00800	0.00582	0.00208	0.00831	0.00151	0.00063	0.00565	0.00028	0.00038	0.00549	0.00020	0.00719
81	0	0.00142	0.00053	0.00074	0.01192	0.00014	0.00074	0.00162	0.00005		0.00014	0.00124	0.00339	0.00027
82	0	0.01242	0.00851	0.01267	0.00317	0.00298	0.03099	0.03598	0.04079	0.06058	0.03213	0.05553	0.01324	0.03219
83	0	0.00296	0.00244	0.00227	0.02316	0.00410	0.00487	0.00352	0.00298	0.00250	0.00163	0.00411	0.00201	0.00246
84	0	0.01135	0.00909	0.00567		0.00043	0.00310	0.00451	0.00407	0.00759	0.00006	0.00287	0.00089	0.00027

Continued

a_{ij} \ b_{ij}	85	86	87	88	89	90	91	92	93	94	95	96	97	98
85	0	0	0	0	0	0	0	0	0	0	0	0	0	0
86	0	0.05929	0.04660	0.07190	0.40026	0.00923	0.07517	0.01273	0.36459	0.00124	0.00049	0.04258	0.01469	0.03111
87	0	0.00103	0.03268	0.00410	0.00075	0.00062	0.00184	0.00232	0.00197	0.00011	0.00011	0.00111	0.00133	0.00522
88	0	0.00015	0.00110	0.06678	0.00003	0.00012	0.00006	0.00021	0	0.00029	0	0.00083	0.00057	0.00039
89	0	0.00516	0.01389	0.00384	0.11726	0.00146	0.09755	0.00452	0.00200	0.00149	0.00035	0.00834	0.00262	0.00528
90	0	0.00492	0.00280	0.00638	0.00544	0.00081	0.00829	0.00657	0.03560	0.00173	0.00107	0.04676	0.02282	0.13344
91	0	0.02499	0.04405	0.02564	0.00308	0.00665	0.01123	0.00400	0.00327	0.00251	0.00067	0.00962	0.01859	0.03496
92	0	0.01113	0.01237	0.01248	0.00395	0.01335	0.00081	0.02094	0.00523	0.00311	0.00307	0.01175	0.02137	0.01022
93	0	0.00072	0.00085	0.00088	0.00005	0.00015	0.00138	0.00104	0.00808	0.00133	0.00037	0.00038	0.00018	0.00009
94	0	0.00281	0.00377	0.00307	0.00033	0.00076	0.00012	0.00139	0.00085	0.00925	0.00041	0.00094	0.00067	0.00067
95	0	0.00016	0.00009	0.00008	0.00013	0.00020	0.00743	0.00034	0.00013	0.00009	0.00320	0.00291	0.00008	0.04188
96	0	0.01247	0.01463	0.01434	0.00151	0.00263	0	0.07937	0.00119	0.12476	0.03520	0.08429	0.01541	0.00482
97	0	0.00001	0.00008	0.00007	0.00003	0.00148	0	0.00165	0.00136	0.00677	0.00008	0.02164	0.06773	0.00005
98	0	0	0.00001	0	0.00001	0	0	0.00222	0.00039	0.00039	0.00017	0.00135	0.00064	0
99	0	0.00376	0.00332	0.01521	0.02904	0.02586	0.00244	0.01705	0.00490	0.01040	0.00197	0.02304	0.00902	0.00856
100	0	0.08862	0.09934	0.08749	0.01938	0.06237	0.03938	0.02698	0.01768	0.08116	0.00763	0.03396	0.02430	0.07488
101	0	0.00353	0.01304	0.00445	0.00838	0.00475	0.00341	0.01847	0.00439	0.00442	0.00782	0.02524	0.01744	0.00746
102	0	0	0	0.00002	0.00003	0.00009	0	0.00443	0.00496	0.00129	0.00058	0.01373	0.00672	0.00001
103	0	0	0	0	0	0.00007	0	0.00558	0.00092	0.00116	0.00047	0.00397	0.00234	0
104	0	0	0	0	0.00003	0.00006	0	0.00141	0.00101	0.00256	0.00002	0.01608	0.00028	0.00001
105	0	0.00006	0.00005	0.00025	0.00048	0.00013	0	0.00209	0.00059	0.00205	0.00241	0.01585	0.00117	0.00022
106	0	0.02954	0.01364	0.01008	0.00958	0.00524	0.02758	0.01904	0.01858	0.03281	0.01485	0.02844	0.05339	0.00592
107	0	0.00679	0.00528	0.00401	0.00373	0.00341	0	0.02527	0.00568	0.01599	0.00590	0.01422	0.01122	0.00632
108	0	0.00020	0.00036	0.00080	0.00015	0.00016	0.00014	0.00245	0.00030	0.00359	0.00128	0.01124	0.00548	0.00479
109	0	0	0	0.00001	0.00002	0	0.00232	0.01360	0.00256	0.00011	0.00086	0.02353	0.00517	0.00001
110	0	0.01618	0.02015	0.01436	0.03656	0.00080	0	0.00221	0.00075	0.00300	0.00026	0.00159	0.00054	0.00862
111	0	0.00013	0.00012	0.00032	0.00060	0.00045	0.01187	0.02214	0.00538	0.00001	0.00379	0.01967	0.00481	0.00063
112	0	0	0	0	0	0	0	0.00031	0.00148	0	0.00001	0.00025	0.00080	0
113	0	0.00042	0.00155	0.00052	0.00098	0.00055	0	0.00012	0	0.00002	0.00046	0	0	0.00088
114	0	0.00267	0.00308	0.00277	0.00305	0.02631	0.00042	0.01119	0.09450	0.01680	0.03904	0.03894	0.01596	0.02861
115	0	0.00141	0.00168	0.00033	0.00051	0.00017	0	0.00188	0.00180	0.00056	0.00007	0.00089	0.00068	0.00005
116	0	0	0	0	0	0	0	0	0	0	0	0	0	0
117	0	0	0	0	0	0	0	0	0	0	0	0	0	0
118	0	0.00147	0.00170	0.00225	0.00344	0.00202	0.00498	0.00791	0.01132	0.00244	0.00386	0.00904	0.00214	0.00601
119	0	0.00002	0.00002	0.00001	0.00039	0.00022	0.00654	0.00072	0.00016	0.00013	0.00053	0.00167	0.00058	0.00988
120	0	0.00018	0.00009	0.00010	0.00145	0.00006	0.00146	0.00002	0.00022	0	0	0.00005	0.00001	0.00062
121	0	0.00545	0.00483	0.00110	0	0.00078	0.00006	0.00158	0.00320	0.00013	0.00006	0.00170	0.00055	0.00004
122	0	0	0	0	0.02092	0	0.00022	0	0	0	0	0	0	0
123	0	0	0	0	0	0	0	0	0	0	0	0	0	0
124	0	0	0	0	0	0.01141	0	0	0	0	0	0	0	0

Continued

a_{ij} \ b_{ij}	99	100	101	102	103	104	105	106	107	108	109	110	111	112
1	0	0.00093	0.04506	0	0	0.00008	0	0	0	0	0.00469	0.00169	0.02006	0.00010
2	0	0.00122	0.00348	0	0	0	0	0	0	0	0.00376	0.00128	0.00076	0.00003
3	0	0.01049	0.12429	0	0	0	0	0	0	0	0.00121	0.00131	0.02742	0.00002
4	0	0.00161	0.17279	0	0	0	0	0	0	0	0	0.00021	0.01816	0.00010
5	0	0	0.00274	0	0.00325	0	0.00028	0	0	0.00131	0.00095	0.00003	0.00123	0
6	0	0.00186	0.00430	0.02622	0	0.00068	0	0.01224	0.00178	0.02257	0.00815	0.01502	0.01702	0.00007
7	0	0	0	0	0.00007	0	0	0	0	0	0.00149	0.00016	0	0
8	0	0	0.00368	0.00003	0	0	0	0	0	0	0.00001	0.00010	0.00012	0
9	0	0	0	0	0	0	0	0	0	0	0	0	0	0
10	0	0	0	0	0	0	0	0	0	0	0	0	0	0
11	0	0	0	0.10726	0	0	0.00001	0	0	0	0.00385	0.00006	0	0
12	0	0.00031	0.00003	0.00052	0.00005	0.00060	0.00172	0	0	0.00598	0.00058	0.00079	0.00081	0.00006
13	0	0.01960	0.05839	0	0	0.00039	0.00022	0	0	0	0.00272	0.00021	0.00664	0
14	0	0.00029	0.00610	0.00409	0	0.00004	0.00333	0	0	0	0.00030	0.00080	0.00025	0.00013
15	0	0.00061	0.10520	0	0	0.00234	0.00040	0	0	0	0.00101	0.00007	0.02875	0.00016
16	0	0.00171	0.04692	0	0	0.00060	0.04018	0	0	0	0.00018	0.00022	0.01429	0.00065
17	0	0.00194	0.04392	0	0	0.00006	0.00356	0	0	0	0.00126	0.00145	0.01325	0.00029
18	0	0.00528	0.09395	0.00040	0	0.00031	0.04195	0	0	0.00140	0.00137	0.06989	0.05531	0.00047
19	0	0.01226	0.04392	0	0	0.00077	0.00032	0.00399	0.00069	0	0.00309	0.00380	0.02939	0.00010
20	0	0.01667	0.05036	0.02193	0	0	0.00011	0	0	0	0.00031	0.00240	0.01686	0.00022
21	0	0.00523	0.00178	0	0	0	0	0	0	0.00136	0.00192	0.10307	0.02141	0.00001
22	0.00098	0.00241	0	0.01256	0.00046	0.00106	0	0.00026	0.00077	0.00059	0.00025	0.01551	0.00332	0
23	0	0.00006	0	0.00506	0.00035	0	0	0.00021	0.00041	0	0	0.00065	0.00014	0.00017
24	0	0.00517	0	0.00020	0	0	0	0	0	0	0	0.02047	0.00286	0.00001
25	0	0.00086	0	0.00063	0	0	0	0	0	0.00048	0.00035	0.00076	0.00873	0
26	0	0	0	0.00101	0.00012	0	0	0	0	0	0.00082	0.00119	0	0.00064
27	0	0	0	0.00873	0.00065	0	0	0	0	0	0	0.01443	0.01957	0.00004
28	0.01659	0.01953	0.00410	0.00512	0.00619	0.00539	0.00348	0.00552	0.00486	0.00626	0.00593	0.00084	0.00167	0.00005
29	0	0.00036	0.00030	0.00280	0.00003	0.00018	0.00156	0.00008	0.00097	0.00032	0.00137	0.01510	0.00114	0.00009
30	0.00078	0.00134	0.00969	0.01087	0.00009	0.00002	0.00005	0.00063	0.00015	0.00253	0.00158	0.05953	0.00913	0.00020
31	0.00874	0.01554	0.00240	0.00337	0.00110	0.00097	0.00032	0.01198	0.01379	0.00397	0.00211	0.02409	0.01144	0.00040
32	0.00619	0.01416	0.00081	0.00155	0.00117	0.00146	0.00114	0.01684	0.01867	0.00884	0.00339	0.00817	0.01020	0.00080
33	0.01480	0.04789	0.00087	0.00263	0.00501	0.00200	0.00207	0.02045	0.04978	0.00623	0.00620	0.00967	0.00849	0.00012
34	0.00634	0.00032	0.00003	0.00247	0.00368	0.00168	0.00272	0.01963	0.02590	0.01259	0.00548	0.00439	0.00115	0.00024
35	0	0.00152	0.00356	0.00035	0.01109	0.00007	0.00012	0.00088	0.00043	0.00108	0.00417	0.01074	0.01817	0
36	0.01378	0.03059	0	0.14736	0.31157	0.37401	0.16223	0.00819	0.01032	0.01026	0.22169	0	0	0
37	0	0	0	0.00006	0	0	0	0	0	0	0.00216	0.00032	0.00008	0
38	0	0	0	0.00099	0.00004	0.00051	0	0	0	0	0.00086	0	0	0
39	0	0	0	0	0	0	0	0	0	0	0.00037	0	0	0
40	0	0.01543	0.00006	0	0	0	0	0	0	0	0.00042	0	0	0
41	0.00078	0.00293	0.00221	0.00431	0.00070	0.00195	0.00001	0.00023	0.00046	0.00113	0.00251	0.02753	0.00117	0.00003
42	0	0.00112	0.00277	0.00290	0.00089	0.00111	0.00082	0.00235	0.00736	0.00333	0.00313	0.09903	0.03223	0.00015

Continued

a_{ij} \ b_{ij}	99	100	101	102	103	104	105	106	107	108	109	110	111	112
43	0.00043	0.00004	0	0.00076	0.00007	0.00021	0.00019	0.00118	0.00196	0.00126	0.00211	0.06073	0.00762	0.00012
44	0.00031	0.00858	0.00061	0.00045	0.00278	0.00052	0.00015	0.00049	0.00041	0.00052	0.00759	0.00112	0.00243	0.00009
45		0	0	0.00041	0	0.00005	0.00011	0	0	0	0	0	0	0
46	0.00204	0.00245	0.00051	0.00424	0.06298	0.00441	0.00275	0.00033	0.00029	0.00026	0.02700	0.00362	0.00339	0.00002
47	0.00428	0.01241	0.00201	0.00282	0.00062	0.00069	0.01093	0.00060	0.00132	0.00067	0.00172	0.00726	0.00784	0.00002
48	0.00250	0.00695	0.00021	0.00225	0.00067	0	0.00006	0.00055	0.00020	0.05836	0.00753	0.00123	0.00063	0.00007
49		0	0	0.01745	0.00043	0	0.00001	0.00019	0.00078	0.00644	0.00813	0.00063	0.00297	0.00003
50	0.00005	0.00849	0.00021	0.00194	0.00076	0.00002	0.00002	0.00125	0.00076	0.09203	0.00651	0.01521	0.00240	0.00006
51	0.00080	0.00102	0.00052	0.00207	0.00068	0.00014	0.00012	0.00097	0.00141	0.00184	0.00128	0.00085	0.00118	0.00001
52	0.00151	0.00201	0.00388	0.00067	0.00012	0.00010	0.00006	0.00045	0.00061	0.00191	0.00059	0.00059	0.00775	0.00002
53		0	0	0.00072	0.00002	0.00003	0.00003	0.00006	0.00019	0.00080	0.00032	0.00291	0.00022	0
54		0	0	0.00116	0.00040	0	0	0.00015	0.00007	0.00008	0.00084	0.00063	0.00015	0.00006
55		0	0	0	0	0	0	0	0	0	0.00015	0	0	0
56		0	0	0	0	0	0	0	0	0	0.00021	0	0	0
57	0.00386	0.00090	0.00040	0.02552	0.00161	0.00007	0.00003	0	0	0.00396	0.00359	0.00001	0.00001	0.00004
58		0	0	0	0	0	0	0	0	0	0.00101	0	0	0
59		0	0	0	0	0	0	0	0	0	0.00015	0.00006	0	0
60		0	0.00013	0.00028	0.00003	0.00001	0	0	0.00198	0.00250	0.00042	0.00023	0	0
61	0.00841	0.00980	0.00367	0.00748	0.00389	0.00089	0.00096	0.00484	0	0.01265	0.01252	0.00431	0.00694	0.00011
62		0.00234	0.00027	0.00456	0.00068	0.00017	0.00003	0	0	0.00178	0.00058	0.00037	0.00195	0
63		0	0	0.00044	0.00350	0.00006	0.00057	0	0	0	0.00462	0.00014	0	0
64	0.05702	0.00932	0.00090	0.03052	0.02285	0.05863	0.06781	0.01555	0.01894	0.01681	0.06824	0.00612	0.01329	0.00059
65		0	0	0	0	0	0	0	0	0	0.00021	0	0	0
66	0.00202	0.02275	0.00052	0.00145	0.00014	0.00057	0.00029	0.00919	0.00320	0.00128	0.00199	0.01201	0.00212	0.00001
67		0	0	0.15863	0	0	0	0	0	0	0.00495	0	0	0
68	0.01449	0.05206	0.00039	0.00432	0.15427	0.00045	0.01093	0.00283	0.00442	0.00748	0.07341	0.00594	0.00308	0.00108
69		0	0	0	0	0.14074	0	0	0	0	0.00001	0	0	0
70		0	0	0	0	0	0.20766	0	0	0	0	0	0	0
71	0.00554	0	0	0.00006	0.00001	0.00041	0.00263	0.00109	0.00287	0	0.00013	0.00002	0.00017	0
72	0.00049	0.00208	0.00009	0.00091	0.00365	0.00140	0.00017	0.00016	0.00899	0.00136	0.00706	0.00131	0.00021	0.00001
73	0.00117	0	0	0.00144	0.00014	0.00119	0.00144	0.00685	0.00143	0.00034	0.00232	0.00007	0.02089	0.00001
74	0.00230	0.01590	0.00137	0.00163	0.00042	0.00119	0.00127	0.00685	0.00197	0.00267	0.00101	0.00786	0.01063	0.00008
75	0.25589	0.04690	0.00081	0.01584	0.00145	0.00127	0.00179	0.00552	0.00125	0.03430	0.01113	0.00176	0.00510	0.00024
76	0.00048	0.00391	0.00045	0.00062	0.00133	0.00230	0.00313	0.03040	0.00931	0.00360	0.00160	0.00071	0.00549	0.00024
77	0.00494	0.03025	0.00076	0.00025	0.00032	0.00006	0.00006	0.00352	0.00631	0.00057	0.00050	0.00184	0.00059	0.00005
78		0.00088	0	0.00306	0.00007	0.00036	0.00047	0.00341	0.00038	0.00023	0.00040	0.00283	0.00305	0.00001
79	0.06120	0.00256	0.00004	0.00215	0.00021	0.00065	0.00057	0.00989	0.00375	0.00430	0.00204	0.00033	0.00237	0.00010
80	0.04036	0.00145	0	0.00160	0.00011	0.00015	0.00017	0.00057	0.00018	0.00015	0.00208	0.00036	0.00621	0.00001
81	0.00045	0.00410	0.00008	0.00066	0.00102	0.00013	0.07833	0.02700	0.00600	0.00505	0.00078	0.03094	0.01196	0.00021
82	0.10041	0.00796	0.00018	0.03836	0.04202	0.09148	0.01180	0.01449	0.01744	0.01342	0.05841	0.00528	0.02399	0.00054
83	0.00378	0.00936	0.00301	0.00374	0.00433	0.00268	0.00012	0.01266	0.01254	0.00632	0.00478	0.01267	0.01101	0.00200
84	0.00031	0.00362	0.00126	0.00331	0.00105	0.00177		0.00706	0.01852	0.02693	0.00786	0.02179		0.00017

Continued

a_{ij} \ b_{ij}	99	100	101	102	103	104	105	106	107	108	109	110	111	112
85	0	0	0	0	0	0	0	0	0	0	0	0	0	0
86	0.05902	0.01575	0.00842	0.04586	0.01417	0.00211	0.00275	0.01177	0.00937	0.01682	0.04084	0.01658	0.06039	0.00282
87	0.00343	0.00242	0.00206	0.00211	0.00148	0.00046	0.00017	0.00224	0.00189	0.00285	0.00242	0.00083	0.00332	0.00009
88	0.00074	0.00055	0.00412	0.00010	0.00102	0.00014	0.00001	0.00044	0.00005	0.00126	0.00308	0.00208	0.00306	0.00003
89	0.00629	0.00304	0.00326	0.00285	0.00666	0.00276	0.00087	0.00321	0.00351	0.00302	0.00530	0.00574	0.01225	0.00047
90	0.06992	0.01040	0.00104	0.11229	0.01569	0.00049	0.00265	0.05070	0.02706	0.16165	0.06381	0.03836	0.07200	0.00189
91	0.00376	0.00142	0.00217	0.00603	0.00227	0.00249	0.00232	0.00092	0.00154	0.00439	0.00315	0.00377	0.00286	0.00027
92	0.00250	0.00342	0.00393	0.01111	0.00891	0.00304	0.00595	0.00162	0.00223	0.00739	0.00584	0.00631	0.00397	0.00061
93	0.00005	0.00011	0.00003	0.00053	0.00112	0.00134	0.00058	0.00003	0.00027	0.00004	0.00082	0.00007	0.00007	0
94	0.00018	0.00028	0.00031	0.00090	0.00139	0.00161	0.00069	0.00011	0.00016	0.00038	0.00112	0.00039	0.00032	0.00001
95	0.00407	0.00022	0.00005	0.00012	0.00031	0.00008	0.00044	0.00112	0.00036	0.00042	0.00026	0.00053	0.00055	0.00016
96	0.00060	0.01292	0.00103	0.00622	0.05352	0.16097	0.08452	0.00138	0.00522	0.00219	0.00307	0.00391	0.00164	0.00202
97	0.00013	0.00402	0.00004	0	0.00119	0.00123	0.00012	0.00032	0.00403	0.00033	0.00009	0.00012	0.00016	0.00003
98	0.00001	0.00277	0.00028	0	0.00085	0.00014	0.00015	0.00939	0.00681	0.00079	0.00110	0.00103	0.00312	0.00098
99	0	0	0	0.00274	0	0.00479	0.00524	0.03214	0.03762	0.01072	0.01207	0.00856	0.03341	0.01132
100	0.03082	0.16741	0.00270	0.03067	0.01910	0.02426	0.02733	0.02434	0.02960	0.03594	0.03143	0.06291	0.04950	0.01992
101	0.00769	0.03366	0.06893	0.00402	0.02960	0.00499	0.02351	0.02417	0.06131	0.02791	0.01482	0.01230	0.01265	0.09990
102	0.00002	0.00672	0.00181	0	0.02198	0.00073	0.00037	0.01140	0.01563	0.00238	0.00315	0.00340	0.00258	0.12867
103	0	0	0	0	0.00360	0.00093	0.00040	0.00507	0.00821	0.00156	0.00446	0.00121	0.00210	0.05966
104	0.00002	0.00295	0	0	0.02451	0.01639	0.00006	0.00691	0.01077	0.00199	0.00328	0.00146	0.00075	0.02649
105	0.00031	0.00419	0	0	0.00260	0.00197	0.01438	0.01798	0.02507	0.00636	0.00409	0.00467	0.00551	0.22234
106	0.01066	0.05599	0.00004	0.02090	0.00246	0.03097	0.03388	0.23985	0.00180	0.18192	0.00580	0.00131	0.03268	0.02135
107	0.00639	0.06675	0.00366	0	0.01881	0.01210	0.01534	0.01029	0.06177	0.00931	0.01953	0.00290	0.01278	0.00379
108	0.00802	0.01076	0.00116	0.00002	0.03752	0.00092	0.00612	0.08728	0.13991	0.02487	0.00519	0.02701	0.01052	0.00674
109	0.00001	0.02672	0.00518	0.00315	0.00683	0.00296	0.00462	0.01383	0.03574	0.01350	0.07552	0.00589	0.00888	0.04932
110	0.01296	0.00193	0	0	0.01682	0.00087	0.00010	0.00335	0.00827	0.00429	0.00585	0.02927	0.01415	0.05480
111	0.00160	0.02125	0.00093	0.01674	0.00301	0.00329	0.01874	0.03708	0.06820	0.00095	0.00919	0.00661	0.04035	0.08973
112	0	0.00734	0.00133	0	0.01495	0.00031	0.00007	0.00069	0.00068	0.00085	0.00096	0.00081	0.00513	0.15631
113	0.00091	0	0	0	0.00019	0.00002	0.00081	0.00175	0.00080	0.00046	0.00019	0.00022	0.00132	0.00226
114	0.09968	0.00392	0.00021	0.00205	0.00012	0.00523	0.07611	0.12146	0.14936	0.05863	0.02044	0.01921	0.02756	0.01938
115	0.00012	0.05310	0.02898	0	0.01012	0.00014	0.00014	0.00064	0.00106	0.00047	0.00267	0.00019	0.00075	0.00074
116	0	0.00006	0.00140	0	0.00154	0	0	0	0	0	0	0	0	0
117	0	0	0	0.00635	0.01233	0.00370	0.00797	0	0	0	0	0	0	0
118	0.00415	0.00489	0.00092	0.00754	0.00453	0.00040	0.00096	0.00737	0.00718	0.00291	0.00813	0.00353	0.00301	0.00062
119	0.02120	0	0	0.00217	0.00114	0	0.00001	0.00436	0.00908	0.00139	0.00332	0.00096	0.00391	0.00564
120	0.00069	0.00025	0	0.00009	0	0.00040	0.00004	0.00024	0.00002	0.00018	0.00009	0.00001	0	0.00002
121	0.00003	0.00458	0	0.00026	0.00048	0	0	0.00185	0.00164	0.00277	0.00284	0.00017	0.00192	0.00027
122	0	0	0	0	0	0	0	0	0	0	0.00019	0	0	0
123	0	0	0	0	0	0	0	0	0	0	0.00047	0	0	0
124	0	0	0	0	0	0	0	0	0	0	0	0	0	0

Continued

a_{ij} \ b_{ij}	113	114	115	116	117	118	119	120	121	122	123	124
1	0.03162	0.00177	0.00232	0.00989	0.02191	0.00294	0.00027	0.00344	0.00099	0.39930	0.00670	0
2	0.00655	0.00038	0.00010	0.00239	0.00034	0.00147	0.00004	0.00383	0.00058	0.01341	0.00055	0
3	0.01301	0.00102	0.00009	0.00242	0.00393	0.00314	0.00001	0.00204	0	0.03507	0.00007	0
4	0.02010	0.00135	0.00006	0.00006	0.00154	0.00060	0.00005	0.00023	0.00019	0.03547	0.00578	0
5	0.00010	0	0.00370	0.00063	0.00039	0.00054	0.00001	0.00003	0.01291	0.00044	0.00064	0
6	0.00629	0.00589	0.00841	0.02636	0.03058	0.03629	0.00245	0.01259	0.00019	0.00513	0.00464	0.00638
7	0	0	0.00012	0	0	0	0	0.00003	0	0	0	0.00006
8	0.00041	0	0.00013	0	0	0	0	0.00021	0	0	0	0
9	0	0	0	0	0	0	0	0.00011	0	0	0.00196	0
10	0	0	0	0	0	0	0	0.01240	0	0	0.01135	0
11	0	0	0	0	0	0	0	0.00002	0	0	0	0
12	0	0	0	0.00063	0.00009	0.00060	0.00011	0.00164	0.00110	0.00025	0.01387	0
13	0.00003	0.00031	0.00022	0.00030	0.00003	0.00806	0.00012	0.00012	0.00097	0.00011	0.00074	0
14	0.01643	0.00025	0.00039	0.00706	0.03059	0.00030	0	0.00288	0.00027	0.10708	0.00020	0
15	0.00102	0.00004	0.00009	0.00056	0.00107	0.00699	0	0.00035	0.00002	0	0	0
16	0.02091	0.00040	0.00027	0.03568	0.02267	0.00041	0.00028	0.00032	0.00011	0.00108	0.00102	0
17	0.01768	0.00031	0.00009	0.00488	0.00475	0.00338	0.00013	0.00053	0.00008	0.00106	0.00059	0
18	0.04378	0.00287	0.00019	0.00799	0.01603	0.00142	0.00034	0.00117	0.00223	0.00087	0.00577	0
19	0.09136	0.00228	0.00574	0.00250	0.00361	0.00189	0.00059	0.00062	0.00077	0.00066	0.00065	0
20	0.09938	0.00267	0.00018	0.02004	0.00350	0.00266	0.00104	0.00071	0.00234	0.00079	0.00170	0
21	0.10002	0.00201	0.00009	0.00083	0.00441	0.00078	0.00119	0.00149	0.00209	0.00025	0.00173	0
22	0.00410	0.00216	0.00876	0.00253	0.00616	0.00007	0.00036	0.00057	0.00255	0.00013	0.00007	0.00957
23	0.00028	0.00050	0.00015	0.00067	0.00141	0.00037	0.00085	0.01897	0.00030	0.00025	0.00026	0.00043
24	0	0.00056	0.00001	0	0.00004	0.00081	0.00016	0.00006	0.00003	0.00004	0.00001	0.00002
25	0.00017	0	0.00002	0.00022	0.00040	0.00030	0.00044	0.00022	0.00037	0.00034	0.00017	0.00001
26	0.00168	0.00033	0.00032	0.00457	0.00692	0.00550	0.00016	0.00040	0.00074	0	0	0.00034
27	0.00421	0.00040	0.00041	0.00581	0.00027	0.00045	0.00024	0.00007	0.00031	0.00466	0.00702	0
28	0.01155	0.00150	0.00288	0.03635	0.01169	0.00326	0.00162	0.00218	0.00604	0.00035	0.00086	0.02333
29	0.00158	0.00757	0.00052	0.00539	0.00072	0.01731	0.00027	0.00033	0.00182	0.00034	0.00041	0.00339
30	0.00073	0.00053	0.00060	0.00235	0.00334	0.03022	0.00088	0.00089	0.00206	0.00250	0.00337	0.00372
31	0.00507	0.00513	0.00515	0.00474	0.00439	0.05369	0.00146	0.00145	0.00681	0.00713	0.00570	0.00977
32	0.00225	0.01542	0.00399	0.00241	0.00665	0.02051	0.24660	0.00580	0.02737	0.00518	0.00592	0.01422
33	0.00683	0.03898	0.00748	0.00640	0.02487	0.00508	0.10276	0.01178	0.01960	0.00550	0.00851	0.06867
34	0.00595	0.06450	0.00194	0.00910	0.02750	0.08667	0.00673	0.00678	0.01965	0.00047	0.00053	0.01389
35	0.01781	0.01295	0.00031	0.07237	0.00261	0.00021	0.00181	0.00058	0.00107	0.00906	0.02720	0.00068
36	0.00664	0.00108	0.00256	0.00678	0.02136	0.00253	0.00222	0.00874	0.01495	0	0.00005	0.02677
37	0	0.01113	0	0	0	0.00019	0	0.00012	0.00011	0.00012	0.00146	0
38	0.00014	0.00032	0.00182	0.00039	0.00008	0.00017	0.00074	0.00638	0.00213	0.00170	0.00016	0.00009
39	0	0	0	0.00376	0.00008	0	0	0.00406	0.00002	0.00790	0.00010	0
40	0	0	0	0.00176	0.00035	0	0	0.00307	0.00045	0.00120	0	0
41	0.00208	0.00888	0.00169	0.00232	0.00420	0.00587	0.00515	0.01088	0.00197	0	0.00077	0.00505
42	0.00786	0.00464	0.00544	0.01045	0.00572	0.00294	0.00138	0.00254	0.00401	0.00106	0.00066	0.00360

Continued

b_{ij} / a_{ij}	113	114	115	116	117	118	119	120	121	122	123	124
43	0.00424	0.01012	0.01658	0.00277	0.00177	0.00833	0.00674	0.00952	0.00319	0.00044	0.00322	0.00059
44	0.00049	0.00092	0.57817	0.01678	0.04123	0.02123	0.00233	0.00680	0.00639	0.02326	0.00339	0.00583
45	0.00273							0.00051		0.00071		
46	0.00102	0.00086	0.00907	0.00207	0.00109	0.00077	0.00031	0.00239	0.00129	0.00020	0.00165	0.00122
47	0.00152	0.00218	0.00469	0.00163	0.00102	0.00176	0.00096	0.00367	0.00212	0.00779	0.00630	0.00064
48	0.00056	0.00420	0.00248	0.00836	0.00919	0.00872	0.00256	0.00246	0.00738	0.00333	0.03174	0.01673
49	0.00044	0.00292	0.00102	0.00103	0.00172	0.00433	0.00018	0.00217	0.00218	0.00043	0.04276	0.00131
50	0.00077	0.00194	0.00096	0.00501	0.00433	0.00826	0.00179	0.00134	0.00243	0.00088	0.00536	0.00621
51	0.00127	0.00154	0.00083	0.00104	0.00073	0.00226	0.00118	0.00269	0.00134	0.00009	0.00041	0.00013
52	0.00199	0.00061	0.00026	0.00043	0.00152	0.00137	0.00014	0.00070	0.00034	0.00019	0.00170	0.00046
53	0.00048	0.00079	0.00003	0.00439	0.00012	0.00030	0.00003	0.00257	0.00062	0.00002	0.00017	0
54	0.00015	0.00126	0.00007	0.00228	0.00002	0.00072	0.00002	0.00257	0.00048	0.00012	0.00024	0
55	0	0	0	0	0	0.00015	0	0.00070	0.00006	0	0	0
56	0	0	0	0	0	0.00033	0	0.00184	0.00005	0	0	0
57	0.00006	0.00127	0.00009	0.00050	0.00050	0.00051	0.00029	0.00377	0.00141	0.00016	0.01664	0
58	0	0.00043	0	0	0	0.00009	0	0.00155	0.00015	0	0	0
59	0	0	0	0	0	0.00007	0	0.00633	0.00005	0	0	0
60	0	0.00024	0.00001	0.00002	0	0.00012	0.00002	0.00298	0.00033	0.00026	0.00156	0
61	0.00517	0.03038	0.00503	0.01257	0.01325	0.01202	0.00295	0.03820	0.01330	0.00210	0.08827	0.02567
62	0	0.00304	0.00079	0.00799	0.00249	0.00191	0.00091	0.00083	0.00330	0.00070	0.00014	0.00066
63	0.00095	0.00080	0.00008	0.00006	0	0.00148	0.00004	0.01903	0.00156	0.00088	0.00108	0.00046
64	0.01275	0.02015	0.01022	0.01925	0.01724	0.01186	0.00783	0.02443	0.03245	0.00759	0.05039	0.03771
65	0	0	0	0.00116	0	0	0	0.00039	0	0.00265	0.00085	0.00251
66	0.00292	0.00218	0.06413	0.00533	0.00331	0.00345	0.00088	0.00966	0.01602	0.00197	0.00655	0.00411
67	0	0	0	0	0	0.00007	0	0.00012	0	0	0.00015	0
68	0.00260	0.03053	0.00143	0.01243	0.00895	0.00646	0.00218	0.00375	0.01993	0.00488	0.00540	0.03089
69	0.00007	0	0	0	0	0.00005	0	0.00014	0	0	0.00137	0
70	0	0	0	0	0	0	0	0.01390	0.00302	0	0	0
71	0.00003	0.00025	0	0.00056	0.00005	0.00004	0.00003	0.00004	0.00014	0.00003	0.00002	
72	0.00004	0.00214	0.00032	0.00283	0.00103	0.00118	0.00105	0.00023	0.00067	0.00145	0.00018	0.00174
73	0.00178	0.00126	0.00014	0.00053	0.00026	0.00229	0.00020	0.00361	0.00064	0.00056	0.00105	0.00064
74	0.01425	0.02731	0.00172	0.01167	0.00701	0.00752	0.00206	0.04312	0.00292	0.00077	0.00092	0.00279
75	0.00850	0.03032	0.00286	0.00826	0.00498	0.00634	0.00632	0.05515	0.01637	0.00050	0.01360	0.00422
76	0.00313	0.20093	0.00170	0.00171	0.01064	0.01761	0.00307	0.02302	0.05064	0.00140	0.00207	0.01047
77	0.00236	0.00155	0.00043	0.00277	0.00108	0.00439	0.00267	0.01042	0.00161	0.00043	0.00047	0.00099
78	0.00143	0.02130	0.00047	0.00047	0.00021	0.00866	0.00233	0.06495	0.00299	0.00002	0.00120	0.00368
79	0.00563	0.03310	0.00081	0.00295	0.00046	0.00618	0.01007	0.15300	0.00799	0.00028	0.00158	0.00201
80	0.00088	0.00482	0.00091	0.00067	0.00007	0.00367	0.00053	0.03211	0.01880	0.00206	0.00178	0.00133
81	0.00439	0.01690	0.00086	0.00045	0.00269	0.00618	0.00366	0.00346	0.00760	0.00129	0.00078	0.00307
82	0.00654	0.01159	0.00844	0.01539	0.01535	0.00864	0.00686	0.00843	0.01520	0.00654	0.03838	0.02311
83	0.00476	0.01159	0.00223	0.00483	0.00441	0.00461	0.00744	0.06685	0.00518	0.00242	0.00349	0.00577
84	0.00690	0.00261	0.00820	0.00817	0.00591	0.00574	0.00319	0.01218	0.00262	0.00023	0.00768	0.00101

Continued

b_{ij} / a_{ij}	113	114	115	116	117	118	119	120	121	122	123	124
85	0	0	0	0	0	0	0	0	0	0	0	0
86	0.02461	0.00541	0.02022	0.05974	0.03301	0.06564	0.01155	0.02098	0.02215	0.01219	0.02585	0.02145
87	0.00144	0.00060	0.00068	0.00623	0.00475	0.00445	0.00151	0.00148	0.00259	0.00198	0.00272	0.00472
88	0.00123	0.00047	0.00102	0.00292	0.00334	0.00351	0.00063	0.00066	0.00122	0.00003	0.00010	0.00093
89	0.00673	0.00339	0.00380	0.01918	0.00810	0.02510	0.00260	0.00399	0.00668	0.00294	0.00489	0.00602
90	0.03562	0.02432	0.03075	0.11613	0.14200	0.14918	0.01687	0.01818	0.06315	0.01944	0.07333	0.06184
91	0.00353	0.00250	0.00276	0.00524	0.00311	0.00489	0.00582	0.00549	0.00279	0.00708	0.00894	0.00178
92	0.00580	0.00353	0.00368	0.00575	0.00312	0.00508	0.00474	0.00439	0.00857	0.01035	0.01849	0.00464
93	0.00004	0.00004	0.00003	0.00023	0.00048	0.00004	0.00001	0.00005	0.00006	0.00003	0.00004	0.00010
94	0.00030	0.00031	0.00016	0.00039	0.00041	0.00039	0.00017	0.00038	0.00027	0.00051	0.00141	0.00024
95	0.00026	0.00165	0.00095	0.00069	0.00045	0.00048	0.00032	0.00146	0.00282	0.00018	0.00024	0.00041
96	0.00115	0.00138	0.00138	0.00242	0.00343	0.00301	0.00183	0.00180	0.00518	0.00337	0.00815	0.00543
97	0.00023	0.00022	0.00001	0.00002		0.00002	0.00209	0.00019		0.00014		
98	0.00128	0.00195	0.00347	0.00318	0.00269	0.00693	0.02284	0.00333	0.00754	0.00238	0.00140	0.01201
99	0.00897	0.01560	0.00952	0.02414	0.02673	0.02703	0.09394	0.01265	0.03480	0.01289	0.01742	0.06948
100	0.05769	0.05615	0.09376	0.04808	0.03982	0.04846	0.06139	0.05316	0.03579	0.03363	0.04293	0.03913
101	0.02692	0.01984	0.00552	0.03964	0.06536	0.02250	0.01685	0.01526	0.04532	0.02906	0.03941	0.05588
102	0.00308	0.00908	0.00242	0.03261	0.00564	0.02510	0.00766	0.02331	0.03053	0.00757	0.01743	0.02746
103	0.00180	0.00434	0.00066	0.00465	0.00363	0.00980	0.00394	0.00413	0.01539	0.00552	0.00963	0.00939
104	0.00223	0.00656	0.00097	0.00255	0.00371	0.01663	0.00883	0.00633	0.01240	0.00627	0.01455	0.01142
105	0.00420	0.01045	0.00204	0.02115	0.01137	0.01575	0.00977	0.01113	0.02225	0.00746	0.00515	0.04068
106	0.03769	0.00840	0.00167	0.00378	0.00254	0.00379	0.02449	0.00527	0.00526	0.01539	0.02548	0.04332
107	0.01154	0.00361	0.00111	0.00746	0.01099	0.00329	0.00379	0.00264	0.01595	0.00518	0.01142	0.01140
108	0.05642	0.02205	0.00208	0.00232	0.06566	0.00604	0.00799	0.00295	0.01463	0.00238	0.00100	0.02090
109	0.00797	0.01541	0.00495	0.01241	0.01958	0.02692	0.01253	0.01368	0.02340	0.00776	0.01708	0.02064
110	0.00376	0.00489	0.00173	0.02366	0.01078	0.01116	0.00597	0.00572	0.01015	0.00331	0.00312	0.00663
111	0.00691	0.01947	0.00364	0.04306	0.03166	0.03592	0.01833	0.02197	0.07968	0.01728	0.01678	0.08790
112	0.00382	0.00045	0.00011	0.00058	0.00268	0.00191	0.00030	0.00018	0.00114	0.00077	0.00070	0.00216
113	0.00048	0.00006	0.00002	0.00054	0.00023	0.00018	0.00087	0.00005	0.00064	0.00009	0.00020	0.00138
114	0.01713	0.03962	0.00152	0.01767	0.01639	0.02105	0.03301	0.01122	0.03188	0.00335	0.04101	0.01776
115	0.00028	0.00041	0.00286	0.00142	0.00274	0.00550	0.00083	0.00179	0.00281	0.00071	0.00171	0.00416
116	0	0	0	0	0	0	0	0	0	0	0	0
117	0	0	0	0	0	0	0	0	0	0	0	0
118	0.00123	0.00313	0.00302	0.01336	0.02798	0.00665	0.00383	0.00559	0.01391	0.00411	0.00723	0.01046
119	0.01095	0.01225	0.00104	0.01041	0.01045	0.02284	0.15230	0.00423	0.01124	0.00292	0.00213	0.01137
120	0.00015	0.01871	0.00022	0.00019	0	0.00805	0.00006	0.01630	0.00549	0.00032	0.00031	0.00052
121	0.00325	0.00099	0.00095	0.00121	0.00030	0.00342	0.00102	0.00785	0.07437	0.00158	0.01107	0.00246
122	0.00283	0	0	0	0	0	0	0	0	0.04622	0	0
123	0	0	0	0	0	0	0	0	0	0	0.10488	0
124	0	0	0	0	0	0	0	0	0	0	0	0

APPENDIX 4

Output Coefficients among Industries in China in 1997

a_{ij} \ b_{ij}	1	2	3	4	5	6	7	8	9	10	11	12	13	14
1	0.14220	0.00106	0.19581	0.01491	0.01756	0	0	0	0	0	0	0	0	0.31395
2	0.01148	0.08326	0.00769	0.00093	0.04664	0.01312	0	0	0.00012	0.00014	0.00001	0.00011	0.05689	0.00085
3	0.02966	0.00026	0.13265	0	0	0	0	0	0	0	0	0	0	0.01498
4	0	0	0	0.13717	0.00316	0	0	0	0	0	0	0	0	0.04191
5	0.12257	0.00487	0.07298	0.00766	0.11605	0.02241	0.00409	0.00005	0.00061	0.00501	0.00105	0.00076	0.00055	0.00093
6	0	0.00034	0.00659	0.00126	0.00205	0.02894	0.00324	0.00031	0.00038	0.00200	0	0.00129	0.00077	0.00340
7	0	0	0.00507	0	0	0	0.00029	0.03683	0.07046	0.00007	0	0.00367	0.00004	0.00008
8	0	0	0	0	0	0.00042	0	0	0	0.00011	0	0	0	0
9	0	0	0	0	0	0	0	0	0	0.00066	0	0	0	0
10	0	0	0	0	0	0	0	0	0	0.19503	0	0	0	0
11	0	0	0.00930	0.00717	0.00203	0.00219	0.00106	0.00004	0.00002	0.00018	0.00848	0	0	0.00207
12	0	0	0	0	0.01127	0.19955	0.00032	0.00010	0.00013	0.00029	0.00022	0.07646	0	0.00010
13	0.00465	0.00307	0.00637	0.00617	0.00141	0	0.00127	0.00011	0.00247	0.00559	0.00008	0.00243	0.00950	0.13482
14	0.05305	0.00054	0.44815	0.08130	0	0	0	0	0	0	0	0	0	0.01163
15	0	0	0	0	0	0	0	0	0	0	0	0	0	0.01507
16	0	0	0	0	0	0	0	0	0	0	0	0	0	0.32602
17	0	0	0	0	0	0	0	0	0	0	0	0	0	0.04152
18	0	0	0	0	0	0	0	0	0	0	0	0	0	0
19	0.00476	0.00030	0.00182	0.00102	0.00161	0.00176	0.00339	0	0	0	0	0	0	0
20	0	0	0	0	0	0	0	0	0	0.00342	0	0	0	0
21	0	0	0	0	0	0	0	0	0	0	0	0	0	0
22	0.00054	0.00004	0.00039	0.00006	0.00168	0.00154	0.00058	0.00006	0.00003	0.00059	0.00002	0.00020	0.00003	0.00164
23	0.00013	0	0.00002	0.00005	0.00156	0.00073	0.00076	0.00004	0.00002	0.00025	0.00001	0.00024	0.00001	0.00012
24	0.03598	0.00155	0.08885	0.00244	0.07226	0.00176	0.00144	0.00010	0.00013	0.00061	0.00031	0.03862	0.00008	0.01270
25	0	0	0	0	0.00460	0.00038	0.00032	0.00002	0.00001	0.00011	0.00001	0.00014	0	0.00017
26	0.00276	0.00059	0.00042	0.00075	0.00338	0.00264	0.00145	0.00023	0.00056	0.00369	0.00005	0.00114	0.00019	0.00057
27	0	0	0	0	0.00763	0	0	0	0	0	0	0	0	0
28	0.00275	0.00088	0.00145	0.00077	0.00921	0.01603	0.01022	0.00092	0.00259	0.00791	0.00060	0.00688	0.00138	0.00625
29	0.00170	0.00011	0.00826	0.00010	0.00081	0.00370	0.00318	0.00022	0.00043	0.00318	0.00012	0.00073	0.00018	0.00050
30	0.00238	0.00098	0.00513	0.00445	0.00227	0.00667	0.00237	0.00022	0.00063	0.00291	0.00010	0.00112	0.00053	0.00018
31	0.01373	0.00045	0.00254	0.00490	0.01062	0.00329	0.00124	0.00018	0.00061	0.00100	0.00009	0.00176	0.00019	0.00028
32	0.00131	0.00020	0.00018	0.00038	0.00287	0.00043	0.00019	0.00001	0.00003	0.00006	0	0.00016	0.00001	0.00115
33	0.00160	0.00037	0.00040	0.00066	0.00189	0.00090	0.00082	0.00006	0.00012	0.00039	0.00002	0.00019	0.00005	0.00044
34	0.02329	0.00296	0.00593	0.00913	0.00357	0.00687	0.00512	0.00042	0.00056	0.00148	0.00015	0.00130	0.00010	0.00074
35	0.00772	0.00201	0.00347	0.00659	0.00222	0.00207	0	0	0	0.00376	0	0.00810	0	0
36	0.03637	0.00172	0.00517	0.02150	0.00283	0.00992	0.00914	0.00147	0.00277	0.00430	0.00033	0.01885	0.00208	0.00294
37	0	0	0	0	0	0	0	0	0.00477	0.00060	0	0	0	0
38	0.00016	0.00002	0.00016	0.00009	0.00012	0.00233	0.00701	0.00084	0.00036	0.01070	0.00030	0.00146	0.00003	0.00975

Continued

a_{ij} \ b_{ij}	1	2	3	4	5	6	7	8	9	10	11	12	13	14
39	0.85779	0.01728	0.00021	0	0	0	0	0	0	0	0	0	0	0.00094
40	0.58079	0.01996	0	0	0	0	0	0	0	0	0	0	0	0
41	0.00063	0.00006	0.00010	0.00011	0.00029	0.00166	0.00378	0.00167	0.00009	0.00157	0.00004	0.00133	0.00007	0.00177
42	0.00178	0.00030	0.00109	0.00047	0.00056	0.00585	0.00527	0.00032	0.00085	0.00441	0.00026	0.00407	0.00026	0.00649
43	0.00020	0.00002	0.00090	0.00013	0.00014	0.00898	0.00558	0.00005	0.00285	0.00720	0.00013	0.02037	0.00011	0.01075
44	0.00148	0.00086	0.02701	0.00462	0.00027	0.00053	0	0	0.00001	0.00014	0.00004	0.00002	0.00004	0.01296
45	0	0	0	0.00420	0.00025	0	0	0	0	0	0	0	0	0
46	0.00656	0.00048	0.00188	0.00134	0.00138	0.03414	0.00474	0.00048	0.00254	0.01112	0.00335	0.00499	0.00641	0.00064
47	0.03728	0.00084	0.00102	0.00393	0.00157	0.00212	0.00088	0.00007	0.00032	0.00261	0.00254	0.01978	0.00013	0.00891
48	0.00357	0.00049	0.00215	0.00105	0.00087	0.00401	0.00171	0.00036	0.00041	0.00141	0.00010	0.00136	0.00012	0.00011
49	0.00703	0.00065	0.00197	0.00245	0.00306	0.00569	0.00124	0.00026	0.00054	0.00166	0.00004	0.00201	0.00002	0.00008
50	0.00430	0.00041	0.00394	0.00139	0.00049	0.00343	0.00341	0.00055	0.00033	0.00183	0.00007	0.00104	0.00017	0.00017
51	0.00118	0.00021	0.00056	0.00045	0.00202	0.00529	0.00193	0.00015	0.00083	0.00125	0.00014	0.00310	0.00032	0.00073
52	0.00028	0.00006	0.00028	0.00009	0.00549	0.00840	0.00262	0.00027	0.00136	0.00210	0.00076	0.00570	0.00051	0.00256
53	0.00021	0.00006	0.00007	0.00007	0.00035	0.00114	0.00060	0.00005	0.00010	0.00010	0.00002	0.00066	0.00001	0.00003
54	0.00014	0.00003	0.00006	0.00005	0.00015	0.00124	0.00039	0.00002	0.00020	0.00049	0.00001	0.00311	0.00002	0.00003
55	0	0	0	0	0	0.00053	0	0.00035	0.00120	0.00021	0	0.00014	0	0
56	0	0	0	0.00006	0	0.01331	0.00273	0.00044	0.00096	0.00200	0.00006	0.00494	0.00002	0.00006
57	0.00030	0.00003	0.00003	0	0.00037	0.02129	0.00177	0.00033	0.00068	0.00096	0.00006	0.00015	0.00004	0
58	0	0	0	0	0	0.00483	0.00462	0.00042	0.00362	0.00311	0.00009	0.00055	0.00002	0
59	0	0	0	0.00003	0	0.00065	0.00035	0.00002	0.00003	0.00114	0	0.00002	0	0
60	0.00008	0.00090	0.00081	0.00104	0.00003	0.00877	0.00127	0.00007	0.00010	0.00052	0.00001	0.00028	0.00001	0.00003
61	0.01130	0	0	0	0.00325	0.01089	0.00266	0.00023	0.00247	0.00550	0.00015	0.00281	0.00109	0.00107
62	0	0	0	0	0	0.00361	0.00225	0.00029	0.00002	0.00075	0.00005	0.00087	0.00033	0.00083
63	0	0	0	0	0	0.01939	0.00276	0.00029	0.00243	0.00943	0.00039	0.00794	0.00016	0.00109
64	0.00865	0.00082	0.00134	0.00512	0.00164	0.01442	0.00642	0.00106	0.00294	0.00613	0.00043	0.00612	0.00138	0.00112
65	0.34440	0.01330	0.04417	0.08644	0.02139	0.00181	0.00025	0.00002	0.00016	0.00078	0.00051	0.00467	0.01419	0
66	0.00211	0.00025	0.00032	0.00097	0.00062	0.03000	0.02405	0.00356	0.02320	0.00768	0.00069	0.05086	0.00142	0.00232
67	0	0	0	0	0	0.00662	0.00301	0.00043	0.00042	0.00048	0.00001	0.00003	0.00016	0
68	0.00119	0.00042	0.00113	0.00061	0.00037	0.00564	0.00570	0.00073	0.00566	0.00264	0.00031	0.01465	0.00243	0.00393
69	0.00451	0.00020	0.00001	0.46997	0.00145	0	0	0	0	0	0	0	0	0
70	0	0	0	0	0	0	0	0	0	0	0	0	0	0
71	0.00300	0.00037	0.00035	0.00096	0.00107	0.00250	0.00127	0.00020	0.00052	0.00280	0.00012	0.00257	0.00030	0.00047
72	0.00536	0.00331	0.00519	0.00840	0.00343	0.00909	0.00888	0.00087	0.00320	0.00402	0.00042	0.00212	0.00032	0.00187
73	0.00686	0.00029	0.00366	0.00173	0.00024	0.00140	0.00064	0.00016	0.00030	0.00143	0.00004	0.00045	0.00008	0.00020
74	0.00217	0.00005	0.00017	0.00257	0.00254	0.01824	0.00511	0.00050	0.00114	0.00144	0.00017	0.00220	0.00034	0.00081
75	0.00129	0.00024	0.00050	0.00021	0.00010	0.00120	0.00115	0.00023	0.00030	0.00040	0.00006	0.00071	0.00015	0.00021
76	0.00011	0.00004	0	0.00007	0.00013	0.00884	0.00070	0.00020	0.00079	0.00069	0.00012	0.00558	0.00021	0.00062
77	0.00084	0.00009	0.00024	0.00063	0.00007	0.00056	0.00172	0.00011	0.00050	0.00134	0.00001	0.00034	0	0.00012
78	0.00001	0	0	0	0	0.00270	0.00491	0.00091	0.00366	0.00119	0.00011	0.00176	0.00016	0.00058
79	0.00058	0.00005	0.00011	0.00080	0.00024	0.00977	0.01701	0.00333	0.00184	0.00648	0.00035	0.00525	0.00033	0.00253
80	0.00042	0.00006	0.00066	0.00031	0.00009	0.00978	0.00281	0.00048	0.00164	0.00195	0.00034	0.00401	0.00089	0.00109
81	0.00309	0.00037	0.00032	0.00044	0.00006	0.00253	0.00070	0.00333	0.00366	0.00648	0.00011	0.00525	0.00089	0.00253
82	0.03541	0.00253	0.00539	0.02240	0.00496	0.01544	0.00413	0.00028	0.00164	0.00491	0.00045	0.00893	0.00046	0.00184

Continued

a_{ij} / b_{ij}	1	2	3	4	5	6	7	8	9	10	11	12	13	14
83	0.03243	0.00230	0.01689	0.00591	0.00298	0.00933	0.00571	0.00038	0.00111	0.00273	0.00032	0.00428	0.00090	0.00951
84	0.00686	0.00009	0.00240	0.00105	0.00034	0.03569	0.00792	0.00095	0.00254	0.00483	0.00070	0.00883	0.00024	0.00329
85	0	0	0	0	0	0.00048	0.00003	0	0.00001	0.00006	0.00001	0.00001	0	0.04644
86	0.03675	0.00118	0.00411	0.00373	0.00530	0.03006	0.01249	0.00166	0.00951	0.01822	0.00176	0.02032	0.00081	0.00676
87	0.0007	0.00003	0.01267	0.00089	0	0.00788	0.00150	0.00027	0.00013	0.00074	0.00058	0.00036	0.00059	0.00200
88	0	0	0	0	0	0.00458	0.00027	0.00002	0.00021	0.00006	0.00006	0.00009	0.00001	0.00041
89	0.00889	0.00031	0.00165	0.00132	0.00076	0.01042	0.00437	0.00042	0.00298	0.00236	0.00046	0.00913	0.00033	0.00415
90	0.02850	0.00199	0.01291	0.00267	0.00152	0.00531	0.00260	0.00021	0.00113	0.00167	0.00020	0.00330	0.00033	0.00116
91	0.03209	0.00189	0.01162	0.00409	0.00334	0.01569	0.00519	0.00043	0.00445	0.01066	0.00041	0.03871	0.00150	0.01147
92	0.05411	0.00844	0.02316	0.01518	0.00601	0.01343	0.00178	0.00019	0.00509	0.02123	0.00154	0.02489	0.00172	0.02128
93	0.01254	0.00124	0.00469	0.00721	0.00097	0.00344	0.00502	0.00288	0.00159	0.00153	0.00011	0.00885	0.00072	0.00108
94	0.02450	0.00104	0.01525	0.00683	0.00219	0.01080	0.00317	0.00033	0.01450	0.00987	0.00078	0.02235	0.00114	0.01353
95	0.00491	0.00052	0.00466	0.00753	0.00173	0.00432	0.00246	0.00032	0.00112	0.00153	0.00021	0.00362	0.00037	0.00190
96	0.01636	0.00078	0.01012	0.00344	0.00147	0.01261	0.00195	0.00021	0.00283	0.01603	0.00074	0.00815	0.00078	0.01459
97	0.10038	0.00364	0.04539	0.03779	0.00609	0.00015	0	0	0	0.00065	0.00004	0.00004	0	0.00081
98	0.02346	0.00264	0.01842	0.00916	0.02262	0.00009	0.00002	0	0	0.00001	0	0	0	0.00003
99	0.00230	0.00039	0.00131	0.00095	0.00022	0.00937	0.00491	0.00031	0.00126	0.00092	0.00009	0.04197	0.00034	0.00233
100	0.02442	0.00154	0.02529	0.00473	0.00356	0.00864	0.00213	0.00024	0.00115	0.00328	0.00031	0.00589	0.00090	0.01655
101	0.00595	0.00082	0.00189	0.00118	0.00100	0.01230	0.00137	0.00015	0.00319	0.00409	0.00058	0.01403	0.00137	0.00587
102	0.01700	0.00126	0.00686	0.00493	0.00154	0.00006	0.00003	0	0.00001	0.00001	0	0.00820	0	0.00001
103	0.06179	0.01074	0.01698	0.01366	0.01538	0	0	0	0	0	0	0.02689	0	0
104	0.06974	0.00342	0.01144	0.03048	0.00653	0.00010	0.00005	0.00002	0.00001	0.00001	0	0.00476	0	0.00002
105	0.00156	0.00018	0.00041	0.00084	0.00017	0.00073	0.00039	0.00002	0.00010	0.00007	0	0.00433	0.00003	0.00018
106	0.02291	0.00241	0.01091	0.00415	0.00291	0.00779	0.00601	0.00040	0.00095	0.00463	0.00044	0.00834	0.00102	0.01002
107	0.02099	0.00224	0.00538	0.01090	0.00162	0.01166	0.01230	0.00090	0.00159	0.00341	0.00043	0.00682	0.00196	0.00438
108	0.00575	0.00133	0.00168	0.00120	0.00055	0.00331	0.00010	0.00002	0.00009	0.00127	0.00003	0.00240	0.00011	0.00179
109	0.04255	0.00764	0.01178	0.00619	0.00603	0.00004	0.00002	0	0.00001	0.00001	0	0.00194	0	0.00001
110	0.03769	0.00446	0.00481	0.00474	0.00080	0.02188	0.00882	0.00066	0.00325	0.00509	0.00160	0.01157	0.00595	0.00828
111	0.01622	0.00145	0.00594	0.00291	0.00475	0.00087	0.00026	0.00002	0.00008	0.00076	0.00001	0.01305	0.00003	0.00020
112	0	0	0	0	0	0	0	0	0	0	0	0.00172	0	0
113	0.00050	0.00010	0.00030	0.00019	0.00013	0.01506	0.00186	0.00020	0.00404	0.00560	0.00079	0.01926	0.00186	0.00807
114	0.01178	0.00129	0.00081	0.00162	0.00086	0.00770	0.00041	0.00004	0.00022	0.00183	0.00019	0.00110	0.00020	0.00406
115	0.01358	0.00346	0.00539	0.00470	0	0.09501	0.00407	0.00028	0.00276	0.00464	0.00023	0.00597	0.00231	0.00191
116	0	0	0	0	0	0	0	0	0	0	0	0	0	0
117	0	0	0	0	0	0	0	0	0	0	0	0	0	0
118	0.03996	0.00442	0.01002	0.00949	0.00268	0.02890	0.00146	0.00018	0.00117	0.00483	0.00063	0.00463	0.00171	0.00290
119	0.01676	0.00186	0.00516	0.00239	0.00173	0.00033	0.00008	0.00001	0	0.00002	0	0.00449	0	0.00012
120	0.00331	0.00106	0.00074	0.00160	0.00026	0.00619	0.00255	0.00034	0.00042	0.00305	0.00004	0.00287	0.00004	0.00290
121	0.04362	0.00409	0.01360	0.01767	0.00317	0.01525	0.00261	0.00042	0.00254	0.00894	0.00013	0.01194	0.00599	0.00428
122	0.58322	0.07642	0.18913	0.09665	0.02310	0.06119	0.03076	0.00220	0.00564	0.01507	0	0.01620	0	0
123	0.13108	0.00550	0.00333	0.00589	0.00227	0	0	0	0	0	0	0	0	0
124	0	0	0	0	0	0	0	0	0	0	0	0	0	0

Continued

a_{ij} / b_{ij}	15	16	17	18	19	20	21	22	23	24	25	26	27	28
1	0.02107	0.00021	0.00002	0.05534	0.05274	0.02437	0.02706	0.02987	0.00003	0.00727	0.00159	0.00011	0.04606	0.00018
2	0.00008	0.00013	0	0.00904	0.00024	0.00044	0.00002	0	0.08719	0.00001	0.00031	0	0.00001	0.00032
3	0	0.34739	0	0.06999	0	0.00233	0	0	0	0.00005	0.04361	0.00227	0.02584	0.00020
4	0	0	0.44484	0.04337	0	0	0	0	0	0	0	0	0	0
5	0.00018	0.00014	0.00163	0.11120	0.00113	0.00890	0.00001	0.00008	0	0.00001	0.01156	0.00002	0.00001	0.00003
6	0.00206	0.00077	0.00016	0.00982	0.00619	0.00082	0.00094	0.00704	0.00199	0.00023	0.00250	0.00318	0.00094	0.00181
7	0.00005													
8	0.00009	0		0.00059	0.00150	0.00009	0.00033	0.00164	0.00021	0	0	0	0	0
9														
10	0	0		0	0	0	0	0	0	0	0	0	0	0
11	0	0		0.06529	0	0	0	0	0	0	0	0	0	0
12	0	0	0	0	0	0	0	0	0	0	0	0	0	0
13	0.00010	0.00005	0.00159	0.00078	0.00053	0.00013	0.00016	0.00031	0.00010	0.00002	0.00007	0.00007	0.00007	0.00026
14	0.00031	0.00092	0.00381	0.12043	0.03549	0.00981	0	0	0	0	0	0	0	0
15	0.11402	0.00278	0.00048	0.47343	0.03879	0.25529	0	0	0	0	0	0	0	0
16	0	0.08603	0.00053	0.08098	0.00012	0.00012	0	0	0	0	0	0	0	0
17	0	0	0.08072	0.05538	0.00027	0.00001	0	0	0	0	0	0	0	0
18	0.00244	0.03432	0.00398	0.35076	0.05082	0.06225	0.00431	0.00451	0.00001	0.00003	0.00005	0.00002	0.00001	0.00002
19	0	0.00071	0.00007	0.00711	0.25356	0.04192	0.00359	0	0	0	0	0	0	0
20	0	0.01269	0	0.02815	0.01231	0.13392	0	0	0	0	0	0	0	0
21	0	0	0	0	0	0	0.47929	0	0	0	0	0	0	0
22	0.00004	0.00002	0.00002	0.00011	0.00007	0.00001	0.00192	0.35092	0.00484	0.00060	0.00502	0.07759	0.01064	0.19254
23	0.00003	0.00002	0.00002	0.00009	0.00008	0.00003	0.00003	0.00934	0.26308	0.00231	0.00143	0.14151	0.00075	0.36058
24	0.00161	0.00006	0.00005	0.00412	0.00266	0.00065	0.00346	0.02621	0.00678	0.05416	0.00272	0.01548	0.00855	0.50289
25	0.00003	0.00001	0.00001	0.00015	0.00018	0	0.00001	0.00426	0.00641	0.00021	0.54643	0.04375	0.00198	0.24112
26	0.00050	0.00007	0.00004	0.00036	0.00054	0.00041	0.00013	0.02054	0.01028	0.00046	0.00244	0.27438	0.00471	0.11323
27	0	0	0	0	0	0	0	0	0.09814	0.01826	0.00556	0.00519	0.11154	0.19014
28	0.00050	0.00095	0.00044	0.00385	0.00260	0.00128	0.00285	0.37402	0.00104	0.00114	0.00515	0.00377	0.00101	0.08535
29	0.00011	0.00047	0.00008	0.00064	0.00067	0.00034	0.00014	0.03105	0.00596	0.00002	0.00014	0.00100	0.00016	0.02046
30	0.00007	0.00006	0.00008	0.00112	0.00068	0.00078	0.00012	0.00655	0.00030	0.00003	0.00008	0.00015	0.00034	0.00034
31	0.00005	0.00025	0.00037	0.00144	0.00143	0.00064	0.00023	0.00082	0.00018	0.00003	0.00012	0.00009	0.00018	0.00086
32	0.00007	0.00305	0.00148	0.02499	0.02161	0.01069	0.02358	0.00090	0.00067	0.00012	0.00023	0.00277	0.00141	0.00912
33	0.00003	0.00051	0.00004	0.02041	0.01626	0.00985	0.03526	0.00370	0.00039	0.00003	0.00021	0.00037	0.00017	0.00096
34	0.00013	0.00026	0.00149	0.00091	0.00168	0.00015	0.00029	0.00043	0.00120	0.00015	0.00038	0.00088	0.00032	0.00223
35	0	0	0	0	0.00001	0	0.00001	0.00140	0.00153	0	0.00003	0.00003	0	0.00011
36	0.00042	0.00006	0.00066	0.00293	0.00108	0.00096	0.00025	0.00003	0.00087	0.00026	0.00026	0.00095	0.00064	0.00165
37	0.00079	0.00052	0	0	0	0	0	0.00195	0	0	0	0	0	0
38	0.00080	0	0.00012	0.01139	0.00263	0.00100	0.00255	0.00454	0.00075	0.00026	0.00083	0.00217	0.00173	0.00335
39	0	0.00011	0	0.00142	0.00010	0.00001	0	0	0	0	0	0	0	0
40	0	0	0	0	0	0	0	0	0	0	0	0	0	0
41	0.00004	0.00007	0.00078	0.00063	0.00040	0.00047	0.00017	0.02501	0.00375	0.00005	0.00251	0.00859	0.00155	0.00229
42	0.00022	0.00041	0.00028	0.01185	0.00460	0.00599	0.01913	0.01894	0.00114	0.00036	0.00159	0.00242	0.00049	0.00310

Continued

a_{ij} / b_{ij}	15	16	17	18	19	20	21	22	23	24	25	26	27	28
43	0.00007	0.00020	0.00050	0.00442	0.00099	0.00272	0.00099	0.00183	0.00005	0.00004	0.00163	0.00018	0.00019	0.00110
44	0.00001	0.00021	0	0.00021	0.00037	0.00080	0.00001	0.00005	0.00001	0	0	0	0.00006	0.17809
45	0.00023	0	0	0	0.00020	0.00039	0.02584	0.21439	0.02784	0.00487	0.07254	0.07811	0.02460	0.01609
46	0.00098	0.00024	0.00026	0.00858	0.00116	0.00089	0.00059	0.00474	0.00043	0.00011	0.00030	0.00078	0.00042	0.01066
47	0.00098	0.00374	0.00061	0.02878	0.00289	0.01166	0.00170	0.00389	0.00102	0.00014	0.00019	0.00226	0.00217	0.01066
48	0.00006	0.00005	0.00003	0.00026	0.00298	0.00006	0.00009	0.00026	0.00011	0.00001	0.00003	0.00008	0.00006	0.00019
49	0.00007	0.00005	0.00003	0.00032	0.00038	0.00004	0.00002	0.00016	0.00011	0.00001	0.00003	0.00005	0.00002	0.00008
50	0.00018	0.00007	0.00003	0.00032	0.00041	0.00007	0.00012	0.00033	0.00011	0.00002	0.00004	0.00010	0.00007	0.00027
51	0.00014	0.00007	0.00012	0.00891	0.09684	0.00485	0.00033	0.00198	0.00022	0.00003	0.00036	0.00047	0.00022	0.00123
52	0.00032	0.00051	0.00025	0.00315	0.00653	0.00516	0.00078	0.00795	0.00088	0.00009	0.00063	0.00193	0.00117	0.00458
53	0.00010	0	0	0.00014	0.00008	0.00002	0	0.00004	0	0	0.00001	0.00019	0	0.00002
54	0.00004	0	0	0.00016	0.00082	0.00002	0.00001	0.00003	0.00007	0	0.00004	0	0.00130	0.00014
55	0	0	0	0	0	0	0	0	0	0	0	0	0	0
56	0	0	0	0.00026	0.00013	0.00006	0.00002	0.00013	0.00004	0.00001	0.00002	0.00002	0.00006	0.00019
57	0.00007	0.00001	0.00001	0	0	0	0	0	0	0	0	0	0	0
58	0	0	0	0	0	0	0	0	0	0	0	0	0	0
59	0	0	0	0	0	0	0	0	0	0	0	0	0	0
60	0.00005	0.00001	0.00003	0.00026	0.00042	0.00136	0.00029	0.00014	0.00015	0.00009	0.00042	0.00001	0.00005	0.00010
61	0.00014	0.00024	0.00060	0.00658	0.00606	0.00838	0.00068	0.00190	0.00068	0.00001	0.00007	0.00046	0.00059	0.00277
62	0.00004	0.00004	0	0.00045	0.00027	0.00003	0.00007	0.00003	0.00007	0.00007	0.00001	0.00001	0.00001	0.00049
63	0.00041	0.00024	0.00019	0.00185	0.00147	0.00134	0.00072	0.00363	0.00097	0.00029	0.00074	0.00086	0.00463	0.00197
64	0.00057	0.00032	0.00023	0.00205	0.00163	0.00060	0.00186	0.00338	0.00064	0.00014	0.00046	0.00062	0.00099	0.00092
65	0	0	0	0	0	0	0	0	0	0	0	0	0	0
66	0.00115	0.00051	0.00034	0.00373	0.00385	0.00079	0.00341	0.06269	0.00203	0.00039	0.00063	0.00110	0.00124	0.00226
67	0	0	0	0	0	0	0	0	0	0	0	0	0	0
68	0.00057	0.00047	0.00013	0.00345	0.00135	0.00059	0.00053	0.00165	0.00028	0.00008	0.00045	0.00051	0.00040	0.00128
69	0	0	0	0	0	0	0	0	0	0	0	0	0	0
70	0	0	0	0	0	0	0	0	0	0	0	0	0	0
71	0	0	0	0	0	0	0	0	0	0	0	0	0	0
72	0.00007	0.00013	0.00005	0.00040	0.00077	0.00013	0.00025	0.00041	0.00007	0.00002	0.00006	0.00017	0.00007	0.00036
73	0.00023	0.00033	0.00040	0.00171	0.00115	0.00026	0.00013	0.01530	0.00309	0.00004	0.00023	0.00056	0.00053	0.00241
74	0.00004	0.00020	0.00004	0.00028	0.00024	0.00013	0.00017	0.00036	0.00020	0.00002	0.00021	0.00014	0.00086	0.00037
75	0.00033	0.00015	0.00012	0.00100	0.00090	0.00029	0.00067	0.00626	0.00053	0.00005	0.00022	0.00032	0.00033	0.00079
76	0.00007	0.00009	0.00001	0.00031	0.00026	0.00011	0.00019	0.00043	0.00006	0.00002	0.00005	0.00015	0.00006	0.00025
77	0.00019	0.00045	0.00014	0.00116	0.00074	0.00060	0.00028	0.00123	0.00015	0.00004	0.00053	0.00061	0.00044	0.00204
78	0.00003	0.00001	0.00002	0.00008	0.00008	0.00002	0.00011	0.00048	0.00008	0.00001	0.00002	0.00007	0.00008	0.00007
79	0.00017	0.00012	0.00009	0.00052	0.00048	0.00018	0.00092	0.00175	0.00054	0.00004	0.00012	0.00036	0.00023	0.00045
80	0.00087	0.00062	0.00061	0.00324	0.00380	0.00152	0.00163	0.00519	0.00084	0.00008	0.00062	0.00136	0.00069	0.00284
81	0.00035	0.00052	0.00008	0.00170	0.00142	0.00063	0.00104	0.00210	0.00034	0.00034	0.00025	0.00088	0.00033	0.00152
82	0.00066	0.00054	0.00061	0.00336	0.00232	0.00090	0.00408	0.00409	0.00049	0.00015	0.00060	0.00076	0.00156	0.00143
83	0.00071	0.00260	0.00148	0.00681	0.00356	0.00205	0.00482	0.01111	0.00255	0.00036	0.00242	0.00283	0.00217	0.00843
84	0.00061	0.00261	0.00403	0.01250	0.00823	0.00273	0.00130	0.03320	0.00430	0.00116	0.00324	0.00580	0.00218	0.04085

Continued

a_{ij} / b_{ij}	15	16	17	18	19	20	21	22	23	24	25	26	27	28
85	0.00016	0.00004	0	0.00061	0.00111	0.00003	0.00005	0.00042	0.00001	0	0.00023	0.00002	0.00015	0.00007
86	0.00158	0.00160	0.00138	0.00923	0.00678	0.00154	0.00102	0.01662	0.00084	0.00078	0.00021	0.00157	0.00068	0.00279
87	0.00213	0.00055	0.00009	0.01863	0.01134	0.00221	0.00145	0.02900	0.00447	0.00026	0.00326	0.00457	0.00149	0.00402
88	0.00009	0.00008	0.00139	0.00396	0.00331	0.00116	0.00217	0.00691	0.00070	0.00003	0.00083	0.00040	0.00195	0.00221
89	0.00079	0.00236	0.00207	0.01143	0.00558	0.00367	0.00106	0.01712	0.00307	0.00076	0.00363	0.00399	0.00128	0.00838
90	0.00027	0.00096	0.00022	0.00197	0.00117	0.00071	0.00073	0.00449	0.00031	0.00007	0.00036	0.00018	0.00043	0.00161
91	0.00132	0.00104	0.00142	0.01290	0.00573	0.00343	0.00154	0.00717	0.00192	0.00032	0.00208	0.00205	0.00275	0.00612
92	0.00257	0.00306	0.00311	0.01454	0.00690	0.00846	0.00280	0.01301	0.00274	0.00086	0.00250	0.02480	0.00410	0.01227
93	0.00019	0.00018	0.00023	0.00108	0.00081	0.00039	0.00012	0.00098	0.00034	0.00009	0.00012	0.00040	0.00023	0.00058
94	0.00138	0.00186	0.00184	0.01140	0.00482	0.00375	0.00186	0.00616	0.00182	0.00041	0.00130	0.00899	0.00224	0.00554
95	0.00019	0.00285	0.00039	0.00395	0.01229	0.00168	0.00068	0.00396	0.00205	0.00009	0.00214	0.00137	0.00071	0.00477
96	0.00144	0.00122	0.00044	0.00655	0.00457	0.00199	0.00274	0.00488	0.00102	0.00046	0.00137	0.00324	0.00265	0.00334
97	0.00018	0.00012	0.00040	0.00570	0.00099	0.00036	0.00110	0.00132	0.00037	0.00001	0.00012	0.00071	0.00013	0.00449
98	0	0	0	0.00004	0.00001	0.00016	0.00002	0.00001	0	0	0.00001	0.00001	0.00004	0.00002
99	0.00137	0.00104	0.00078	0.00494	0.00241	0.00055	0.00163	0.00478	0.00237	0.00011	0.00248	0.01063	0.00041	0.01050
100	0.00126	0.01522	0.00547	0.01527	0.00811	0.00466	0.00558	0.02352	0.00909	0.00077	0.00949	0.00768	0.00524	0.02195
101	0.00058	0.00142	0.00084	0.00669	0.00465	0.00317	0.00131	0.03501	0.00155	0.00058	0.00221	0.00406	0.00344	0.00871
102	0.00001	0.00001	0	0.00003	0.00002	0	0.00001	0.00003	0.00001	0	0.00002	0.00207	0	0.00007
103	0	0	0	0	0	0	0	0	0	0	0	0.00678	0	0
104	0.00001	0.00001	0.00001	0.00005	0.00003	0.00001	0.00002	0.00005	0.00002	0	0.00003	0.00121	0	0.00011
105	0.00011	0.00008	0.00006	0.00039	0.00019	0.00004	0.00013	0.00038	0.00019	0	0.00019	0.00113	0.00003	0.00082
106	0.00139	0.00187	0.00137	0.00753	0.00391	0.00147	0.00169	0.01226	0.00358	0.00055	0.00414	0.00551	0.00310	0.00714
107	0.00110	0.00086	0.00059	0.00601	0.00421	0.00228	0.00221	0.01182	0.00199	0.00047	0.00140	0.00325	0.00300	0.01088
108	0.00015	0.00310	0.00023	0.00377	0.00127	0.00312	0.00055	0.00191	0.00050	0.00017	0.00076	0.00122	0.00562	0.00848
109	0	0.00001	0	0.00002	0.00001	0	0.00001	0.00002	0.00001	0	0.00001	0.00049	0	0.00004
110	0.00136	0.00161	0.00058	0.00746	0.00441	0.00154	0.00279	0.01453	0.00136	0.00019	0.00098	0.00143	0.00132	0.00770
111	0.00011	0.00035	0.00004	0.00196	0.00045	0.00008	0.00016	0.00032	0.00019	0	0.00014	0.00331	0.00008	0.00107
112	0	0	0	0	0	0	0	0	0	0	0	0.00043	0	0
113	0.00079	0.00195	0.00115	0.00917	0.00637	0.00391	0.00180	0.04832	0.00212	0.00078	0.00303	0.00559	0.00473	0.01188
114	0.00031	0.00133	0.00041	0.03006	0.01845	0.01435	0.00128	0.00775	0.00211	0.00006	0.00070	0.00318	0.00050	0.01620
115	0.00107	0.00612	0.00054	0.00116	0.00685	0.00150	0.00140	0.00466	0.00066	0.00014	0.00034	0.00035	0.00047	0.00294
116	0	0	0	0	0	0	0	0	0	0	0	0	0	0
117	0	0	0	0	0	0	0	0	0	0	0	0	0	0
118	0.00077	0.00088	0.00036	0.00401	0.00353	0.00103	0.00147	0.00514	0.00191	0.00030	0.00258	0.00168	0.00138	0.00668
119	0.00001	0	0	0.00016	0.00004	0.00058	0.00006	0.00006	0.00001	0	0.00004	0.00115	0.00016	0.00006
120	0.00029	0.00013	0.00018	0.00316	0.00410	0.00243	0.00185	0.00072	0.00032	0.00003	0.00060	0.00026	0.00045	0.00221
121	0.00034	0.00117	0.00101	0.01061	0.00545	0.01753	0.00102	0.00404	0.00089	0.00003	0.00075	0.00538	0.00089	0.00489
122	0	0	0	0	0	0	0	0	0	0	0	0	0	0
123	0	0	0	0	0	0	0	0	0	0	0	0	0	0
124	0	0	0	0	0	0	0	0	0	0	0	0	0	0

Continued

b_{ij} \ a_{ij}	29	30	31	32	33	34	35	36	37	38	39	40	41	42
1	0.00001	0.00005	0.00118	0.01234	0	0	0	0	0	0	0	0	0.00071	0.00037
2	0.00009	0.09331	0.00883	0.02752	0	0.00006	0.01409	0	0.00004	0.00001	0.00003	0.00006	0.00077	0.00060
3	0.07477	0	0	0	0	0.00014	0.00004	0	0	0	0	0	0	0.00051
4	0	0	0	0	0	0	0	0	0	0	0	0	0.05097	0.00001
5	0.00127	0.03321	0.02817	0.09517	0.00075	0.00246	0.00629	0.00843	0.04986	0.01877	0.00005	0.00041	0.00192	0.00492
6	0.00081	0.00478	0.00118	0.01144	0	0.00016	0.00075	0.81862	0.00608	0.00218	0.04172	0.00325	0.01841	0.00285
7	0	0	0	0	0.00027	0	0.00001	0.02198	0.00276	0.01556	0.01963	0.00018	0.02744	0.00042
8	0	0	0	0.00058	0	0	0	0	0	0	0.37982	0.00130	0.16134	0.00034
9	0	0	0	0	0	0	0	0	0	0	0	0	0	0
10	0	0	0	0	0	0	0	0	0	0.03434	0.00414	0.00038	0.00350	0.00237
11	0	0	0	0	0	0	0	0	0	0.64679	0.11508	0.00902	0.02120	0.00512
12	0	0	0	0	0	0	0.00046	0.00076	0.00060	0.02640	0.06587	0.00111	0.00157	0.00056
13	0.00007	0.13965	0.01495	0.28704	0	0.00004	0.00007	0	0	0.00030	0.00051	0.00012	0.00053	0.00888
14	0	0	0	0.00022	0	0	0	0	0	0	0	0.00067	0.00643	0.00334
15	0	0	0	0	0	0	0	0	0	0.00021	0.00006	0.00010	0.00125	0.04039
16	0.44034	0	0	0	0	0	0	0	0	0	0	0	0.00094	0
17	0	0	0	0	0	0	0	0	0	0	0	0	0	0
18	0.00016	0.00012	0	0	0	0	0	0	0	0	0	0	0.00069	0.00124
19	0	0	0	0	0	0	0	0	0	0	0	0.01983	0.04590	0.01085
20	0	0	0	0	0	0	0	0	0	0	0.00186	0.00286	0	0
21	0	0	0	0	0	0	0	0	0	0	0	0	0	0
22	0.04275	0.00580	0.02536	0.00815	0.00056	0.00375	0.00844	0.00014	0.00002	0.00161	0.00236	0.00005	0.00240	0.00055
23	0.00804	0.00138	0.01074	0.04429	0.00016	0.00001	0.03688	0.00009	0.00001	0.00008	0.00569	0.00001	0.00028	0.00003
24	0.01178	0.00005	0.01595	0.00217	0.00019	0.00001	0.00048	0.00028	0.00001	0.00031	0.00080	0.00004	0.00058	0.00010
25	0.04049	0.00001	0.00173	0.02181	0.00057	0.00010	0.00180	0	0.00001	0.00009	0.00006	0.00001	0.00016	0.00002
26	0.10499	0.00006	0.09702	0.00401	0.00197	0.00033	0.02572	0.00054	0.00004	0.00072	0.00065	0.00011	0.00060	0.00013
27	0.01625	0	0.01458	0	0	0.00001	0.01575	0	0	0	0	0	0	0
28	0.03382	0.00768	0.00403	0.01098	0.00262	0.00039	0.04158	0.00701	0.00102	0.00523	0.00576	0.00102	0.00582	0.00134
29	0.72269	0.01186	0.04974	0.00093	0.00043	0.00019	0.02125	0.00256	0.00023	0.00058	0.00086	0.00087	0.00087	0.00065
30	0.00046	0.17823	0.31148	0.00786	0.00020	0.00642	0.01124	0.00094	0.00013	0.00049	0.00052	0.00023	0.00079	0.00091
31	0.00041	0.00561	0.13737	0.01492	0.00091	0.00104	0.00302	0.00122	0.00024	0.00056	0.00042	0.00020	0.00127	0.00024
32	0.00541	0.00554	0.00691	0.22021	0.12311	0.01299	0.00690	0.00023	0.00004	0.00045	0.00016	0.00221	0.00184	0.01848
33	0.00097	0.00008	0.00049	0.03491	0.06000	0.03491	0.00274	0.00049	0.00022	0.00017	0.00171	0.00094	0.00074	0.00487
34	0.00077	0.00112	0.00044	0.00259	0.00438	0.01978	0.00106	0.00134	0	0.00058	0.00065	0.00023	0.00081	0.00074
35	0	0	0	0.00045	0.00005	0.00009	0.38942	0.00001	0	0.00009	0.00005	0	0.00029	0
36	0.00125	0.00130	0.00150	0.00485	0.00138	0.00015	0.00124	0.04381	0.00071	0.00389	0.00713	0	0.03121	0.00618
37	0	0	0	0	0	0	0	0	0.00945	0.02026	0.05411	0.00097	0.01119	0
38	0.00970	0.00304	0.00059	0.06024	0.00099	0.00099	0.00200	0.00793	0.00031	0.12416	0.03619	0.00066	0.12004	0.04458
39	0	0.00278	0	0.00152	0	0	0	0	0	0.00663	0.09311	0.04511	0.00292	0.00007
40	0	0	0	0	0	0	0	0	0	0	0.00157	0.16885	0.00381	0.01427
41	0.00748	0.01492	0.00837	0.01046	0.01202	0.00146	0.00401	0.00768	0.00005	0.01756	0.00182	0.02396	0.27213	0.01979
42	0.00291	0.00267	0.00203	0.00394	0.00647	0.00520	0.00612	0.00340	0.00042	0.00341	0.01867	0.00501	0.05879	0.21269

Continued

a_{ij} / b_{ij}	29	30	31	32	33	34	35	36	37	38	39	40	41	42
43	0.00408	0.00184	0.00162	0.00197	0.00482	0.00317	0.00504	0.00853	0.00027	0.00204	0.00078	0.00180	0.02803	0.00135
44	0.00001	0	0.00002	0.00048	0.00007	0	0	0.00002	0	0.00003	0.00109	0.00043	0.00358	0.00075
45	0.00365	0.00012	0.00306	0.01229	0.00041	0.00016	0.02798	0	0.00149	0.00089	0.00120	0.00009	0.00053	0.00062
46	0.03405	0.00101	0.00203	0.00667	0.00430	0.00133	0.02923	0.00522	0.00012	0.00186	0.00435	0.00155	0.00181	0.00987
47	0.01118	0.00160	0.00685	0.00755	0.00706	0.00338	0.02106	0.00130	0.00011	0.06659	0.06094	0.00412	0.00301	0.01644
48	0.00006	0.00170	0.00019	0.00074	0.00010	0.00001	0.00015	0.00175	0.00011	0.00087	0.00050	0.00020	0.00209	0.00015
49	0.00002	0.00194	0.00054	0.00065	0.00004	0	0.00025	0.00246	0.00009	0.00091	0.00033	0.00022	0.00304	0.00078
50	0.00009	0.00053	0.00025	0.00193	0.00014	0.00002	0.00014	0.00213	0.00030	0.00366	0.00192	0.00064	0.00494	0.00084
51	0.00092	0.00046	0.00340	0.00392	0.00169	0.00220	0.00142	0.00166	0.00071	0.00169	0.00127	0.00749	0.00586	0.00512
52	0.00182	0.00132	0.00378	0.00457	0.00606	0.00039	0.00650	0.00411	0.01042	0.00275	0.00211	0.00071	0.00479	0.00082
53	0.00008	0.00205	0.00438	0.00050	0.00012	0.00003	0.00026	0.00223	0.00001	0.00682	0.00074	0.00103	0.00359	0.00041
54	0.00038	0.00261	0.00037	0.00106	0.00004	0.00044	0.00154	0.00717	0	0.01702	0.00109	0.00083	0.01315	0.00046
55	0	0	0	0	0	0	0.00063	0	0	0.00004	0.00002	0	0.00046	0.00002
56	0	0	0	0	0	0	0	0	0	0	0.00026	0.00002	0.00014	0.00001
57	0.00004	0.00010	0.00979	0.00040	0.00010	0.00009	0.00169	0.00212	0.00009	0.00018	0.00050	0.00001	0.00053	0.00004
58	0	0	0	0	0	0	0.00125	0	0	0.00030	0.00168	0.00009	0.00109	0.00015
59	0	0	0	0	0	0	0.00775	0	0.00005	0.00086	0.00083	0.00061	0.00171	0.00048
60	0.00014	0.00007	0.00492	0.00325	0.00088	0.00236	0.00630	0.00056	0.00015	0.00196	0.00107	0.00007	0.00075	0.00103
61	0.00379	0.00475	0.01542	0.00416	0.00139	0.00249	0.00120	0.00191	0.00003	0.00080	0.00131	0.00115	0.01194	0.00116
62	0.00129	0.00020	0.00001	0.00045	0.00003	0	0.00840	0.00085	0.00061	0.00151	0.00040	0.00018	0.00026	0.00008
63	0.00356	0.00087	0.00404	0.00431	0.00280	0.00013	0.00082	0.00123	0.00219	0.00036	0.00286	0.00073	0.00472	0.00054
64	0.00107	0.00097	0.00213	0.00682	0.00170	0.00010	0.00164	0.00748		0.00346	0.00870	0.00070	0.00467	0.00055
65	0	0	0	0	0	0	0	0	0	0.00321	0	0	0	0
66	0.00214	0.00317	0.00132	0.01388	0.00371	0.00015	0.00088	0.01588	0.00543	0.00892	0.01224	0.00167	0.00953	0.00094
67	0	0	0	0	0	0	0	0	0	0	0	0	0	0
68	0.00093	0.00129	0.00183	0.00743	0.00191	0.00023	0.00072	0.00463	0.00034	0.00235	0.00249	0.00104	0.00332	0.00168
69	0	0	0	0	0	0	0	0	0	0	0	0	0	0
70	0	0	0	0	0	0	0	0	0	0	0	0	0	0
71	0	0	0	0	0	0	0	0	0	0	0	0	0	0
72	0.00009	0.00015	0.00044	0.00070	0.00070	0.00002	0.00005	0.00138	0.00070	0.00044	0.00064	0.00019	0.00180	0.00018
73	0.00088	0.00040	0.00062	0.00594	0.00158	0.00006	0.00526	0.00291	0.00017	0.00253	0.00186	0.00041	0.00220	0.00032
74	0.00006	0.00008	0.00023	0.00125	0.00022	0.00002	0.00043	0.00077	0.00006	0.00032	0.00046	0.00017	0.00037	0.00007
75	0.00054	0.00073	0.00047	0.00332	0.00105	0.00005	0.00095	0.00647	0.00025	0.00237	0.00300	0.00049	0.00220	0.00031
76	0.00010	0.00011	0.00037	0.00058	0.00041	0.00002	0.00004	0.00141	0.00010	0.00031	0.00056	0.00013	0.00048	0.00012
77	0.00131	0.00062	0.00170	0.00356	0.00310	0.00013	0.00280	0.00096	0.00048	0.00202	0.00106	0.00041	0.00181	0.00042
78	0.00009	0.00005	0.00014	0.00023	0.00005	0.00001	0.01740	0.00047	0.00002	0.00014	0.00026	0.00005	0.00023	0.00004
79	0.00023	0.00027	0.00054	0.00455	0.00125	0.00008	0.00042	0.00562	0.00056	0.00072	0.00116	0.00036	0.00097	0.00023
80	0.00168	0.00094	0.00151	0.00684	0.00314	0.00024	0.00070	0.01090	0.00076	0.00867	0.01077	0.00197	0.00995	0.00131
81	0.00040	0.00066	0.00226	0.00326	0.00903	0.00012	0.00032	0.00464	0.00086	0.00173	0.00303	0.00072	0.00225	0.00069
82	0.00139	0.00088	0.00157	0.00360	0.00368	0.00020	0.00044	0.01633	0.00123	0.00359	0.00574	0.00072	0.00886	0.00129
83	0.00397	0.00263	0.00422	0.00704	0.00284	0.00066	0.00210	0.01120	0.00102	0.00400	0.00422	0.00129	0.00704	0.00266
84	0.01052	0.00248	0.01227	0.05301	0.00965	0.00073	0.00819	0.00638	0.00111	0.00551	0.00627	0.00233	0.00812	0.00233

Continued

b_{ij} \ a_{ij}	29	30	31	32	33	34	35	36	37	38	39	40	41	42
85	0.00020	0.00094	0.00006	0.17041	0.00005	0	0.00014	0.00313	0	0.00128	0.00170	0.00053	0.00026	0.00509
86	0.00141	0.00435	0.00445	0.02360	0.00270	0.00041	0.00214	0.01094	0.00585	0.04627	0.02640	0.00299	0.01880	0.00156
87	0.00139	0.00346	0.00172	0.04466	0.00169	0.00049	0.00043	0.02371	0.00981	0.03259	0.01951	0.00564	0.04127	0.00301
88	0.00038	0.00109	0.00008	0.00179	0.00053	0.00016	0.00014	0.00061	0.01625	0.01381	0.01508	0.00187	0.01367	0.03248
89	0.00261	0.00328	0.00510	0.02117	0.00297	0.00045	0.00154	0.01480	0.00129	0.01214	0.01044	0.00928	0.01731	0.00226
90	0.00170	0.00035	0.00080	0.00153	0.00200	0.00022	0.00049	0.00301	0.00014	0.00123	0.00197	0.00039	0.00219	0.00099
91	0.00452	0.00591	0.00524	0.02106	0.00366	0.00044	0.00248	0.01172	0.01034	0.01728	0.02022	0.00629	0.02181	0.00429
92	0.00758	0.00695	0.00882	0.01347	0.00417	0.00127	0.00552	0.00517	0.00326	0.01076	0.01493	0.00298	0.01384	0.00585
93	0.00051	0.00058	0.00057	0.00190	0.00058	0.00007	0.00046	0.49087	0.00403	0.00408	0.03687	0.00082	0.03860	0.00264
94	0.00481	0.00502	0.00623	0.01077	0.00246	0.00061	0.00287	0.07560	0.00662	0.01515	0.01827	0.00265	0.01638	0.00433
95	0.01276	0.00103	0.00296	0.00383	0.00183	0.00133	0.00428	0.00334	0.00029	0.00161	0.00384	0.00147	0.00369	0.00268
96	0.00278	0.00258	0.00362	0.00634	0.00146	0.00035	0.00103	0.01903	0.00731	0.00785	0.01162	0.00154	0.00829	0.00228
97	0.00005	0.00008	0.00143	0.00118	0.00060	0.00035	0.00042	0.00110	0.00005	0.00069	0.00052	0.00002	0.00439	0.00105
98	0	0	0.00003	0.00016	0.00002	0	0	0.00002	0	0.00002	0.00004	0	0.00003	0.00001
99	0.01027	0.00262	0.01032	0.00271	0.00124	0.00040	0.00324	0.00754	0.00019	0.00275	0.00205	0.00157	0.00455	0.00195
100	0.02192	0.00793	0.01267	0.01744	0.00825	0.00153	0.00512	0.01115	0.00368	0.00554	0.00858	0.00224	0.01084	0.00398
101	0.00336	0.00275	0.00584	0.01830	0.00484	0.00155	0.00214	0.00278	0.00184	0.00331	0.00475	0.00301	0.00891	0.00329
102	0.00006	0.00002	0.00019	0.00002	0.00001	0	0.00002	0.00005	0	0.00002	0.00001	0.00001	0.00003	0.00001
103	0	0	0.00060	0	0	0	0	0	0	0	0	0	0	0
104	0.00011	0.00003	0.00012	0.00003	0.00001	0	0.00003	0.00008	0.00001	0.00003	0.00002	0.00002	0.00005	0.00002
105	0.00081	0.00021	0.00018	0.00021	0.00010	0.00003	0.00025	0.00059	0.00099	0.00022	0.00016	0.00012	0.00036	0.00015
106	0.00400	0.00315	0.00411	0.00947	0.00313	0.00054	0.00182	0.01007	0.00099	0.00539	0.00876	0.00143	0.01051	0.00179
107	0.00267	0.00295	0.00441	0.01295	0.00406	0.00070	0.00162	0.01252	0.00138	0.00382	0.00608	0.00204	0.00967	0.00253
108	0.00336	0.00133	0.00907	0.00213	0.00273	0.00036	0.00468	0.00082	0.00001	0.00292	0.00035	0.00029	0.00155	0.00295
109	0.00004	0.00001	0.00005	0.00001	0.00001	0	0.00001	0.00003	0	0.00001	0.00001	0.00001	0.00002	0.00001
110	0.00239	0.00264	0.00357	0.01071	0.00417	0.00062	0.00180	0.02235	0.00278	0.00928	0.01126	0.00277	0.01400	0.00274
111	0.00075	0.00016	0.00051	0.00057	0.00025	0.00003	0.00022	0.00059	0.00001	0.00017	0.00012	0.00010	0.00054	0.00099
112	0	0	0.00004	0	0	0	0	0	0	0	0	0	0	0
113	0.00457	0.00376	0.00803	0.02517	0.00661	0.00212	0.00293	0.00378	0.00254	0.00453	0.00650	0.00413	0.01221	0.00452
114	0.00554	0.00118	0.00949	0.00452	0.00155	0.00055	0.00135	0.00262	0.00015	0.00099	0.00159	0.00232	0.00301	0.01910
115	0.00335	0.00274	0.00105	0.03291	0.00092	0.00005	0.00025	0.00569	0.00081	0.00319	0.00341	0.00120	0.00390	0.00040
116	0	0	0	0	0	0	0	0	0	0	0	0	0	0
117	0	0	0	0	0	0	0	0	0	0	0	0	0	0
118	0.00231	0.00335	0.00186	0.01101	0.00359	0.00061	0.00204	0.00443	0.00065	0.00370	0.00447	0.00235	0.00494	0.00206
119	0.00001	0.00001	0.00020	0.00058	0.00007	0	0.00008	0.00008	0	0.00006	0.00016	0	0.00012	0.00003
120	0.00273	0.00113	0.00053	0.00884	0.00117	0.00017	0.00043	0.02270	0.00040	0.00337	0.00189	0.00217	0.01050	0.00786
121	0.01458	0.00174	0.00622	0.03052	0.00287	0.00054	0.00231	0.01532	0.00053	0.00618	0.00428	0.00441	0.01784	0.00441
122	0	0	0	0	0	0	0	0	0	0	0	0	0	0
123	0	0	0	0	0	0	0	0	0	0	0	0	0	0
124	0	0	0	0	0	0	0	0	0	0	0	0	0	0

Continued

a_{ij} \ b_{ij}	43	44	45	46	47	48	49	50	51	52	53	54	55	56
1	0.00063	0.00821	0.00077	0.00006	0	0	0	0.00207	0	0	0	0	0	0
2	0.16965	0.00055	0	0.29475	0.00001	0	0.00157	0.00290	0.00051	0.00018	0.00047	0.00123	0	0
3	0	0.02389	0	0	0	0	0	0	0	0	0	0	0	0
4	0.00118	0.00218	0	0	0	0	0	0	0	0	0	0	0	0
5	0.19890	0.04684	0.00334	0.00666	0.00262	0.07427	0.00214	0.00061	0.00243	0.00061	0.00034	0.00012	0.00844	0.00643
6	0.01201	0.00160	0.00230	0.00422	0.00048	0.00017	0.00543	0.05678	0.01034	0.00615	0.01117	0.00424	0.00143	0.00124
7	0.03426	0.00030	0.00553	0.00043	0.00025	0.00080	0.00001	0.00024	0.00800	0.00065	0.00011	0.00057	0.00050	0.00178
8	0.05199	0.00127	0	0.00036	0	0	0.00003	0.00162	0.02959	0.03063	0.00053	0.02143	0.16534	0.07498
9	0	0	0	0	0	0	0	0.00012	0	0	0	0.00097	0.00075	0.01178
10	0.00918	0	0	0	0	0	0	0.00211	0	0	0	0.00398	0	0
11	0.03233	0.00248	0	0	0	0.00691	0.00044	0.06520	0.00217	0.00059	0.00300	0.00077	0.01690	0.00478
12	0.00468	0.00017	0	0	0	0.09692	0.04367	0.12840	0.03170	0.09637	0.00072	0.02932	0.00043	0.00037
13	0.00052	0.00023	0	0	0	0.00076	0.00050	0.00055	0.00086	0	0.03622	0	0	0
14	0.00094	0.00602	0	0	0	0	0	0	0	0	0	0	0	0
15	0.00142	0.04018	0	0	0	0	0	0	0	0	0	0	0	0
16	0.00010	0.00527	0	0	0	0	0	0	0	0	0	0	0	0
17	0.00011	0.01878	0	0	0	0	0	0	0	0	0	0	0	0
18	0.00236	0.04878	0.00196	0.00010	0.00045	0.00370	0	0	0	0	0	0.00151	0	0
19	0.01799	0.04948	0.00071	0.00114	0.00046	0	0	0	0.00074	0.00526	0.00246	0	0	0
20	0	0	0	0	0	0	0	0	0	0	0	0	0	0
21	0	0	0	0	0	0	0	0	0	0	0	0	0	0
22	0.00757	0.00192	0.00537	0.06210	0.03182	0.00386	0.00388	0.00083	0.00122	0.00320	0.00037	0.00330	0.00008	0.00005
23	0.00011	0.00091	0.00214	0.00076	0.01941	0.00386	0.00980	0.00153	0.00012	0.00091	0.00070	0.00019	0.00009	0.00005
24	0.00033	0.00025	0.00168	0.00065	0.00086	0.00276	0.00045	0.00057	0.00159	0.00117	0.00006	0.00022	0.00021	0.00013
25	0.00006	0.00007	0.00168	0.00058	0.00010	0.00011	0.00010	0.00007	0.00010	0.00091	0.00005	0.00003	0.00005	0.00002
26	0.00048	0.00208	0.00114	0.08930	0.07793	0.00093	0.00031	0.00039	0.00033	0.00027	0.00496	0.00013	0.00038	0.00029
27	0	0.00510	0.06434	0.01262	0.01187	0	0.00296	0.00229	0	0	0.00078	0.00104	0	0
28	0.00484	0.00382	0.00161	0.00447	0.00598	0.01378	0.00257	0.01253	0.00300	0.00171	0.00433	0.00141	0.00200	0.00291
29	0.00104	0.00108	0.00027	0.01530	0.00503	0.00435	0.00050	0.00370	0.00070	0.00046	0.00490	0.00043	0.00030	0.00039
30	0.00126	0.00037	0.00024	0.00030	0.00055	0.00093	0.00104	0.00239	0.00413	0.00129	0.00082	0.00035	0.00055	0.00044
31	0.00103	0.00097	0.00029	0.00047	0.00225	0.00089	0.00049	0.00826	0.00465	0.00359	0.00163	0.00659	0.00025	0.00018
32	0.00522	0.01826	0.00764	0.00392	0.01496	0.07034	0.00028	0.05548	0.00711	0.00969	0.00937	0.00097	0.00003	0.00016
33	0.00073	0.02143	0.00018	0.00032	0.00116	0.00072	0.00006	0.00031	0.00014	0.00113	0.00011	0.00090	0.00019	0.00007
34	0.00064	0.00389	0.00036	0.00151	0.00182	0.00146	0.00089	0.00165	0.00120	0.00499	0.00012	0.00041	0.00058	0.00044
35	0.00002	0	0	0	0.00200	0.00017	0.00001	0.00004	0	0.00052	0	0.00127	0	0
36	0.02009	0.00065	0.00434	0.00335	0.00573	0.06626	0.00197	0.03438	0.01152	0.01070	0.00200	0.00445	0.00192	0.00334
37	0.00827	0	0	0.00259	0	0	0	0.01188	0	0	0.00378	0.07533	0.35905	0.06818
38	0.09454	0.03383	0.01238	0.02123	0.01070	0.00245	0.00066	0.04603	0.03843	0.00851	0.00134	0.00623	0.00033	0.00201
39	0.01198	0.00071	0	0	0	0	0	0	0	0	0	0	0	0
40	0.00091	0.00070	0	0	0	0	0	0	0	0	0	0	0	0
41	0.06746	0.02866	0.05097	0.00936	0.08556	0.00181	0.00082	0.00369	0.00257	0.00660	0.00882	0.00125	0.00012	0.00023
42	0.04290	0.01205	0.00740	0.01121	0.01640	0.00783	0.00096	0.07164	0.00212	0.00103	0.00241	0.01886	0.00606	0.00068

Continued

a_{ij} \ b_{ij}	43	44	45	46	47	48	49	50	51	52	53	54	55	56
43	0.07061	0.00430	0.10190	0.02970	0.27821	0.00246	0.00029	0.03508	0.00391	0.00116	0.00089	0.00376	0.00330	0.00043
44	0.00033	0.26400	0.00003	0.00001	0.00004	0.00002	0	0.00003	0.00002	0.00003	0	0	0.00001	0.00001
45	0.00138	0.00107	0.15193	0.02364	0.00814	0.00036	0.00031	0.00430	0.00010	0.00010	0.00007	0.00207	0	0
46	0.00886	0.01079	0.00164	0.08746	0.00801	0.01143	0.00208	0.00644	0.00292	0.00149	0.00305	0.00147	0.00168	0.00194
47	0.01078	0.00716	0.00376	0.00234	0.21222	0.03559	0.00101	0.00945	0.00316	0.00228	0.00204	0.00193	0.00022	0.00032
48	0.00098	0.00037	0.00025	0.00028	0.00036	0.02836	0.12104	0.02646	0.00038	0.00251	0.00279	0.00175	0.00031	0.00034
49	0.00106	0.00034	0.00047	0.00035	0.00026	0.01151	0.01849	0.00671	0.00037	0.00272	0.00165	0.00132	0.00045	0.00047
50	0.00153	0.00101	0.00077	0.00025	0.00044	0.00855	0.00467	0.09911	0.00194	0.00184	0.00312	0.00225	0.00075	0.00173
51	0.00785	0.02744	0.00055	0.00094	0.00252	0.00274	0.00110	0.00776	0.07856	0.00122	0.00188	0.00136	0.00048	0.00072
52	0.00360	0.00374	0.00129	0.00249	0.00453	0.01123	0.00176	0.00807	0.00470	0.02808	0.01493	0.00230	0.00799	0.00142
53	0.00134	0.00023	0.00062	0.00002	0.00008	0.06039	0.01457	0.03426	0.02603	0.01417	0.04787	0.00632	0.02569	0.05418
54	0.00559	0.00083	0.00106	0.00043	0.00484	0.02422	0.11288	0.03586	0.01355	0.00862	0.06774	0.19581	0.00148	0.01057
55	0.00002	0.00001	0	0.00088	0.00029	0.00937	0.00006	0.00017	0.00005	0.00005	0.00002	0.00006	0.00642	0.61462
56	0.00010	0.00009	0.00007	0.00008	0.00004	0.00428	0.00396	0.00160	0.00010	0.00012	0.00001	0.02846	0.01511	0.04185
57	0.00033	0.00088	0.00017	0.00197	0.00031	0.00865	0.01899	0.01314	0.00017	0.00026	0.00029	0.00018	0.00044	0.00159
58	0.00055	0.00041	0.00119	0.00082	0.00073	0.00735	0.00176	0.00157	0.00036	0.00119	0.00044	0.00178	0.00160	0.08351
59	0.00761	0.00008	0.00007	0.00020	0.00008	0.00008	0.00005	0.00064	0.00128	0.00020	0.00005	0.06274	0.00003	0.02807
60	0.00298	0.00053	0.00024	0.00120	0.00079	0.00174	0.00126	0.00165	0.00076	0.00039	0.00069	0.00126	0.00074	0.00071
61	0.00356	0.00222	0.00060	0.00655	0.00387	0.02896	0.01853	0.00534	0.00562	0.00201	0.00110	0.00145	0.00044	0.01456
62	0.00100	0.00114	0.00006	0.00006	0.00001	0.00020	0.00002	0.00008	0.00039	0.00007	0.00002	0.00008	0.00003	0.00011
63	0.00338	0.00156	0.00164	0.01754	0.00769	0.02380	0.00225	0.01168	0.00282	0.00709	0.00130	0.00316	0.00480	0.01815
64	0.00490	0.00176	0.00150	0.00130	0.00319	0.04660	0.00234	0.00278	0.00149	0.00255	0.00117	0.03421	0.00171	0.00352
65	0	0	0	0	0	0	0	0	0	0	0	0	0	0
66	0.01116	0.00282	0.00319	0.00283	0.00465	0.01945	0.00240	0.00339	0.00386	0.00375	0.00277	0.00404	0.00401	0.00779
67	0	0	0	0	0	0	0	0	0	0	0	0	0.00055	0.00196
68	0.00305	0.00153	0.00133	0.00142	0.00376	0.00493	0.00107	0.00226	0.00145	0.00162	0.00141	0.00083	0.00158	0.00237
69	0	0	0	0	0	0	0	0	0	0	0	0	0	0
70	0	0	0	0	0	0	0	0	0	0	0	0	0	0
71	0	0	0	0	0	0	0	0	0	0	0	0	0	0
72	0.00083	0.00038	0.00026	0.00033	0.00065	0.00136	0.00022	0.00166	0.00033	0.00028	0.00013	0.00022	0.00101	0.00036
73	0.00460	0.00192	0.00123	0.00112	0.01218	0.00639	0.00109	0.00144	0.00130	0.00040	0.00069	0.01029	0.00103	0.00197
74	0.00038	0.00025	0.00016	0.00014	0.00041	0.00129	0.00016	0.00040	0.00020	0.00051	0.00007	0.00016	0.00011	0.00025
75	0.00286	0.00090	0.00077	0.00086	0.00194	0.00439	0.00087	0.00169	0.00102	0.00125	0.00095	0.00189	0.00115	0.00164
76	0.00052	0.00069	0.00023	0.00020	0.00054	0.00111	0.00018	0.00030	0.00029	0.00016	0.00011	0.00012	0.00013	0.00031
77	0.00160	0.00121	0.00067	0.00166	0.00428	0.00355	0.00064	0.00575	0.00258	0.00183	0.00131	0.00167	0.00069	0.00103
78	0.00034	0.00020	0.00015	0.00013	0.00027	0.00116	0.00019	0.00012	0.00028	0.00024	0.00005	0.00015	0.00036	0.00014
79	0.00093	0.00106	0.00072	0.00050	0.00126	0.00252	0.00077	0.00057	0.00049	0.00083	0.00022	0.00041	0.00131	0.00080
80	0.01040	0.00565	0.00350	0.00381	0.00847	0.01216	0.00144	0.00579	0.00329	0.00279	0.00205	0.00242	0.00229	0.00249
81	0.00261	0.00192	0.00128	0.00177	0.00336	0.00610	0.00108	0.00163	0.00167	0.00126	0.00066	0.00074	0.00056	0.00109
82	0.00835	0.00387	0.00263	0.00197	0.00485	0.00496	0.00402	0.00384	0.00133	0.00199	0.00089	0.00166	0.00133	0.00285
83	0.00604	0.00454	0.00346	0.00376	0.00856	0.00735	0.00394	0.00821	0.00216	0.00282	0.00172	0.00250	0.00204	0.00258
84	0.00838	0.00867	0.00368	0.00743	0.00853	0.01500	0.00980	0.02380	0.00808	0.00826	0.00476	0.00387	0.00330	0.01243

Continued

a_{ij} / b_{ij}	43	44	45	46	47	48	49	50	51	52	53	54	55	56
85	0.00062	0.00002	0.00022	0.00122	0.00724	0.03202	0.00098	0.01084	0.00662	0.00003	0.00067	0.00371	0.09798	0.11624
86	0.02159	0.00912	0.00898	0.00680	0.01713	0.05372	0.00731	0.02141	0.00608	0.00529	0.00708	0.00872	0.00657	0.01784
87	0.03289	0.01511	0.00939	0.01962	0.00395	0.00155	0.00186	0.00135	0.00161	0.00530	0.00136	0.00214	0.00192	0.00146
88	0.01578	0.00226	0.00058	0.00934	0.00724	0.00121	0.00040	0.00220	0.02885	0.02272	0.00107	0.00108	0.00255	0.01041
89	0.01151	0.00914	0.00464	0.00472	0.00952	0.00596	0.00744	0.00619	0.00326	0.00445	0.00134	0.00302	0.00263	0.00853
90	0.00235	0.00137	0.00065	0.00062	0.00128	0.00186	0.00035	0.00213	0.00046	0.00063	0.00037	0.00069	0.00029	0.00040
91	0.01296	0.00484	0.00448	0.00581	0.01627	0.03383	0.02202	0.04829	0.00775	0.01521	0.00812	0.00817	0.01711	0.00728
92	0.01157	0.00756	0.00467	0.00775	0.01619	0.03418	0.01605	0.02587	0.00806	0.01073	0.00642	0.00952	0.01292	0.00584
93	0.03055	0.00084	0.00360	0.00153	0.00302	0.00289	0.00091	0.01256	0.00637	0.00635	0.00106	0.00331	0.00265	0.00245
94	0.01317	0.00385	0.00355	0.00457	0.01110	0.02782	0.01181	0.03062	0.00765	0.01257	0.00644	0.00702	0.03520	0.01680
95	0.00356	0.01481	0.00160	0.00181	0.01214	0.00625	0.00258	0.00352	0.01022	0.00122	0.00148	0.00271	0.00054	0.00098
96	0.00688	0.00278	0.00268	0.00259	0.00648	0.02271	0.00605	0.01862	0.00446	0.00519	0.00432	0.00423	0.00566	0.00569
97	0.00110	0.00124	0.00051	0.00114	0.00097	0.00111	0.00003	0.00117	0.00117	0.00011	0.01293	0.00022	0.00010	0.00032
98	0.00002	0.00019	0.00001	0.00001	0.00011	0.00003	0.00001	0.00006	0.00019	0.00001	0.00005	0.00002	0.00006	
99	0.01292	0.00285	0.00176	0.00299	0.01148	0.00137	0.00186	0.00380	0.00171	0.00118	0.00747	0.00546	0.00127	0.00083
100	0.00952	0.01102	0.00599	0.00769	0.01440	0.01699	0.00856	0.01710	0.00487	0.00482	0.00407	0.00460	0.00374	0.00287
101	0.00637	0.00743	0.00260	0.00418	0.01056	0.01019	0.00973	0.00990	0.00331	0.00343	0.00397	0.00807	0.00121	0.00257
102	0.00018	0.00002	0.00001	0.00002	0.00007	0.00001	0.00001	0	0.00001	0.00001	0.00005	0.00003	0	0.00001
103	0.00048	0	0	0	0	0	0	0	0	0	0	0	0	0
104	0.00014	0.00003	0.00002	0.00003	0.00012	0.00001	0.00002	0.00003	0.00002	0.00001	0.00008	0.00006	0	0.00001
105	0.00050	0.00022	0.00014	0.00023	0.00090	0.00011	0.00015	0.01363	0.00013	0.00009	0.00059	0.00043	0.00001	0.00007
106	0.00830	0.00529	0.00512	0.00396	0.01070	0.01262	0.00324	0.00559	0.00382	0.00488	0.00258	0.00531	0.00216	0.00274
107	0.00778	0.00626	0.00311	0.00480	0.00982	0.00660	0.00313	0.00373	0.00299	0.00208	0.00173	0.00251	0.00130	0.00189
108	0.00228	0.00449	0.00082	0.00262	0.00790	0.00069	0.00052	0	0.00114	0.00292	0.00101	0.00039	0.00012	0.00027
109	0.00006	0.00001	0.00001	0.00001	0.00005	0.00001	0.00001	0	0.00001	0	0.00003	0.00002	0	0
110	0.01304	0.00493	0.00374	0.00395	0.00726	0.01151	0.00428	0.00805	0.00464	0.00474	0.00149	0.00390	0.00314	0.00605
111	0.00075	0.00133	0.00012	0.00045	0.00065	0.00063	0.00069	0.01292	0.00012	0.00085	0.00061	0.00058	0.00002	0.00006
112	0.00003	0	0	0	0	0	0	0	0	0	0	0	0	0
113	0.00873	0.01020	0.00357	0.00563	0.01442	0.01401	0.01336	0.01346	0.00452	0.00471	0.00540	0.01098	0.00165	0.00354
114	0.00272	0.02289	0.00042	0.00158	0.00426	0.00286	0.00070	0.00490	0.00153	0.00173	0.00224	0.00095	0.00016	0.00042
115	0.00291	0.00158	0.00041	0.00281	0.00509	0.00501	0.00179	0.00794	0.00145	0.00108	0.00084	0.00207	0.00287	0.00101
116	0	0	0	0	0	0	0	0	0	0	0	0	0	0
117	0	0	0	0	0	0	0	0	0	0	0	0	0	0
118	0.00476	0.00422	0.00146	0.00421	0.00459	0.00824	0.00349	0.00440	0.00188	0.00194	0.00254	0.00233	0.00138	0.00189
119	0.00014	0.00069	0.00004	0.00004	0.00041	0.00010	0.00005	0.00024	0.00070	0.00002	0.00020	0.00006	0.00021	0.00001
120	0.00719	0.01186	0.00112	0.00216	0.01589	0.00105	0.00023	0.00074	0.00187	0.00192	0.00058	0.00209	0.00015	0.00072
121	0.01292	0.00858	0.00360	0.00531	0.00997	0.01541	0.00245	0.00467	0.00128	0.00351	0.00233	0.00164	0.00137	0.00293
122	0	0	0	0	0	0	0	0	0	0	0	0	0	0
123	0	0	0	0	0	0	0	0	0	0	0	0	0	0
124	0	0	0	0	0	0	0	0	0	0	0	0	0	0

Continued

a_{ij} / b_{ij}	57	58	59	60	.61	62	63	64	65	66	67	68	69	70
1	0	0	0	0	0	0	0	0	0	0	0	0	0	0
2	0.00453	0	0	0	0.00312	0	0	0.00045	0.00022	0.00138	0	0.00269	0.00039	0
3	0	0	0	0	0	0	0	0	0	0	0	0	0	0
4	0	0	0	0	0	0	0	0	0	0	0	0	0	0
5	0	0.00182	0.01090	0.00127	0.00299	0.00112	0.00104	0	0.00155	0.00323	0.00112	0.00405	0.00013	0.00020
6	0.05504	0.00007	0.00105	0.00009	0.00579	0.00004	0.00017	0.00904	0.00006	0.00480	0.00004	0.00014	0.00002	0.00002
7	0.01344	0.00004	0.00409	0.00094	0.00010	0.00554	0.00013	0.00019	0.00100	0.00020	0.00147	0.01141	0.00004	0.00026
8	0.04952	0.05554	0.00940	0.00036	0.00405	0.00650	0.00118	0.00771	0.00243	0.00704	0.00153	0.00591	0.00065	0.00016
9	0.36479	0.07145	0.45386	0.09933	0.15400	0.00472	0.00049	0.01330	0.00032	0.00911	0.00031	0.00073	0.00010	0.00013
10	0.01352	0	0	0	0.01025	0	0	0.03104	0	0.00346	0	0	0	0
11	0	0.00304	0.00268	0.00018	0	0	0	0	0	0	0	0	0	0
12	0.01913	0.00304	0.00268	0.00018	0.01212	0.00085	0.00067	0.00306	0.00046	0.00076	0.00063	0.00066	0.00050	0.00002
13	0.00203	0.00030	0.00156	0.00037	0.00311	0.00093	0.00161	0.00188	0.00018	0.00109	0.00127	0.00061	0.00170	0.00010
14	0	0	0	0	0	0	0	0	0	0	0	0	0	0
15	0	0	0	0	0	0	0	0	0	0	0	0	0	0
16	0	0	0	0	0	0	0	0	0	0	0	0	0	0
17	0	0	0	0	0	0	0	0	0	0	0	0	0	0
18	0	0	0	0	0	0	0	0	0	0	0	0	0	0
19	0	0	0	0	0	0	0	0	0	0	0	0	0	0
20	0	0	0	0	0	0	0	0	0	0	0	0	0	0
21	0	0	0	0	0	0	0	0	0	0	0	0	0	0
22	0.00086	0.00002	0.00035	0.00048	0.00269	0.00046	0.00045	0.00319	0.00044	0.00744	0.00011	0.00367	0.00085	0.00003
23	0.00044	0.00001	0.00015	0.00054	0.00217	0.00054	0.00036	0.00183	0.00005	0.00183	0.00008	0.01008	0.00085	0.00003
24	0.00107	0.00030	0.00035	0.00033	0.01831	0.00007	0.00006	0.00060	0.00016	0.00055	0.00005	0.00093	0.00073	0.00007
25	0.00022	0	0.00008	0.00060	0.00128	0.00002	0.00002	0.00130	0.00002	0.00013	0.00001	0.00112	0.00006	0.00031
26	0.00168	0.00019	0.00069	0.00046	0.00265	0.00031	0.00016	0.00313	0.00023	0.00116	0.00020	0.01442	0.00021	0.00155
27	0	0	0	0	0	0	0	0	0	0.00987	0	0.00348	0.00019	0.00002
28	0.01408	0.00092	0.00495	0.00230	0.01469	0.00272	0.00134	0.01357	0.00266	0.00901	0.00125	0.00834	0.00143	0.00054
29	0.00205	0.00030	0.00052	0.00048	0.00289	0.00064	0.00030	0.00442	0.00405	0.00283	0.00010	0.01255	0.00017	0.00004
30	0.00326	0.00071	0.00105	0.00059	0.01353	0.00102	0.00105	0.00460	0.00060	0.00488	0.00159	0.00749	0.00146	0.00008
31	0.00105	0.00044	0.00156	0.00204	0.06733	0.00254	0.00221	0.01061	0.00151	0.01015	0.00024	0.00706	0.00060	0.00017
32	0.00072	0.00003	0.00034	0.00128	0.01277	0.00044	0.00014	0.00942	0.00015	0.00292	0.00003	0.00379	0.00002	0.00003
33	0.00067	0.00019	0.00019	0.00039	0.00209	0.00036	0.00016	0.00220	0.00032	0.00171	0.00007	0.00171	0.00008	0.00002
34	0.00324	0.00013	0.00076	0.00263	0.00357	0.00055	0.00043	0.00318	0.00273	0.00279	0.00024	0.00261	0.00034	0.00021
35	0.00005	0.00020	0.00032	0.00003	0.00090	0.00005	0.00004	0.00005	0.00002	0.00009	0.00018	0.00035	0.00004	0
36	0.02579	0.00041	0.00571	0.00325	0.00904	0.00228	0.00179	0.00859	0.00195	0.00643	0.00091	0.00754	0.00099	0.00035
37	0.18278	0.02246	0.01727	0.00172	0.02833	0.00728	0.00748	0.02373	0.00463	0.00838	0.00169	0.00626	0.00085	0.00003
38	0.00879	0.01292	0.02962	0.00526	0.01222	0.00110	0.00064	0.00625	0.00079	0.00407	0.00132	0.00836	0.00131	0.00343
39	0	0	0	0	0	0	0	0	0	0	0	0	0	0
40	0	0	0	0	0	0	0	0	0	0	0	0	0	0
41	0.00294	0.00059	0.00110	0.00145	0.01948	0.00093	0.00120	0.05821	0.00321	0.00382	0.00114	0.01157	0.00242	0.00036
42	0.00647	0.00018	0.00347	0.00096	0.00735	0.00125	0.00062	0.00779	0.00161	0.00700	0.00027	0.00349	0.00021	0.00012

Continued

a_{ij} / b_{ij}	57	58	59	60	61	62	63	64	65	66	67	68	69	70
43	0.00294	0.00023	0.00321	0.00013	0.00517	0.00040	0.00029	0.00265	0.00046	0.00396	0.00069	0.00220	0.00026	0.00035
44	0.00008	0	0.00002	0	0.00002	0.00001	0.00003	0.00005	0.00001	0.00012	0	0.00007	0.00013	0
45	0	0	0	0	0.01966	0.00320	0.00010	0.00034	0.00004	0.00405	0.00010	0.00022	0.00010	0.00002
46	0.01140	0.00040	0.00274	0.00089	0.02384	0.00384	0.00375	0.04663	0.04171	0.02299	0.00302	0.09452	0.00176	0.00051
47	0.00245	0.00037	0.00117	0.00073	0.01014	0.00115	0.00408	0.02985	0.00101	0.00731	0.00066	0.02078	0.00022	0.00010
48	0.00212	0.00227	0.00126	0.00033	0.00244	0.00372	0.00012	0.00074	0.00015	0.00086	0.00012	0.00088	0.00014	0.00003
49	0.00228	0.00033	0.00072	0.00045	0.00140	0.00022	0.00004	0.00062	0.00006	0.00061	0.00080	0.00044	0.00009	0.00001
50	0.00414	0.00062	0.00116	0.00027	0.00832	0.00057	0.00022	0.00112	0.00021	0.00135	0.00026	0.00117	0.00043	0.00004
51	0.00475	0.00031	0.00220	0.00165	0.03701	0.00143	0.00726	0.00682	0.00248	0.00475	0.00050	0.03022	0.00237	0.00025
52	0.00817	0.00061	0.00233	0.00138	0.02249	0.00317	0.00129	0.01142	0.00233	0.00805	0.00051	0.00582	0.00108	0.00025
53	0.12679	0.00239	0.00690	0.00151	0.01190	0.01112	0.06273	0.01060	0.00097	0.00656	0.00202	0.01268	0.00390	0.00007
54	0.01989	0.00597	0.01777	0.00112	0.03871	0.00147	0.00143	0.00934	0.00238	0.01026	0.00114	0.01386	0.00110	0.00173
55	0.13034	0.00018	0.00022	0.00006	0.10662	0.00749	0.00682	0.04226	0.02538	0.01361	0.00164	0.00692	0.00604	0.00001
56	0.39012	0.00158	0.00113	0.00023	0.12087	0.00603	0.02581	0.06859	0.00960	0.05612	0.00996	0.10984	0.00203	0.00011
57	0.15964	0.00270	0.00049	0.00048	0.21797	0.01590	0.00893	0.06181	0.01745	0.07184	0.00642	0.02978	0.00873	0.00049
58	0.34737	0.10761	0.01265	0.00303	0.08853	0.03610	0.01874	0.11208	0.00544	0.04055	0.00311	0.01089	0.00208	0.00012
59	0.01685	0.01301	0.09933	0.30490	0.12822	0.00454	0.01051	0.02356	0.00645	0.01395	0.00059	0.01505	0.00163	0.00025
60	0.00594	0.00028	0.00604	0.03324	0.12125	0.00618	0.00244	0.02317	0.01358	0.02759	0.00225	0.01592	0.03617	0.00165
61	0.01245	0.00064	0.00348	0.00151	0.15365	0.00392	0.02232	0.03728	0.05433	0.02369	0.00167	0.00879	0.00181	0.00028
62	0.00071	0.00002	0.00048	0.00002	0.00034	0.22179	0.00114	0.05550	0.13842	0.02291	0.00560	0.10350	0.04798	0.00275
63	0.07035	0.00085	0.00435	0.00270	0.08824	0.03612	0.09774	0.05550	0.08068	0.04681	0.00410	0.03574	0.00290	0.00060
64	0.02340	0.00079	0.00398	0.00125	0.01126	0.02383	0.02774	0.13070	0.02917	0.05052	0.00478	0.02977	0.00958	0.00175
65	0	0	0	0	0	0	0.00089	0.00815	0.42625	0.00608	0	0	0.00275	0
66	0.05885	0.00193	0.00840	0.00314	0.01322	0.00173	0.00106	0.01064	0.00382	0.12168	0.00030	0.00536	0.01110	0.00064
67	0.01599	0.00003	0.00191	0.00004	0.00050	0.00016	0.00588	0.00536	0.00262	0.00012	0.28163	0.00209	0.00001	0
68	0.01476	0.00072	0.00194	0.00103	0.00930	0.00432	0.00085	0.01178	0.00689	0.00588	0.00094	0.47151	0.00200	0.00035
69	0	0	0	0	0	0	0	0	0	0	0	0	0.18868	0
70	0	0	0	0	0	0	0	0	0	0	0	0	0	0.21979
71	0	0	0	0	0	0	0	0	0	0	0	0	0	0
72	0.00168	0.00012	0.00083	0.00038	0.00461	0.00201	0.00017	0.00149	0.00126	0.00683	0.00279	0.00723	0.00822	0.00069
73	0.00861	0.00012	0.00159	0.00264	0.02694	0.02471	0.01513	0.22471	0.00910	0.14773	0.01919	0.02836	0.00931	0.00031
74	0.00116	0.00007	0.00036	0.00013	0.00314	0.00274	0.00029	0.00432	0.00052	0.01673	0.00392	0.01639	0.00281	0.00023
75	0.01413	0.00044	0.00216	0.00077	0.00559	0.00548	0.01430	0.01856	0.00613	0.01300	0.00438	0.01110	0.00408	0.00044
76	0.00167	0.00005	0.00040	0.00017	0.00131	0.00040	0.00085	0.00107	0.00028	0.00442	0.00016	0.00265	0.00012	0.00030
77	0.00688	0.00044	0.00141	0.00088	0.01032	0.00190	0.00126	0.00721	0.00171	0.00479	0.00046	0.00944	0.00085	0.00031
78	0.00115	0.00009	0.00047	0.00041	0.00127	0.00062	0.00333	0.03425	0.00019	0.02854	0.00061	0.00158	0.00012	0.00037
79	0.00531	0.00014	0.00102	0.00102	0.00611	0.00066	0.00067	0.00254	0.00052	0.02067	0.00026	0.00189	0.01187	0.00152
80	0.01638	0.00071	0.00370	0.00189	0.01557	0.00870	0.00782	0.01472	0.00370	0.01649	0.00223	0.02376	0.00675	0.00372
81	0.00626	0.00025	0.00175	0.00078	0.00768	0.00209	0.00086	0.00558	0.00201	0.00505	0.00092	0.00305	0.00042	0.00063
82	0.02052	0.00049	0.00509	0.00164	0.00847	0.00145	0.00080	0.00515	0.00101	0.00544	0.00103	0.00493	0.00086	0.00015
83	0.01565	0.00120	0.00468	0.00340	0.02139	0.00299	0.00211	0.01266	0.00330	0.00698	0.00097	0.01196	0.00129	0.00036
84	0.05493	0.00206	0.00815	0.00348	0.04171	0.00282	0.00722	0.03137	0.00318	0.01108	0.00260	0.01822	0.00191	0.00040

Continued

a_{ij} / b_{ij}	57	58	59	60	61	62	63	64	65	66	67	68	69	70
85	0.16868	0.00290	0.21359	0.00520	0.01413	0.00152	0.00119	0.05142	0.00142	0.00651	0.00133	0.00230	0.00002	0.00001
86	0.04320	0.01004	0.02659	0.01094	0.04856	0.00273	0.00216	0.01663	0.00208	0.01170	0.00289	0.00217	0.00002	0.00001
87	0.01563	0.00098	0.00633	0.00150	0.01379	0.00209	0.00215	0.00564	0.00171	0.00879	0.00212	0.00859	0.00120	0.00036
88	0.09978	0.00472	0.00535	0.00667	0.01604	0.00436	0.00132	0.01236	0.00200	0.01092	0.00238	0.00738	0.00120	0.00097
89	0.02209	0.00131	0.00512	0.00305	0.02166	0.00337	0.00154	0.01228	0.00155	0.00908	0.00118	0.00784	0.00070	0.00034
90	0.00345	0.00015	0.00162	0.00050	0.00426	0.00116	0.00088	0.00351	0.00100	0.00644	0.00053	0.00314	0.00021	0.00021
91	0.04893	0.00497	0.01132	0.00435	0.03145	0.00308	0.00261	0.01672	0.00323	0.01355	0.00175	0.01009	0.00127	0.00035
92	0.02214	0.00356	0.01109	0.00371	0.03912	0.00316	0.00233	0.01903	0.00401	0.01463	0.00096	0.01506	0.00100	0.00033
93	0.02189	0.00055	0.00293	0.00123	0.00455	0.00118	0.00073	0.00363	0.00079	0.00280	0.00044	0.00345	0.00038	0.00015
94	0.08721	0.01130	0.00825	0.00183	0.04836	0.00320	0.00194	0.01447	0.00287	0.01135	0.00105	0.01527	0.00104	0.00025
95	0.00827	0.00035	0.00152	0.00117	0.01707	0.00255	0.00241	0.01350	0.00339	0.01103	0.00091	0.03695	0.00206	0.00054
96	0.02510	0.00265	0.00760	0.00335	0.03419	0.00223	0.00173	0.00934	0.00207	0.00742	0.00070	0.00589	0.00096	0.00020
97	0.00255	0.00013	0.00030	0.00005	0.00090	0.00429	0.00024	0.00144	0.00030	0.00160	0.00026	0.00389	0.00001	0.00014
98	0.00001	0	0.00001	0.00001	0.00015	0.00013	0.00001	0.00008	0	0.00017	0	0.00005	0	0
99	0.01472	0.00197	0.02209	0.00113	0.06185	0.00876	0.00195	0.01781	0.01168	0.01442	0.00259	0.01777	0.00179	0.00285
100	0.01717	0.00130	0.00463	0.00278	0.02125	0.00240	0.00178	0.01170	0.00312	0.00870	0.00092	0.01184	0.00121	0.00030
101	0.01011	0.00131	0.00295	0.00406	0.03170	0.00461	0.00364	0.02615	0.00279	0.01489	0.00088	0.00672	0.00231	0.00028
102	0.00009	0.00001	0.00424	0.00001	0.00057	0.00005	0.00001	0.00011	0.00018	0.00009	0.00002	0.00011	0.00001	0.00002
103	0	0	0.01381	0	0.00091	0	0	0	0.00052	0	0	0	0	0
104	0.00015	0.00002	0.00250	0.00001	0.00066	0.00009	0.00002	0.00019	0.00013	0.00015	0.00003	0.00018	0.00002	0.00003
105	0.00116	0.00015	0.00264	0.00009	0.00388	0.00069	0.00015	0.00140	0.00036	0.00113	0.00020	0.00139	0.00014	0.00022
106	0.02092	0.00153	0.00747	0.00564	0.07756	0.00387	0.00439	0.01922	0.00221	0.01484	0.00067	0.01175	0.00155	0.00069
107	0.01196	0.00095	0.00438	0.00342	0.01580	0.00242	0.00226	0.01157	0.00212	0.01002	0.00074	0.00637	0.00113	0.00049
108	0.00144	0.00015	0.00065	0.00103	0.01494	0.00087	0.00060	0.00480	0.00046	0.00510	0.00005	0.00306	0.00009	0.00011
109	0.00006	0.00001	0.00102	0	0.00026	0.00004	0.00001	0.00007	0.00005	0.00006	0.00001	0.00007	0.00001	0.00001
110	0.04812	0.00222	0.00708	0.00275	0.03994	0.00487	0.00233	0.01202	0.00248	0.01246	0.00308	0.01142	0.00156	0.00093
111	0.00261	0.00019	0.00710	0.00017	0.00657	0.00067	0.00013	0.00635	0.00053	0.00523	0.00014	0.00121	0.00012	0.00017
112	0	0	0.00088	0	0.00006	0	0	0	0.00003	0	0	0	0	0
113	0.01381	0.00175	0.00400	0.00558	0.04302	0.00630	0.00500	0.03564	0.00375	0.02035	0.00120	0.00901	0.00318	0.00037
114	0.00353	0.00010	0.00078	0.00123	0.01620	0.00204	0.00124	0.01177	0.00169	0.01176	0.00042	0.00747	0.00023	0.00018
115	0.02496	0.00186	0.03083	0.00268	0.01433	0.00419	0.00235	0.03598	0.00140	0.07006	0.00791	0.00263	0.01069	0.00009
116	0	0	0	0	0	0	0	0	0	0	0	0	0	0
117	0	0	0	0	0	0	0	0	0	0	0	0	0	0
118	0.01217	0.00077	0.00311	0.00247	0.03425	0.00519	0.00255	0.01683	0.00312	0.01237	0.00228	0.00927	0.00219	0.00078
119	0.00004	0.00001	0.00234	0.00005	0.00070	0.00047	0.00002	0.00029	0.00010	0.00062	0.00002	0.00020	0.00001	0.00001
120	0.01132	0.00010	0.00149	0.00073	0.01040	0.00422	0.00114	0.03600	0.00258	0.01408	0.00170	0.03244	0.00031	0.00083
121	0.02239	0.00108	0.01038	0.00549	0.04139	0.00568	0.00346	0.02037	0.00274	0.02001	0.00145	0.01312	0.00458	0.00108
122	0	0	0	0	0	0	0	0	0	0	0	0	0	0
123	0	0	0	0	0	0	0	0	0	0	0	0	0	0
124	0	0	0	0	0	0	0	0	0	0	0	0	0	0

Continued

b_{ij} \ a_{ij}	71	72	73	74	75	76	77	78	79	80	81	82	83	84
1	0	0	0	0	0	0	0	0	0	0	0	0	0.00008	0.00019
2	0	0	0	0	0	0	0	0	0	0	0	0.00380	0.04761	0.00001
3	0	0	0	0	0	0	0	0	0	0	0	0	0.02715	0.02621
4	0	0	0	0	0	0	0	0	0	0	0	0	0.00173	0.00004
5	0.00027	0.00047	0.00070	0.00019	0.00230	0	0	0.00067	0.00017	0.00035	0	0	0.00364	0.00016
6	0	0.00001	0.00144	0.00002	0.00007	0.00002	0.00019	0.00001	0	0	0.00014	0.00193	0.00184	0.00892
7	0.00003	0.00092	0.00001	0.00043	0.00820	0	0	0.01499	0.00027	0.00009	0	0.00023	0	0
8	0.00019	0.00165	0.00504	0.00123	0.04520	0	0	0.00043	0.00004	0.00056	0	0.00053	0	0.00038
9	0.00419	0.00186	0.00590	0.00986	0.00952	0	0	0.00114	0.00009	0.00017	0	0.00125	0	0.00057
10	0	0	0.00119	0	0	0	0	0	0	0	0	0.00013	0	0.00005
11	0	0	0	0	0	0	0	0.00185	0	0	0	0	0	0
12	0.00002	0.00013	0.00028	0.00002	0.00109	0	0	0.00022	0.00001	0.00031	0.00001	0.00024	0.00391	0.00025
13	0	0.00016	0.00140	0.00022	0.00163	0	0	0	0.00026	0.00024	0.00002	0.00184	0.00307	0.00095
14														
15														
16														
17														0.04550
18										0.00075				
19														0.00070
20														
21														
22	0.00001	0.00034	0.00042	0.00113	0.00101	0.00001	0.00002	0.00016	0.00016	0.00081	0.00001	0.00303	0.01726	0.01939
23	0.00001	0.00004	0.00002	0.00004	0.00014	0.00001	0.00002	0.00011	0.00001	0.00018	0.00001	0.00161	0.01135	0.00164
24	0.00013	0.00006	0.00002	0.00007	0.00043	0.00002	0.00004	0.00019	0.00002	0.00013	0.00002	0.00040	0.01217	0.00475
25	0	0.00001	0.00001	0.00002	0.00028	0.00001	0.00001	0.00007	0.00001	0.00016	0.00001	0.00091	0.02661	0.00947
26	0.00003	0.00189	0.00022	0.00047	0.00151	0.00002	0.00016	0.00038	0.00009	0.00053	0.00003	0.00113	0.00618	0.00594
27	0.00097	0	0	0	0.00075	0	0	0.00022	0	0	0	0.00147	0.00754	0.01048
28	0.00047	0.00168	0.00174	0.00301	0.00999	0.00045	0.00094	0.00196	0.00142	0.00134	0.00036	0.00342	0.00389	0.00347
29	0.00023	0.00083	0.00027	0.00123	0.00302	0.00015	0.00023	0.00043	0.00049	0.00228	0.00033	0.00101	0.01713	0.00076
30	0.00021	0.00141	0.00044	0.00052	0.00448	0.00025	0.00078	0.00393	0.00050	0.00132	0.00003	0.00193	0.00806	0.00178
31	0.00013	0.00372	0.00238	0.01405	0.02434	0.00058	0.00113	0.00124	0.00162	0.00388	0.00020	0.00056	0.00700	0.01058
32	0.00090	0.00133	0.00041	0.01834	0.02170	0.00160	0.00550	0.00699	0.00177	0.00223	0.00173	0.00008	0.03390	0.00408
33	0.00015	0.00062	0.00020	0.00597	0.00276	0.00054	0.00363	0.00656	0.00075	0.00063	0.00001	0.00014	0.00157	0.00016
34	0.00030	0.00129	0.00039	0.00136	0.00594	0.01326	0.00060	0.00651	0.00067	0.00144	0.00069	0.00065	0.00104	0.00179
35	0.00082	0	0	0.00003	0	0	0.00010	0.00148	0	0	0	0.00020	0.00146	0.00002
36	0.00026	0.00092	0.00114	0.00271	0.00828	0.00025	0.00048	0.00280	0.00060	0.00112	0.00012	0.00265	0.00328	0.00219
37	0.00009	0.00032	0.00219	0.00018	0.00056	0	0	0.00014	0.00001	0.00033	0	0.00018	0	0.00479
38	0.00048	0.00061	0.00042	0.02146	0.00895	0.00004	0.00048	0.01056	0.00366	0.00043	0.00010	0.00115	0.00202	0.00567
39	0	0	0	0	0	0	0	0	0	0	0	0	0	0
40	0	0	0	0	0	0	0	0	0	0	0	0	0	0
41	0.00060	0.00121	0.00114	0.01267	0.03392	0.00003	0.00041	0.00809	0.00127	0.00062	0.01129	0.00279	0.00879	0.00420
42	0.00037	0.00061	0.00056	0.00233	0.00618	0.00027	0.00097	0.00133	0.00154	0.00755	0.00010	0.00113	0.01211	0.00403

Continued

a_{ij} / b_{ij}	71	72	73	74	75	76	77	78	79	80	81	82	83	84
43	0.00011	0.00072	0.00063	0.00792	0.03728	0.00496	0.01086	0.03776	0.00158	0.00080	0.00199	0.00159	0.00792	0.12159
44	0	0	0	0.00001	0.00001	0	0	0.00002	0.00001	0.00005	0	0	0	0
45	0.00003	0.00009	0.00017	0.00044	0.01737	0.00002	0.00416	0.00051	0.00068	0.00005	0.00003	0.00015	0.00552	0.03562
46	0.00619	0.06517	0.00377	0.01587	0.01853	0.00378	0.00557	0.01140	0.02244	0.00581	0.00140	0.00455	0.00166	0.01818
47	0.00096	0.00438	0.00236	0.05708	0.10045	0.01099	0.02578	0.01726	0.00957	0.00549	0.00548	0.00081	0.00894	0.01131
48	0.00002	0.00018	0.00005	0.00019	0.00091	0.00002	0.00004	0.00009	0.00007	0.00005	0.00001	0.00044	0.00020	0.00141
49	0.00003	0.00018	0.00003	0.00013	0.00085	0	0	0.00006	0.00001	0.00005	0.00002	0.00030	0.00099	0.00070
50	0.00003	0.00019	0.00008	0.00019	0.00274	0.00002	0.00005	0.00011	0.00009	0.00009	0.00001	0.00067	0.00023	0.00093
51	0.00029	0.00187	0.00098	0.00542	0.06481	0.01106	0.01159	0.22527	0.01573	0.01989	0.01499	0.00529	0.00092	0.01019
52	0.00048	0.00195	0.00222	0.00433	0.03977	0.00163	0.00090	0.00789	0.00185	0.00185	0.00019	0.00186	0.00140	0.01190
53	0.00001	0.00034	0.00115	0.00139	0.03389	0	0.00014	0.00424	0.00004	0.00063	0	0.00137	0.00008	0.00419
54	0.00007	0.00038	0.00248	0.00191	0.09391	0	0.00084	0.02195	0.00771	0.00070	0.00018	0.00107	0.00055	0.01023
55	0.00002	0.00032	0.00300	0.00028	0.00796	0	0	0.00053	0.00026	0.00052	0	0.00060	0	0.00228
56	0.00072	0.00422	0.00513	0.00690	0.03882	0.00003	0.00003	0.00040	0.00033	0.00174	0	0.01156	0	0.00444
57	0.01224	0.01079	0.00962	0.01331	0.01751	0.00106	0.00031	0.00171	0.00100	0.00593	0.00235	0.00423	0.00004	0.00213
58	0.00047	0.00077	0.00971	0.00327	0.00936	0.00806	0.00201	0.00931	0.00031	0.00075	0.00220	0.00331	0	0.01475
59	0.00023	0.00260	0.00642	0.00576	0.18698	0.00004	0.00020	0.00412	0.00219	0.00227	0.00015	0.00057	0.03106	0.00240
60	0.00180	0.00646	0.01361	0.03526	0.43590	0.00119	0.00052	0.02103	0.01038	0.00695	0.00039	0.00173	0.02328	0.00669
61	0.00853	0.00868	0.00492	0.03132	0.02780	0.00326	0.00481	0.01209	0.00560	0.00499	0.00251	0.00966	0.01653	0.00875
62	0.00028	0.24096	0.04449	0.00003	0.00040	0.00005	0.00005	0.00002	0.00003	0.00004	0.00189	0.00450	0.00007	0.00015
63	0.00074	0.01305	0.02391	0.00949	0.02369	0.00320	0.00508	0.08492	0.00624	0.00540	0.02060	0.01120	0.00145	0.00233
64	0.00060	0.04374	0.01243	0.02267	0.01661	0.00021	0.00197	0.00145	0.00188	0.05770	0.00022	0.00815	0.00079	0.00110
65	0	0.00236	0	0	0	0	0	0	0	0	0	0.00246	0	0
66	0.00028	0.00058	0.00117	0.00676	0.00351	0.00056	0.00024	0.00153	0.00109	0.00081	0.00005	0.00297	0.00166	0.00115
67	0	0	0.00431	0	0	0	0	0	0	0	0	0.02395	0	0
68	0.00025	0.00119	0.00034	0.00127	0.00426	0.00120	0.00040	0.00129	0.00100	0.00037	0.00031	0.04037	0.00046	0.00105
69	0	0.00451	0	0	0	0	0	0	0	0	0	0.01068	0	0
70	0	0.00420	0	0	0	0	0	0	0	0	0	0.13850	0	0
71	0.93022	0.00305	0.00021	0.00038	0.00075	0.00001	0.00007	0.00028	0.00020	0.00019	0.00013	0.00137	0.00037	0.00399
72	0.06043	0.71195	0.07769	0.14141	0.01843	0.00017	0.00316	0.01336	0.00974	0.00394	0.00062	0.00959	0.00191	0.00130
73	0.00052	0.01322	0.00101	0.30351	0.02055	0.02532	0.01091	0.00017	0.00018	0.00025	0.00007	0.00124	0.00022	0.00129
74	0.00001	0.00014	0.01792	0.02946	0.09093	0.01685	0.05946	0.00298	0.04559	0.00845	0.00716	0.00476	0.00038	0.00856
75	0.00011	0.01001	0.00012	0.00037	0.00181	0.28042	0.00070	0.02409	0.01611	0.00240	0.00013	0.00422	0.00030	0.00012
76	0.00002	0.00010	0.00175	0.01146	0.03190	0.00101	0.12967	0.00508	0.00166	0.00175	0.00021	0.00185	0.00059	0.00214
77	0.00040	0.00103	0.00454	0.05793	0.02697	0.18375	0.24070	0.14788	0.13323	0.03358	0.02414	0.00083	0.00005	0.00126
78	0.00001	0.00326	0.00030	0.00117	0.01198	0.03565	0.01727	0.03282	0.46837	0.00995	0.00173	0.00128	0.00063	0.00115
79	0.00007	0.00030	0.00412	0.00492	0.05366	0.00389	0.00204	0.00530	0.00449	0.09353	0.00082	0.00852	0.00138	0.00281
80	0.00044	0.00623	0.00063	0.00194	0.00450	0.00040	0.00024	0.00133	0.00061	0.00082	0.04786	0.00434	0.00103	0.00052
81	0.00012	0.00146	0.00055	0.00175	0.00790	0.00040	0.00054	0.00200	0.00128	0.00093	0.00031	0.05019	0.00057	0.00112
82	0.00028	0.00058	0.00058	0.00566	0.01419	0.00318	0.00379	0.00464	0.00400	0.00222	0.00084	0.00233	0.00145	0.00313
83	0.00092	0.00322	0.00181	0.01133	0.03641	0.00188	0.00305	0.00946	0.00612	0.00889	0.00082	0.00057	0.17177	0.00823
84	0.00216	0.00419	0.00260	0.01133	0.03641	0.00188	0.00305	0.00946	0.00612	0.00889	0.00868	0.00372	0.00823	0.09717

a_{ij} / b_{ij}	71	72	73	74	75	76	77	78	79	80	81	82	83	84
85	0.00003	0.00006	0.00211	0.00029	0.00104	0	0.00003	0.00852	0.00003	0.00037	0.00001	0.00007	0.00049	0.00505
86	0.00086	0.00154	0.00182	0.00244	0.01053	0.00039	0.00067	0.00578	0.00130	0.00176	0.00023	0.00247	0.00215	0.00527
87	0.00110	0.00130	0.00139	0.00087	0.00494	0.00044	0.00028	0.00338	0.00070	0.00099	0.00006	0.00263	0.00119	0.00066
88	0.00246	0.00108	0.00064	0.00082	0.02452	0	0.00017	0.00962	0.00010	0.00169	0.00009	0.00217	0.00038	0.00054
89	0.00194	0.00139	0.00149	0.00518	0.01141	0.00106	0.00098	0.00566	0.00152	0.00063	0.00057	0.00390	0.00309	0.00241
90	0.00025	0.00039	0.00061	0.00261	0.00296	0.00018	0.00036	0.00090	0.00053	0.00063	0.00020	0.00242	0.00108	0.00094
91	0.00131	0.00244	0.00385	0.00395	0.01259	0.00244	0.00195	0.00604	0.00320	0.00163	0.00084	0.00322	0.01390	0.00703
92	0.00219	0.00374	0.00271	0.00710	0.01879	0.00236	0.00449	0.00582	0.00286	0.00266	0.00141	0.00275	0.00513	0.00488
93	0.00010	0.00039	0.00047	0.00112	0.00392	0.00009	0.00017	0.00195	0.00024	0.00040	0.00013	0.00109	0.00120	0.00083
94	0.00100	0.00237	0.00291	0.00437	0.01723	0.00278	0.00282	0.00473	0.00216	0.00164	0.00081	0.00272	0.00528	0.00425
95	0.00070	0.00419	0.00221	0.01724	0.02412	0.03559	0.02100	0.04116	0.03439	0.00744	0.00409	0.00470	0.00232	0.00321
96	0.00089	0.00167	0.00126	0.00247	0.00807	0.00056	0.00109	0.00210	0.00106	0.00134	0.00034	0.00142	0.01827	0.00296
97	0.00012	0.00038	0.00030	0.00302	0.00270	0.00009	0.00159	0.00126	0.00043	0.00022	0.00003	0.00026	0.00032	0.00031
98	0	0.00001	0	0.00002	0.00013	0.00002	0	0.00002	0.00002	0.00002	0	0.00002	0.00001	0
99	0.00028	0.00450	0.00213	0.00870	0.01147	0.00136	0.00211	0.00349	0.00296	0.00687	0.00034	0.00202	0.00844	0.00478
100	0.00230	0.03362	0.00177	0.00802	0.01707	0.00509	0.00621	0.00909	0.00740	0.00273	0.00148	0.00212	0.00847	0.00632
101	0.00072	0.00190	0.00322	0.00623	0.03433	0.00113	0.00109	0.00322	0.00462	0.00329	0.00034	0.00430	0.00453	0.00533
102	0	0.00003	0.00001	0.00005	0.00007	0.00001	0.00001	0.00002	0.00002	0.00004	0	0.00001	0.00005	0.00003
103	0	0	0	0	0	0	0	0	0	0	0	0	0	0
104	0	0.00005	0.00002	0.00009	0.00012	0.00001	0.00002	0.00004	0.00003	0.00007	0	0.00002	0.00009	0.00005
105	0.00002	0.00035	0.00017	0.00068	0.00090	0.00011	0.00017	0.00027	0.00023	0.00054	0.00003	0.00016	0.00066	0.00037
106	0.00211	0.00239	0.00261	0.00526	0.01468	0.00188	0.00273	0.00525	0.00297	0.00429	0.00060	0.00288	0.00372	0.00384
107	0.00047	0.00154	0.00171	0.00381	0.01658	0.00124	0.00244	0.00422	0.00166	0.00208	0.00102	0.00351	0.00482	0.00353
108	0.00220	0.00069	0.00076	0.00727	0.00895	0.00362	0.00328	0.00730	0.00370	0.00243	0.00071	0.00094	0.00313	0.00431
109	0	0.00002	0.00001	0.00004	0.00005	0.00001	0.00001	0.00001	0.00001	0.00003	0	0.00001	0.00004	0.00002
110	0.00071	0.00180	0.00208	0.00889	0.02430	0.00199	0.00241	0.00442	0.00258	0.00266	0.00061	0.00508	0.00143	0.00249
111	0.00002	0.00026	0.00026	0.00239	0.00169	0.00019	0.00020	0.00051	0.00096	0.00050	0.00004	0.00048	0.00068	0.00096
112	0	0	0	0	0	0	0	0	0	0	0	0	0	0
113	0.00099	0.00257	0.00439	0.00850	0.04721	0.00155	0.00149	0.00441	0.00619	0.00446	0.00047	0.00591	0.00607	0.00732
114	0.00083	0.00311	0.00077	0.02387	0.01056	0.00232	0.00692	0.00217	0.00548	0.00157	0.00048	0.00205	0.00249	0.00805
115	0.00022	0.00043	0.00360	0.00142	0.01163	0.00012	0.00027	0.00039	0.00043	0.00132	0.00009	0.00204	0.00312	0.01527
116	0	0	0	0	0	0	0	0	0	0	0	0	0	0
117	0	0	0	0	0	0	0	0	0	0	0	0	0	0
118	0.00066	0.00180	0.00446	0.00205	0.01175	0.00058	0.00284	0.00257	0.00380	0.00294	0.00070	0.00478	0.00285	0.00263
119	0.00001	0.00004	0.00002	0.00004	0.00047	0.00007	0.00001	0.00006	0.00006	0.00009	0	0.00006	0.00002	0
120	0.00039	0.00454	0.00274	0.02614	0.01561	0.01208	0.00482	0.01441	0.01525	0.00266	0.00276	0.00132	0.00101	0.00455
121	0.00069	0.00288	0.00242	0.00597	0.01778	0.00136	0.00388	0.00575	0.00545	0.00349	0.00156	0.00373	0.00266	0.00890
122	0	0	0	0	0	0	0	0	0	0	0	0	0	0
123	0	0	0	0	0	0	0	0	0	0	0	0	0	0
124	0	0	0	0	0	0	0	0	0	0	0	0	0	0

Continued

a_{ij} / b_{ij}	85	86	87	88	89	90	91	92	93	94	95	96	97	98
1	0	0	0	0	0	0	0	0	0	0	0	0	0.00136	0
2	0	0.00057	0	0	0	0.05516	0	0	0	0	0	0	0.00001	0
3	0	0	0	0	0	0	0	0	0	0	0	0	0	0
4	0	0.00018	0	0	0	0.05165	0	0	0	0	0	0	0.00001	0
5	0	0.30585	0.01582	0.01565	0.00039	0.00488	0.00745	0.00177	0.00005	0.00002	0	0	0.00015	0
6	0	0.03873	0.00279	0.00219	0.00006	0	0	0	0	0	0.00006	0.00092	0.00003	0
7	0	0.03131	0.00029	0.00149	0.00018	0	0	0	0	0	0	0.00191	0	0
8	0	0.00095	0.00008	0	0	0	0	0	0	0	0	0	0	0
9	0	0.00837	0.00002	0	0	0	0	0	0	0	0	0	0	0
10	0	0.00126	0.00049	0	0.00014	0.24958	0	0.00008	0	0.00002	0	0	0	0
11	0	0.00011	0.00002	0	0.00006	0.24571	0.00277	0	0	0	0	0.00414	0.00002	0
12	0	0.00173	0.00008	0.00008	0.00031	0	0.00044	0	0.00002	0	0	0.00085	0.00004	0
13	0	0	0	0	0	0	0	0	0	0	0	0.00003	0.00492	0
14	0	0	0	0	0	0	0	0	0	0	0	0	0	0
15	0	0	0	0	0	0	0.00067	0	0	0.00001	0	0	0	0
16	0	0	0	0	0	0	0	0	0	0.00013	0	0.00030	0.00002	0
17	0	0	0	0	0	0	0	0	0	0.00019	0	0.00040	0.00004	0
18	0	0	0	0	0	0	0.00032	0	0	0.00030	0	0.00382	0.00001	0
19	0	0	0	0	0	0.02146	0	0	0	0.00019	0	0.00070	0.00014	0
20	0	0	0	0	0	0	0.00270	0	0	0.00022	0	0.00504	0.00004	0
21	0	0.00038	0.00001	0.00020	0.00001	0.00916	0.00051	0.00011	0	0.00001	0.00001	0.00129	0.00008	0.00003
22	0	0.00431	0.00001	0.00002	0.00001	0.00049	0.00028	0.00003	0	0	0.00001	0.00026	0	0
23	0	0.00062	0.00003	0.00002	0.00002	0.00790	0.00063	0.00017	0	0	0	0.00012	0.00278	0
24	0	0.00019	0	0.00002	0.00001	0.00096	0.00002	0	0	0	0	0.00091	0	0
25	0	0.00133	0.00003	0.00002	0.00008	0	0	0	0	0	0	0	0	0
26	0	0	0	0	0	0.00088	0.00047	0.00324	0	0	0	0.00248	0.00226	0
27	0	0	0	0	0	0.03534	0.00010	0	0	0	0	0	0	0
28	0	0.01672	0.00053	0.00075	0.00234	0.00016	0.00307	0.00521	0.00011	0.00134	0.00026	0.00872	0.00029	0.00735
29	0	0.00287	0.00006	0.00011	0.00029	0.30427	0.00160	0.00213	0.00003	0.00003	0.00003	0.00137	0.00002	0
30	0	0.00247	0.00011	0.00009	0.00037	0.15405	0.00384	0.00080	0	0.00002	0	0.00050	0.00004	0.00004
31	0	0.00219	0.00004	0.00009	0.00041	0.00278	0.00156	0.00213	0.00004	0.00003	0.00004	0.00328	0.00055	0.00242
32	0	0.00031	0	0.00003	0.00001	0.00347	0.00021	0.00068	0.00001	0.00001	0.00001	0.00059	0.00002	0.00158
33	0	0.00100	0.00002	0.00007	0.00019	0.00837	0.00074	0.00225	0.00004	0.00017	0.00021	0.00134	0.00008	0.02568
34	0	0.00356	0.00009	0.00018	0.00034	0	0.00399	0.01312	0.00025	0.00136	0.00095	0.00615	0.00141	0.03925
35	0	0.00075	0	0	0	0.15868	0.00238	0.00331	0.00019	0.00041	0.00012	0.00296	0	0
36	0	0.06423	0.00213	0.00399	0.00072	0.02344	0.02284	0.06961	0.00123	0.01662	0.00523	0.00654	0.00077	0.00096
37	0	0.00041	0.00001	0.00187	0	0.02365	0.00008	0.01454	0	0	0	0	0	0
38	0	0.00576	0.00070	0.00104	0.00904	0	0.00027	0.00008	0.00003	0.00001	0	0.00013	0.00002	0
39	0	0	0	0	0	0	0	0	0	0	0	0	0	0
40	0	0	0	0	0	0	0	0	0	0	0	0	0.00012	0
41	0	0.00121	0.00008	0.00019	0.00022	0.03977	0.00051	0.00017	0.00001	0.00035	0	0.00076	0.00001	0.00001
42	0	0.00565	0.00011	0.00025	0.00709	0.00393	0.00121	0.00234	0.00003	0.00024	0.00015	0.00379	0.00011	0

Continued

b_{ij} \ a_{ij}	85	86	87	88	89	90	91	92	93	94	95	96	97	98
43	0	0.00166	0.00008	0.00014	0.00048	0.02889	0.00009	0.00014	0.00002	0.00004	0	0.00013	0.00001	0.00001
44	0	0.00023	0.00001	0	0.00001	0.00550	0.00014	0.00067	0.00003	0.00037	0.00001	0.00072	0.00005	0.00004
45	0	0	0	0	0	0.03042	0.00002	0.00006	0.00001	0.00001	0	0.00051	0.00001	
46	0	0.00562	0.00027	0.00020	0.00037	0.05074	0.00134	0.06388	0.00012	0.00018	0.00006	0.00334	0.00009	0.00028
47	0	0.00152	0.00006	0.00007	0.00033	0.02039	0.00012	0.00024	0.00001	0.00012	0.00001	0.00045	0.00006	0.00014
48	0	0.00275	0.00018	0.00008	0.00040	0.68996	0.00042	0.00027	0.00004	0.00003	0	0.00184	0.00001	0.00006
49	0	0.00357	0.00008	0.00005	0.00038	0.85804	0.00561	0.00055	0.00004	0	0	0.00101	0.00003	0
50	0	0.00309	0.00014	0.00019	0.00054	0.73647	0.00030	0.00013	0.00002	0.00001	0	0.00127	0.00001	
51	0	0.00506	0.00016	0.00031	0.00085	0.16550	0.00033	0.00093	0.00013	0.00002	0.00001	0.00021	0.00001	0.00023
52	0	0.00950	0.00024	0.00009	0.00137	0.59176	0.00025	0.00006	0.00001	0.00023	0	0.00020	0	0.00051
53	0	0.00424	0.00020	0	0.00005	0.35359	0.00060	0.00022	0.00003	0.00001	0	0.00137	0.00003	
54	0	0.00308	0.00003	0	0.00003	0.18248	0.00075	0.00014	0	0.00001	0	0.00090	0	0
55	0	0	0	0	0	0	0	0	0	0	0	0	0	0
56	0	0	0	0.00009	0	0	0	0	0	0	0	0	0	0
57	0	0.00151	0.00007	0	0.00012	0.21540	0.00216	0.00011	0.00004	0.00001	0	0.00032	0.00021	0.00003
58	0	0	0	0	0	0.00946	0	0	0	0	0	0	0	0
59	0	0	0	0.00003	0	0	0	0	0	0	0	0	0	0
60	0	0.00162	0.00004	0.00013	0.00011	0.08725	0.00012	0.00025	0.00001	0.00001	0.00002	0.00001	0.00043	0.00026
61	0	0.00412	0.00015	0.00009	0.00257	0.24814	0.00060	0.00120	0.00002	0.00009	0	0.00105	0.00003	0
62	0	0.03816	0.00110	0.00044	0.00004	0	0.00197	0.00075	0.00003	0.00029	0.00065	0.00092	0.00003	0
63	0	0.01244	0.00027	0.00108	0.00059	0	0.00040	0.01339	0	0.00025	0.00130	0.00200	0	0.00068
64	0	0.03368	0.00075	0	0.00169	0.08487	0.00424	0.00809	0.00029	0.00264	0	0.00735	0.00023	0
65	0	0	0	0.00079	0	0	0	0	0	0	0.00004	0	0	0.00170
66	0	0.01619	0.00052	0	0.00031	0.12935	0.00029	0.00042	0.00014	0.00008	0	0.00125	0.00002	0
67	0	0	0	0.00051	0	0	0.42093	0	0	0	0.00002	0.00086	0.00012	0.00120
68	0	0.00830	0.00090	0	0.00115	0.00276	0.00063	0.05119	0.00004	0.00011	0	0.00205	0.00013	0
69	0	0	0	0	0	0	0	0	0	0.12754	0.07375	0.00561	0	0
70	0	0	0	0	0	0	0	0	0	0	0	0.00042	0	0.01071
71	0	0	0	0.00009	0	0	0.00022	0	0	0	0.00117	0.00033	0.00011	0.00009
72	0	0.00412	0.00028	0.00009	0.00044	0.00515	0.00100	0.01558	0.00016	0.00257	0.00001	0.01025	0.00002	0.00015
73	0	0.01140	0.00026	0.00010	0.00092	0.00341	0.00062	0.00073	0.00005	0.00032	0.00049	0.00216	0.00001	0.00048
74	0	0.00322	0.00004	0.00021	0.00023	0.17180	0.00072	0.00057	0.00003	0.00005	0.00002	0.00143	0.00002	0.00118
75	0	0.05492	0.00046	0.00008	0.00125	0.24912	0.00155	0.00035	0.00003	0.00003	0.00026	0.00111	0.00004	0.00005
76	0	0.00513	0.00006	0.00014	0.00044	0.00295	0.00073	0.00074	0.00004	0.00009	0.00105	0.00078	0.00001	0.00196
77	0	0.00486	0.00015	0.00002	0.00071	0.06117	0.00048	0.00196	0	0.00019	0.00002	0.00108		0
78	0	0.00227	0.00003	0.00016	0.00015	0	0.00030	0.00014	0.00002	0	0.00005	0.00015	0.00001	0.00070
79	0	0.01612	0.00042	0.00123	0.00059	0.00443	0.00195	0.00060	0.00022	0.00016	0.00012	0.00590	0.00002	0.00173
80	0	0.12166	0.00152	0.00046	0.00440	0.21124	0.00097	0.00095	0.00001	0.00009	0.00013	0.00329	0.00103	0.00019
81	0	0.01813	0.00029	0.00177	0.00239	0.01060	0.00140	0.00724	0.00106	0.00023	0.00685	0.00217	0.00091	0.00512
82	0	0.03598	0.00107	0.00047	0.00310	0.05020	0.01316	0.03629	0.00011	0.01137	0.00052	0.02202	0.00020	0.00058
83	0	0.01271	0.00045	0.00063	0.00122	0.10218	0.00306	0.00525	0.00008	0.00069	0.00001	0.00242	0.00005	0.00003
84	0	0.02631	0.00092		0.00481	0.00580	0.00105	0.00364		0.00114		0.00091		

Continued

b_{ij} \ a_{ij}	85	86	87	88	89	90	91	92	93	94	95	96	97	98
85	0	0	0	0	0	0	0	0	0	0	0	0	0	0
86	0	0.03611	0.00123	0.00211	0.02186	0.03264	0.00671	0.00270	0.00199	0.00005	0.00002	0.00355	0.00021	0.00104
87	0	0.01751	0.02408	0.00335	0.00114	0.06102	0.00457	0.01373	0.00030	0.00012	0.00014	0.00259	0.00054	0.00487
88	0	0.00531	0.00169	0.11333	0.00010	0.02495	0.00030	0.00255	0	0.00066	0.00001	0.00401	0.00048	0.00075
89	0	0.03829	0.00448	0.00137	0.07800	0.06306	0.02964	0.01168	0.00013	0.00023	0.00016	0.00847	0.00046	0.00215
90	0	0.01021	0.00025	0.00064	0.00101	0.00981	0.00307	0.00474	0.00066	0.00041	0.00013	0.01327	0.00113	0.01519
91	0	0.06312	0.00483	0.00312	0.00070	0.09761	0.00246	0.00352	0.00007	0.00030	0.00034	0.00332	0.00112	0.00485
92	0	0.01665	0.00080	0.00090	0.00053	0.11604	0.00769	0.01090	0.00007	0.00557	0.00175	0.00241	0.00076	0.00084
93	0	0.04646	0.00238	0.00273	0.00031	0.05486	0.00323	0.02333	0.00467	0.00957	0.00048	0.00336	0.00027	0.00032
94	0	0.04486	0.00261	0.00236	0.00047	0.07028	0.00091	0.00774	0.00012	0.00028	0.01172	0.00205	0.00025	0.00059
95	0	0.00804	0.00020	0.00018	0.00058	0.05784	0.00388	0.00589	0.00006	0.02880	0.00923	0.01982	0.00009	0.11439
96	0	0.04444	0.00226	0.00246	0.00048	0.05453		0.09846	0.00004	0.01225	0.00017	0.04111	0.00131	0.00094
97	0	0.00023	0.00010	0.00009	0.00007	0.24082		0.01605	0.00034	0.00057	0.00028	0.08276	0.04503	0.00008
98	0	0.00005	0.00001	0	0.00001	0.00051		0.01755	0.00008	0.00097	0.00021	0.00420	0.00034	0
99	0	0.00542	0.00021	0.00106	0.00375	0.21647	0.00051	0.00855	0.00006	0.00154	0.00016	0.00454	0.00031	0.00068
100	0	0.02589	0.00126	0.00123	0.00585	0.10585	0.00169	0.00274	0.00005	0.00051	0.00103	0.00136	0.00017	0.00120
101	0	0.00633	0.00101	0.00038	0.00135	0.04943	0.00089	0.01152	0.00007	0.00071	0.00036	0.00619	0.00074	0.00073
102	0	0.00003		0.00001	0.00002	0.00425		0.01314	0.00038	0.00125	0.00057	0.01601	0.00136	0.00001
103	0	0	0	0	0	0.00696		0.03222	0.00014	0.00249	0.00002	0.00903	0.00092	0
104	0	0.00006	0.00002	0.00001	0.00004	0.00519		0.00739	0.00014	0.00091	0.00122	0.03309	0.00010	0.00001
105	0	0.00042	0.00056	0.00008	0.00029	0.00525		0.00500	0.00004	0.00199	0.00103	0.01490	0.00019	0.00008
106	0	0.02769	0.00116	0.00046	0.00081	0.02854	0.00379	0.00622	0.00016	0.00526	0.00221	0.00365	0.00119	0.00031
107	0	0.03450	0.00006	0.00098	0.00170	0.10059		0.04470	0.00026	0.00036	0.00036	0.00989	0.00136	0.00176
108	0	0.00076		0.00015	0.00005	0.00357	0.00008	0.00325	0.00001	0.00180	0.00049	0.00584	0.00050	0.00100
109	0	0.00002			0.00002	0.00012	0.00263	0.03660	0.00018	0.00002	0.00006	0.02490	0.00095	0
110	0	0.05476	0.00296	0.00234	0.01109	0.01567		0.00260	0.00002	0.00070	0.00101	0.00074	0.00004	0.00160
111	0	0.00048	0.00002	0.00005	0.00019	0.00944	0.00628	0.02780	0.00017	0.00002	0.00003	0.00971	0.00041	0.00012
112	0	0	0	0	0	0	0	0.00370	0.00045	0.00003	0.00071	0.00115	0.00006	0
113	0	0.00870	0.00140	0.00052	0.00184	0.06652		0.00086		0.00125	0.00330	0.00129	0	0.00101
114	0	0.00307	0.00015	0.00015	0.00031	0.17569	0.00007	0.00448	0.00097	0.00107	0.00015	0.00613	0.00044	0.00181
115	0	0.04141	0.00214	0.00047	0.00135	0.02832		0.01922	0.00047			0.00357	0.00047	0.00008
116	0	0	0	0	0	0	0	0	0	0	0	0	0	0
117	0	0	0	0	0	0	0.35685	0	0	0	0	0	0	0
118	0	0.01304	0.00066	0.00096	0.00273	0.10411	0.00849	0.02443	0.00090	0.00140	0.00252	0.01098	0.00045	0.00293
119	0	0.00020	0.00001	0.00005	0.00005	0.01581	0.00261	0.00308	0.00002	0.00011	0.00047	0.00280	0.00017	0.00663
120	0	0.00565	0.00011	0.00015	0.00107	0.01033	0.00025	0.00020	0.00006	0.00008	0.00001	0.00022		0.00104
121	0	0.05362	0.00206	0.00052	0.00128	0.04483	0.00032	0.00543	0.00028	0.00008	0.00004	0.00228	0.00013	0.00002
122	0	0	0	0	0	0	0	0	0	0	0	0	0	0
123	0	0	0	0	0.01689	0.59646	0	0	0	0	0	0	0	0
124	0	0	0	0	0	0	0	0	0	0	0	0	0	0

Continued

b_{ij}＼a_{ij}	99	100	101	102	103	104	105	106	107	108	109	110	111	112
1	0	0.00061	0.00801	0	0	0	0	0	0	0	0.00032	0.00015	0.00074	0
2	0	0.00957	0.00732	0	0	0	0	0	0	0	0.00305	0.00135	0.00033	0.00001
3	0	0.01978	0.06293	0	0	0	0	0	0	0	0.00024	0.00033	0.00289	0
4	0	0.00854	0.24637	0	0.00023	0	0	0	0	0	0.00081	0.00015	0.00540	0.00003
5	0		0.00607	0.00176	0	0.00005	0.00003	0	0.00019	0.00089	0.00212	0.00004	0.00057	0
6	0	0.00468	0.00290	0	0.00012	0	0	0.00677	0	0.00469	0.00044	0.00510	0.00240	0.00001
7	0	0	0	0.00004	0	0	0	0	0	0	0.00005	0.00006	0	0
8	0	0	0.06553	0	0	0	0	0	0	0	0	0.00092	0.00045	0
9	0	0	0	0	0	0	0	0	0	0	0	0	0	0
10	0	0	0	0	0	0	0	0	0	0	0	0	0	0
11	0	0	0	0.01187	0.00002	0	0.00001	0	0	0	0	0.00045	0	0
12	0			0.00015		0	0.00014	0	0	0	0	0.00044	0	0
13	0	0.00348	0.00010	0	0	0.00018	0.00024	0	0	0.00552	0.00165	0.00032	0.00051	0.00001
14	0	0.03800	0.03040	0.00275		0.00002	0.00141	0	0	0	0.00067	0.00021	0.00072	0.00001
15	0	0.00727	0.04118	0	0	0.00003	0.00051	0	0	0	0.00055	0.00025	0.00035	0.00008
16	0	0.00605	0.27935	0	0	0.00062	0.03216	0	0	0	0.00079	0.00126	0.01591	0.00027
17	0	0.05042	0.37207	0	0	0.00048	0.00159	0	0	0	0.00103	0.00087	0.02361	0.00070
18	0	0.03633	0.22056	0.00020	0	0.00003	0.04338	0	0	0	0.00056	0.00367	0.01386	0.00017
19	0	0.05524	0.26381	0	0	0.00009	0.00020	0.00916	0.00031	0.00121	0.00243	0.09857	0.03237	0.00066
20	0	0.29622	0.28496	0.01415	0	0.00050	0.00002	0	0	0	0.00148	0.01239	0.03973	0.00008
21	0	0.24740	0.20077	0	0	0	0.00002	0	0	0	0.00773	0.02139	0.01401	0.00002
22	0.00020	0.00804	0.00073	0.00052	0.00002	0.00004	0	0.00009	0.00005	0.00017	0.00047	0.00985	0.00184	0
23		0.01133		0.00064	0.00005		0	0.00022	0.00008	0.00023	0.00031	0.00369	0.00088	0
24	0.00001	0.00232		0.00222	0		0	0	0	0	0.00012	0.01155	0.00033	0.00004
25		0.02163		0.00007	0.00013		0	0	0	0		0.00377	0.00067	0.00002
26	0	0.03179		0.00099	0.00009		0	0	0	0.00145	0.00133	0.00082	0.01796	0
27	0.01991	0		0.00118	0.00158		0	0	0	0	0.00043		0	0
28		0.17716	0.00999	0.00124	0.00001	0.00132	0.00135	0.01101	0.00186	0.00469	0.00557	0.01765	0.00994	0.00034
29	0.00062	0.00242	0	0.00050	0.00002	0.00003	0.00045	0.00011	0.00028	0.00017	0.00095	0.00076	0.00062	0.00002
30	0.01040	0.00802	0.00049	0.00174	0.00028		0.00001	0.00083	0.00004	0.00125	0.00098	0.01221	0.00038	0.00002
31	0.00175	0.13975	0.02341	0.00081	0.00007	0.00023	0.00012	0.02368	0.00524	0.00295	0.00196	0.07222	0.00459	0.00005
32	0.01069	0.03031	0.00138	0.00009	0.00077	0.00008	0.00010	0.00792	0.00169	0.00156	0.00075	0.00695	0.00137	0.00002
33	0.02287	0.26165	0.00119	0.00038	0.00284	0.00029	0.00048	0.02455	0.01149	0.00281	0.00351	0.00312	0.00312	0.00013
34		0.00872	0.00641	0.00180	0.02954	0.00124	0.00318	0.11770	0.02986	0.02837	0.01549	0.03561	0.01297	0.00126
35	0.00319	0.14348	0.00076	0.00089	0.01541	0.00017	0.00047	0.01830	0.00171	0.00844	0.04076	0.05590	0.00607	0.00063
36	0	0.05360	0.00167	0.00690		0.01771	0.01217	0.00315	0.00076	0.00148	0.04020	0.00254	0.00178	0.00002
37	0	0		0.00004	0		0	0	0	0	0.00487			0
38	0	0		0.00012	0	0.00006	0	0	0	0	0.00040	0.00019	0.00002	0
39	0	0.20359	0.00021	0	0.00004	0	0	0	0	0	0.00013	0	0	0
40	0.00022	0.00630	0.00128	0.00025	0.00028	0.00011	0	0	0	0	0.00058	0	0	0
41	0	0.01215	0.00809	0.00084		0.00033	0	0.00011	0.00004	0.00020		0.00798	0.00014	0
42							0.00038	0.00562	0.00339	0.00300	0.00352	0.14553	0.01965	0.00009

Continued

a_{ij} / b_{ij}	99	100	101	102	103	104	105	106	107	108	109	110	111	112
43	0.00011	0.00007	0	0.00004	0	0.00001	0.00002	0.00050	0.00016	0.00020	0.00042	0.01592	0.00083	0.00001
44	0.00020	0.04060	0.00078	0.00006	0.00037	0.00007	0.00003	0.00050	0.00008	0.00020	0.00372	0.00071	0.00064	0.00002
45	0	0	0	0.00004	0	0.00001	0.00002	0	0	0	0	0	0	0
46	0.00144	0.01306	0.00073	0.00060	0.00949	0.00064	0.00063	0.00039	0.00007	0.00012	0.01491	0.00261	0.00101	0.00001
47	0.00120	0.02638	0.00115	0.00016	0.00004	0.00004	0.00018	0.00028	0.00012	0.00012	0.00038	0.00208	0.00093	0
48	0.00086	0.01804	0.00015	0.00016	0.00005	0	0.00001	0.00032	0.00002	0.01250	0.00202	0.00043	0.00009	0.00001
49	0	0	0	0.00235	0.00006	0	0	0.00021	0.00017	0.01929	0.00425	0.00043	0.00084	0.00001
50	0.00002	0.02155	0.00015	0.00013	0.00005	0	0.00005	0.00070	0.00008	0.00268	0.00171	0.00521	0.00034	0.00001
51	0.00091	0.00885	0.00121	0.00048	0.00017	0	0.00003	0.00184	0.00051	0.00131	0.00114	0.00099	0.00057	0
52	0.00189	0.01903	0.00989	0.00017	0.00003	0.00004	0.00001	0.00093	0.00025	0.00149	0.00058	0.00075	0.00411	0.00001
53	0	0	0	0.00019	0.00001	0.00003	0	0.00012	0.00008	0.00066	0.00033	0.00389	0.00012	0
54	0	0	0	0.00022	0.00008	0	0	0.00023	0.00002	0.00005	0.00062	0.00061	0.00006	0.00002
55	0	0	0	0	0	0	0	0	0	0	0.00007	0	0	0
56	0.00061	0.00106	0.00013	0.00081	0.00005	0	0	0	0	0	0.00016	0	0	0
57	0	0	0	0	0	0	0	0	0	0.00039	0.00044	0	0	0
58	0	0	0	0	0	0	0	0	0	0	0.00196	0	0	0
59	0	0	0	0	0	0	0	0	0	0	0.00005	0.00003	0	0
60	0	0	0.00018	0.00004	0	0	0	0	0	0.00105	0.00022	0.00016	0	0
61	0.00142	0.01255	0.00126	0.00026	0.00014	0.00003	0.00005	0.00136	0.00011	0.00134	0.00166	0.00074	0.00050	0.00001
62	0	0.02158	0.00067	0.00112	0.00018	0.00004	0.00001	0	0	0.00135	0.00056	0.00046	0.00101	0
63	0	0	0	0.00017	0.00145	0.00002	0.00036	0	0	0	0.00027	0.00027	0	0
64	0.01219	0.01506	0.00039	0.00132	0.00104	0.00256	0.00469	0.00552	0.00129	0.00224	0.01141	0.00133	0.00120	0.00006
65	0	0	0	0	0	0	0	0	0	0	0.00033	0	0	0
66	0.00115	0.09806	0.00060	0.00017	0.00002	0.00007	0.00005	0.00871	0.00058	0.00045	0.00089	0.00698	0.00051	0
67	0	0.12710	0	0.19506	0	0	0	0	0	0	0.02360	0	0	0
68	0.00468	0	0.00026	0.00028	0.01063	0.00003	0.00114	0.00152	0.00046	0.00151	0.01855	0.00196	0.00042	0.00015
69	0	0	0	0	0	0	0.53025	0	0	0	0.00005	0	0	0
70	0	0	0	0	0	0.18307	0	0	0	0	0	0	0	0
71	0.03815	0	0	0.00009	0.00002	0.00016	0.00164	0.00347	0.00631	0.00164	0.00070	0.00014	0.00001	0.00001
72	0.00094	0.03027	0.00034	0.00035	0.00150	0.00049	0.00010	0.00045	0.00553	0.00036	0.01065	0.00258	0.00014	0.00001
73	0.00201	0.18280	0.00422	0.00050	0.00005	0.00037	0.00071	0.01731	0.00079	0.00253	0.00312	0.00012	0.00015	0.00005
74	0.00350	0.09044	0.00042	0.00050	0.00014	0.00006	0.00011	0.00234	0.00095	0.00546	0.00120	0.01219	0.01344	0.00003
75	0.06526	0.03307	0.00102	0.00082	0.00008	0.00053	0.00065	0.05643	0.00010	0.00251	0.00222	0.00046	0.00115	0.00003
76	0.00054	0.52159	0.00350	0.00014	0.00032	0.00003	0.00231	0.01333	0.00332	0.00081	0.00140	0.00081	0.00241	0.00012
77	0.01128	0.00246	0.00010	0.00012	0.00016	0.00003	0.00001	0.00210	0.00460	0.00005	0.00012	0.00427	0.00530	0.00005
78	0	0.02370	0	0.00053	0.00005	0.00016	0.00019	0.02015	0.00004	0.00329	0.00195	0.00012	0.00009	0.00005
79	0.07508	0.01612	0	0.00047	0.00003	0.00004	0.00027	0.00138	0.00147	0.00012	0.00240	0.00041	0.00158	0.00005
80	0.05932	0.13318	0	0.00057	0.00003	0.00012	0.00024	0.19289	0.00008	0.00014	0.00261	0.00054	0.00147	0.00005
81	0.00194	0.05856	0.00070	0.00754	0.00094	0.01818	0.02466	0.02342	0.00824	0.01355	0.04446	0.13566	0.01129	0.00039
82	0.09768	0.10199	0.00035	0.00109	0.00873	0.00079	0.00550	0.03030	0.00542	0.00815	0.00539	0.00524	0.00492	0.00023
83	0.00545		0.00881		0.00133		0.00003		0.00577	0.00569		0.01862	0.01463	0.00126
84	0.00024	0.02133	0.00199	0.00052	0.00017	0.00028	0.00003	0.00913	0.00461	0.01309	0.00479	0.01730	0.00363	0.00006

Continued

b_{ij} \ a_{ij}	99	100	101	102	103	104	105	106	107	108	109	110	111	112
85	0	0	0	0	0	0	0	0	0	0	0	0	0	0
86	0.01207	0.02436	0.00350	0.00189	0.00062	0.00009	0.00018	0.00400	0.00061	0.00215	0.00654	0.00346	0.00523	0.00025
87	0.01953	0.10405	0.02383	0.00243	0.00180	0.00053	0.00031	0.02117	0.00343	0.01015	0.01081	0.00481	0.00801	0.00022
88	0.00872	0.04905	0.09889	0.00025	0.00257	0.00035	0.00004	0.00873	0.00020	0.00931	0.02848	0.02505	0.01531	0.00013
89	0.01567	0.05724	0.01651	0.00143	0.00354	0.00141	0.00070	0.01329	0.00279	0.00469	0.01032	0.01459	0.01291	0.00051
90	0.04868	0.05475	0.00148	0.01579	0.00233	0.00007	0.00060	0.05864	0.00602	0.07026	0.03476	0.02724	0.02121	0.00057
91	0.00319	0.00914	0.00373	0.00103	0.00041	0.00043	0.00064	0.00130	0.00042	0.00232	0.00209	0.00326	0.00103	0.00010
92	0.00126	0.01299	0.00400	0.00113	0.00096	0.00031	0.00097	0.00136	0.00036	0.00232	0.00230	0.00323	0.00084	0.00013
93	0.00107	0.01809	0.00131	0.00233	0.00517	0.00593	0.00408	0.00106	0.00189	0.00051	0.01386	0.00145	0.00068	0.00001
94	0.00095	0.01118	0.00341	0.00097	0.00159	0.00177	0.00120	0.00098	0.00027	0.00127	0.00468	0.00212	0.00072	0.00002
95	0.06798	0.02759	0.00173	0.00040	0.00112	0.00028	0.00239	0.03094	0.00193	0.00435	0.00339	0.00903	0.00391	0.00114
96	0.00072	0.11688	0.00250	0.00150	0.01367	0.03934	0.03273	0.00274	0.00199	0.00164	0.00287	0.00477	0.00083	0.00105
97	0.00122	0.28527	0.00067	0	0.00239	0.00237	0.00037	0.00499	0.01208	0.00194	0.00068	0.00111	0.00064	0.00012
98	0.00010	0.15967	0.00441	0	0.00139	0.00022	0.00036	0.11901	0.01660	0.00378	0.00654	0.00801	0.01007	0.00326
99	0	0.04711	0.00265	0.00027	0.00197	0.00047	0.00082	0.02583	0.00581	0.00324	0.00457	0.00422	0.00684	0.00239
100	0.00302	0.12416	0.01373	0.00061	0.00062	0.00049	0.00087	0.00397	0.00093	0.00220	0.00241	0.00629	0.00205	0.00085
101	0.00463	0.15303	0.00221	0.00049	0.00282	0.00061	0.00457	0.02414	0.01177	0.01048	0.00697	0.00754	0.00322	0.02620
102	0.00007	0.14522	0.00001	0	0.00220	0.00043	0.00035	0.05417	0.01427	0.00425	0.00706	0.00993	0.00313	0.16047
103	0.00002	0.12434	0	0	0.02915	0.00106	0.00072	0.04696	0.01460	0.00542	0.01944	0.00687	0.00496	0.14502
104	0.00011	0.15980	0.00002	0	0.00280	0.01690	0.00011	0.05796	0.01737	0.00628	0.01297	0.00754	0.00161	0.05832
105	0.00071	0.10448	0.00017	0.00133	0.00121	0.00093	0.01073	0.06889	0.01846	0.00916	0.00738	0.01098	0.00537	0.22359
106	0.00336	0.15888	0.00234	0	0.00126	0.00199	0.00345	0.12543	0.00018	0.03575	0.00143	0.00042	0.00435	0.00293
107	0.01090	0.13884	0.00402	0	0.01366	0.00422	0.00847	0.02915	0.03365	0.00991	0.02607	0.00505	0.00923	0.00282
108	0.01023	0.25771	0.01342	0.00157	0.00186	0.00024	0.02253	0.18497	0.05700	0.01981	0.00518	0.03515	0.00568	0.00375
109	0.00004	0.03792	0.00001	0.00165	0.00932	0.00157	0.00389	0.05967	0.02963	0.02187	0.15343	0.01559	0.00976	0.05584
110	0.01473	0.18253	0.00215	0	0.00073	0.00020	0.00004	0.00632	0.00300	0.00304	0.00521	0.03392	0.00681	0.02716
111	0.00193	0.06724	0.00328	0.00410	0.00387	0.00081	0.00735	0.07461	0.02638	0.00753	0.00871	0.00817	0.02068	0.04740
112	0	0	0	0	0.00046	0.00072	0.00025	0.01300	0.00248	0.00604	0.00855	0.00938	0.02472	0.77667
113	0.00633	0.20702	0.00304	0	0.00018	0.00003	0.00183	0.02027	0.00178	0.00202	0.00103	0.00156	0.00391	0.00689
114	0.03847	0.15492	0.02271	0.00016	0.00083	0.00041	0.00950	0.07787	0.01841	0.01413	0.00617	0.00756	0.00450	0.00326
115	0.00121	0.00462	0.02802	0	0.00323	0.00029	0.00046	0.01040	0.00335	0.00290	0.02058	0.00190	0.00311	0.00316
116	0	0	0	0	0	0	0	0	0	0	0	0	0	0
117	0	0	0	0.21064	0.43251	0	0	0	0	0	0	0	0	0
118	0.01235	0.11016	0.00556	0.00454	0.00288	0.00225	0.00768	0.03648	0.00683	0.00541	0.01895	0.01073	0.00379	0.00081
119	0.08697	0	0	0.00180	0.00100	0.00034	0.00128	0.02970	0.01189	0.00357	0.01067	0.00403	0.00678	0.01009
120	0.00715	0.01972	0.00001	0.00020	0.00001	0	0.00003	0.00403	0.00006	0.00117	0.00069	0.00005	0.00001	0.00009
121	0.00010	0.11433	0	0.00017	0.00034	0.00027	0.00004	0.01016	0.00173	0.00572	0.00734	0.00058	0.00268	0.00038
122	0	0	0	0	0	0	0	0	0	0	0.00039	0	0	0
123	0	0	0	0	0	0	0	0	0	0	0.00112	0	0	0
124	0	0	0	0	0	0	0	0	0	0	0	0	0	0

Continued

a_{ij} \ b_{ij}	113	114	115	116	117	118	119	120	121	122	123	124
1	0.00055	0.00029	0.00033	0.00002	0.00013	0.00038	0.00001	0.00007	0.00003	0.00869	0.00020	0
2	0.00133	0.00075	0.00017	0.00006	0.00002	0.00224	0.00002	0.00094	0.00018	0.00345	0.00019	0
3	0.00064	0.00048	0.00004	0.00002	0.00007	0.00115	0	0.00012	0	0.00217	0.00001	0
4	0.00278	0.00180	0.00007		0.00007	0.00062	0.00002	0.00004		0.00619	0.00136	0
5	0.00002	0.00001	0.00647	0.00002	0.00003	0.00087	0.00001	0.00001	0.00006	0.00012	0.00023	0
6	0.00041	0.00371	0.00449	0.00022	0.00069	0.01782	0.00042	0.00099	0.00127	0.00042	0.00052	0.00721
7	0	0	0.00007	0	0	0	0	0	0.00002	0	0	0.00008
8	0.00071		0.00181		0	0	0	0.00045		0		0.00009
9	0		0	0	0	0	0	0.00004		0	0.00097	0
10	0		0	0	0	0	0	0.00240		0	0.00311	0
11	0		0	0.00001	0	0	0	0.00003		0		0
12	0		0	0.00001	0		0	0.00021	0.00018	0.00003	0.00255	0
13	0.00001	0.00088	0.00051	0.00005	0.00053	0.00130	0.00008	0.00004	0.00042	0.00004	0.00037	0
14	0.00083	0.00012	0.00016	0.00005	0.00024	0.00305	0.00002	0.00018	0.00002	0.00682	0.00002	0
15	0.00067	0.00023	0.00050	0.00119	0.00201	0.00150	0	0.00028	0.00002	0		0
16	0.00538	0.00100	0.00056	0.00048	0.00126	0.01349	0.00019	0.00010	0.00004	0.00105		0
17	0.01360	0.00228	0.00055	0.00050	0.00269	0.00237	0.00026	0.00050	0.00009	0.00065	0.00133	0
18	0.02131	0.01344	0.00076	0.00009	0.00034	0.01233	0.00042	0.00069	0.00162	0.00030	0.00049	0
19	0.02487	0.00596	0.01274	0.00163	0.00076	0.00291	0.00042	0.00020	0.00031	0.00053	0.00268	0
20	0.06250	0.01614	0.00092	0.00004	0.00059	0.00891	0.00170	0.00053	0.00221	0.00038	0.00070	0
21	0.03865	0.00747	0.00029	0.00001	0.00009		0.00120	0.00070	0.00121	0.00001	0.00112	0
22	0.00016	0.00083	0.00286	0.00001	0.00006	0.00080	0.00004	0.00003	0.00015	0.00002	0.00012	0.00662
23	0.00003	0.00059	0.00015		0.00002	0.00072	0.00027	0.00280	0.00006	0.00035	0.00001	0.00091
24		0.00589	0.00008	0	0.00001	0.00061	0.00046	0.00007	0.00005	0.00001	0.00049	0.00036
25	0.00002	0.00035	0.00002	0.00056	0.00228	0.00030	0.00012	0.00003	0.00006	0.00041		0.00002
26	0.00161	0.00368	0.00246	0.00010	0.00001	0.00583	0.00041	0.00047	0.00106		0.00028	0.00563
27	0.00055	0.00190	0.00044	0.00111	0.00095	0.00030	0.00008	0.00001	0.00006	0.00006		0
28	0.00273	0.01720	0.00554	0.00012	0.00004	0.00974	0.00099	0.00062	0.00214	0.00139	0.00283	0.09514
29	0.00028	0.00089	0.00073	0.00005	0.00018	0.00059	0.00012	0.00007	0.00047	0.00008	0.00026	0.01020
30	0.00011	0.00770	0.00077	0.00014	0.00035	0.00382	0.00035	0.00017	0.00048	0.00007	0.00011	0.01003
31	0.00119	0.03473	0.00983	0.00002	0.00013	0.03040	0.00089	0.00041	0.00239	0.00074	0.00135	0.03950
32	0.00013	0.02089	0.00181	0.00012	0.00122	0.01262	0.03565	0.00039	0.00229	0.00050	0.00054	0.01368
33	0.00097	0.08825	0.00867	0.00083	0.00673	0.05726	0.03793	0.00202	0.00418	0.00093	0.00144	0.16867
34	0.00423	0.08851	0.01122	0.02294	0.00221	0.10926	0.01240	0.00581	0.02093	0.00493	0.01032	0.17040
35	0.04375	0.02546	0.00616	0.00004	0.00034	0.09367	0.01151	0.00171	0.00396	0.00147	0.00223	0.02902
36	0.00030	0.00488	0.00095	0	0	0.00297	0.00026	0.00048	0.00102	0.00052	0.00212	0.02109
37	0	0	0	0.00001	0	0.00090	0	0.00008	0.00010	0	0.00005	0
38	0.00002	0.00036	0.00173	0.00004		0.00222	0.00023	0.00090	0.00037	0.00002	0.00029	0.00018
39	0	0	0	0.00008		0.00013	0	0.00043	0	0.00019	0.00002	0
40	0	0	0	0.00002	0.00004	0.00043	0	0.00127	0.00023	0.00342	0.00006	0
41	0.00012	0.00478	0.00077		0.00008	0.00246	0.00075	0.00073	0.00017	0.00008	0.00007	0.00488
42	0.00223	0.01266	0.01258	0.00038	0.00056	0.00625	0.00102	0.00087	0.00171	0.00038	0.00032	0.01762

Continued

a_{ij} / b_{ij}	113	114	115	116	117	118	119	120	121	122	123	124
43	0.00021	0.00493	0.00683	0.00002	0.00003	0.00316	0.00088	0.00058	0.00024	0.00003	0.00028	0.00052
44	0.00006	0.00109	0.58039	0.00027	0.00175	0.01962	0.00075	0.00101	0.00118	0.00362	0.00071	0.01242
45	0.00027	0	0	0	0	0	0	0.00006	0	0.00009	0	0
46	0.00014	0.00115	0.01026	0.00001	0.00005	0.00080	0.00011	0.00040	0.00027	0.00003	0.00039	0.00293
47	0.00008	0.00116	0.00211	0.00001	0.00002	0.00073	0.00014	0.00024	0.00018	0.00054	0.00060	0.00061
48	0.00004	0.00273	0.00136	0.00007	0.00021	0.00442	0.00045	0.00020	0.00075	0.00028	0.00365	0.01950
49	0.00006	0.00369	0.00109	0.00002	0.00008	0.00427	0.00006	0.00034	0.00043	0.00007	0.00958	0.00297
50	0.00005	0.00123	0.00052	0.00004	0.00010	0.00409	0.00031	0.00011	0.00024	0.00007	0.00060	0.00709
51	0.00028	0.00334	0.00152	0.00003	0.00006	0.00381	0.00069	0.00073	0.00045	0.00002	0.00016	0.00049
52	0.00049	0.00144	0.00052	0.00001	0.00013	0.00254	0.00009	0.00021	0.00013	0.00006	0.00072	0.00195
53	0.00012	0.00197	0.00005	0.00015	0.00001	0.00058	0.00002	0.00080	0.00024	0.00001	0.00007	0
54	0.00003	0.00225	0.00011	0.00005	0	0.00100	0.00001	0.00057	0.00013	0.00003	0.00008	0
55	0	0	0	0	0	0.00014	0	0.00011	0.00001	0	0	0
56	0	0	0	0	0	0.00047	0	0.00042	0.00002	0	0	0
57	0	0.00038	0.00002	0	0	0.00012	0.00002	0.00014	0.00007	0.00001	0.00088	0
58	0	0.00201	0	0	0.00001	0.00032	0	0.00091	0.00011	0	0	0
59	0	0	0.00001	0	0	0.00005	0	0.00065	0.00001	0	0	0
60	0	0.00030		0	0	0.00012	0.00001	0.00048	0.00007	0.00004	0.00035	0
61	0.00017	0.00974	0.00137	0.00005	0.00015	0.00300	0.00026	0.00154	0.00066	0.00009	0.00502	0.01477
62		0.00701	0.00154	0.00025	0.00021	0.00344	0.00057	0.00024	0.00119	0.00021	0.00006	0.00274
63	0.00036	0.00294	0.00024		0	0.00422	0.00004	0.00874	0.00089	0.00042	0.00070	0.00305
64	0.00054	0.00815	0.00350	0.00010	0.00025	0.00374	0.00085	0.00124	0.00205	0.00040	0.00361	0.02739
65		0	0	0.00006	0	0	0	0.00018	0	0.00128	0.00056	0.01659
66	0.00033	0.00236	0.05864	0.00008	0.00013	0.00290	0.00026	0.00131	0.00270	0.00028	0.00125	0.00797
67		0	0	0	0	0.00066	0	0.00017	0	0	0.00031	0
68	0.00017	0.01867	0.00074	0.00010	0.00020	0.00308	0.00036	0.00029	0.00190	0.00039	0.00059	0.03390
69	0.00009	0	0	0	0	0.00050	0	0.00021	0	0	0.00294	0
70		0	0	0	0	0.00001	0	0.02604	0.00704	0	0	0
71	0.00003	0.00321	0.00002	0.00010	0.00002	0.00040	0.00009	0.00006	0.00028	0.00006	0.00005	0.01139
72	0.00001	0.00781	0.00097	0.00014	0.00013	0.00336	0.00104	0.00011	0.00032	0.00069	0.00012	0.00374
73	0.00060	0.00408	0.00038	0.00002	0.00003	0.00580	0.00018	0.00147	0.00131	0.00024	0.00060	0.01443
74	0.00426	0.07862	0.00420	0.00045	0.00072	0.01687	0.00160	0.01556	0.00123	0.00029	0.00047	0.00366
75	0.00043	0.01464	0.00117	0.00005	0.00009	0.00239	0.00082	0.00334	0.01670	0.00003	0.00116	0.03975
76	0.00069	0.42508	0.00305	0.00016	0.00081	0.02904	0.00175	0.00611	0.00109	0.00039	0.00078	0.00765
77	0.00106	0.00668	0.00159	0	0.00017	0.01477	0.00311	0.00564	0.00033	0.00024	0.00036	0.00462
78	0.00490	0.01490	0.00028		0.00001	0.00472	0.00044	0.00570	0	0	0.00015	0.00839
79	0.00136	0.07687	0.00160	0.00009	0.00004	0.01118	0.00631	0.04455	0.00289	0.00008	0.00065	0.00667
80	0.00025	0.01343	0.00214	0.00003	0.00001	0.00797	0.00040	0.01120	0.00815	0.00075	0.00088	0.04488
81	0.00371	0.13760	0.00591	0.00005	0.00079	0.03921	0.00804	0.00354	0.00964	0.00138	0.00112	0.07644
82	0.00125	0.02616	0.01316	0.00038	0.00101	0.01240	0.00341	0.00195	0.00436	0.00158	0.01253	0.02827
83	0.00135	0.03163	0.00516	0.00018	0.00043	0.00981	0.00548	0.00234	0.00220	0.00087	0.00169	0.00266
84	0.00106	0.00384	0.01024	0.00016	0.00031	0.00660	0.00127	0.00225	0.00060	0.00005	0.00201	

Continued

b_{ij} \ a_{ij}	113	114	115	116	117	118	119	120	121	122	123	124
85	0	0	0	0	0	0	0	0	0	0	0	0
86	0.00099	0.00210	0.00663	0.00031	0.00046	0.01981	0.00121	0.00102	0.00134	0.00062	0.00177	0.01491
87	0.00162	0.00652	0.00619	0.00090	0.00184	0.03746	0.00440	0.00200	0.00436	0.00281	0.00521	0.09134
88	0.00287	0.01046	0.01931	0.00088	0.00268	0.06122	0.00379	0.00185	0.00426	0.00010	0.00038	0.03739
89	0.00330	0.01597	0.01517	0.00121	0.00137	0.09228	0.00600	0.00300	0.00491	0.00182	0.00409	0.05096
90	0.00488	0.03206	0.03433	0.00205	0.00671	0.15328	0.00252	0.00111	0.01297	0.00336	0.01714	0.14635
91	0.00059	0.00401	0.00375	0.00011	0.00018	0.00612	0.00122	0.00052	0.00070	0.00149	0.00255	0.00513
92	0.00057	0.00336	0.00296	0.00007	0.00011	0.00376	0.00015	0.00023	0.00127	0.00129	0.00312	0.00792
93	0.00018	0.00171	0.00019	0.00013	0.00070	0.00120	0.00048	0.00048	0.00037	0.00018	0.00079	0.00717
94	0.00032	0.00316	0.00140	0.00005	0.00015	0.00309	0.00274	0.00581	0.00043	0.00068	0.00254	0.00444
95	0.00087	0.05226	0.02553	0.00029	0.00051	0.01189	0.00112	0.00051	0.01391	0.00075	0.00133	0.02317
96	0.00027	0.00312	0.00264	0.00007	0.00028	0.00532	0.01001	0.00043	0.00183	0.00100	0.00327	0.02210
97	0.00042	0.00387	0.00015	0.00001		0.00024		0.00603		0.00034		
98	0.00192	0.02819	0.04243	0.00062	0.00139	0.07803	0.08899	0.00145	0.01698	0.00452	0.00358	0.31153
99	0.00085	0.01429	0.00738	0.00030	0.00088	0.01930	0.02321	0.00124	0.00497	0.00155	0.00283	0.11426
100	0.00111	0.01043	0.01475	0.00012	0.00026	0.00702	0.00308	0.00218	0.00104	0.00082	0.00141	0.01305
101	0.00174	0.02259	0.00532	0.00061	0.00267	0.01997	0.00517	0.01582	0.00804	0.00434	0.00796	0.11422
102	0.00198	0.04915	0.01110	0.00237	0.00110	0.10594	0.01119	0.00547	0.02576	0.00538	0.01673	0.26694
103	0.00222	0.04575	0.00590	0.00066	0.00137	0.08059	0.01122	0.00758	0.02531	0.00765	0.01802	0.17789
104	0.00190	0.06275	0.00784	0.00033	0.00127	0.12394	0.02276	0.00609	0.01847	0.00787	0.02466	0.19594
105	0.00234	0.04562	0.00754	0.00124	0.00178	0.05360	0.01150	0.00039	0.01514	0.00427	0.00398	0.31882
106	0.00387	0.00500	0.00084	0.00003	0.00005	0.00176	0.00394	0.00107	0.00049	0.00120	0.00269	0.04636
107	0.01417	0.01167	0.00305	0.00032	0.00127	0.00829	0.00330	0.00089	0.00803	0.00220	0.00654	0.06613
108	0.00407	0.05325	0.00426	0.00008	0.00568	0.01136	0.00521	0.00843	0.00551	0.00076	0.00043	0.09062
109	0.00084	0.07576	0.02063	0.00082	0.00345	0.10314	0.01661	0.00154	0.01793	0.00501	0.01489	0.18218
110	0.00165	0.01052	0.00315	0.00068	0.00083	0.01872	0.00346	0.00632	0.00340	0.00093	0.00119	0.02562
111	0.00858	0.04465	0.00707	0.00133	0.00260	0.06422	0.01134	0.00049	0.02848	0.00520	0.00682	0.36191
112	0.00066	0.00967	0.00200	0.00017	0.00207	0.03217	0.00177	0.00008	0.00383	0.00219	0.00269	0.08351
113	0.00130	0.00082	0.00019	0.00009	0.00011	0.00186	0.00312	0.00103	0.00131	0.00016	0.00046	0.03273
114	0.00054	0.02895	0.00094	0.00017	0.00043	0.01199	0.00651	0.00419	0.00363	0.00032	0.00531	0.02329
115	0	0.00773	0.04512	0.00036	0.00183	0.07985	0.00417	0	0.00815	0.00175	0.00565	0.13914
116	0	0	0	0	0	0	0	0	0	0	0	0
117	0.00072	0	0	0	0	0	0	0.00395	0	0	0	0
118	0.00884	0.01763	0.01443	0.00101	0.00565	0.02924	0.00583	0.00412	0.01222	0.00304	0.00723	0.10587
119	0.00030	0.09518	0.00685	0.00109	0.00291	0.13828	0.31917	0.03988	0.01360	0.00298	0.00294	0.15857
120	0.00212	0.36510	0.00368	0.00005	0	0.12251	0.00034	0.00616	0.01671	0.00081	0.00108	0.01805
121	0.00144	0.00618	0.00506	0.00010	0.00007	0.01667	0.00173	0	0.07252	0.00130	0.01228	0.02769
122	0	0	0	0	0	0	0	0	0	0.02965	0	0
123	0	0	0	0	0	0	0	0	0	0	0.10640	0
124	0	0	0	0	0	0	0	0	0	0	0	0

Index

T - #0265 - 101024 - C0 - 254/178/13 [15] - CB - 9781420089196 - Gloss Lamination